Histopathology of Seed-Borne Infections

Dalbir Singh • S.B. Mathur

CRC Press
Taylor & Francis Group
Boca Raton London New York

CRC Press is an imprint of the
Taylor & Francis Group, an **informa** business

Cover Photograph: Section of chickpea (*Cicer arietinum*) cotyledon showing inter- and intracellular mycelium of *Ascochyta rabiei*, cause of blight in chickpea. (From Maden, S. et al. 1975. *Seed Sci. Technol.* 3: 667–681. With permission.)

Published 2004 by CRC Press
Taylor & Francis Group
6000 Broken Sound Parkway NW, Suite 300
Boca Raton, FL 33487-2742

© 2004 by Taylor & Francis Group, LLC
CRC Press is an imprint of Taylor & Francis Group, an Informa business

First issued in paperback 2019

No claim to original U.S. Government works

ISBN-13: 978-0-367-45435-7 (pbk)
ISBN-13: 978-0-8493-2823-7 (hbk)

**Visit the Taylor & Francis Web site at
http://www.taylorandfrancis.com**

**and the CRC Press Web site at
http://www.crcpress.com**

Library of Congress Cataloging-in-Publication Data

Singh, Dalbir, 1932-
Histopathology of seed-borne infections / Dalbir Singh, S.B. Mathur.
p. cm.
Includes bibliographical references and index.
ISBN 0-8493-2823-3 (alk. paper)
1. Seed-borne phytopathogens. 2. Seed-borne plant diseases. 3. Histology, Pathological.
I. Mathur, S. B. II. Title.

SB732.8.S56 2004
632.3—dc22

2004041407

Library of Congress Card Number 2004041407

Preface

The book deals with only one aspect of seed-borne infection — the histopathology. Since the publication of the late Dr. Paul Neergaard's book, *Seed Pathology*, which still remains an invaluable guide, phenomenal progress has taken place in the subject. Recent information on histopathology of seeds infected by different groups of microorganisms is scattered in numerous research periodicals. An attempt has therefore been made to consolidate this scattered information and present a coordinated and coherent account. Information on flower and development of anther and ovule leading to the formation of seed, and variability in seed structure of crop plants, relevant to studies in seed pathology has also been provided. Much of the information is based on the material used by the authors for their teaching and incorporates important developments in histopathology. A large number of the illustrations used are from the studies and publications of the authors and their collaborators.

Up-to-date scientific names are used for pathogens based on the following publications:

Farr, D.P., Ellis. G.F., Chamunis, G.P., and Rossman, A.Y. 1989. *Fungi on Plants and Plant Products in United States*. APS Press, St. Paul, MN.
Fauquet, C.M. and Martelli, G.P. 1995. Updated ICTV list of names and abbreviations of viruses, viroides and satellites infecting plants. *Arch. Virol.* 140: 393–413.
Fauquet, C.M. and Mayo, M.A. 1999. Abbreviations of plant virus names — 1999. *Arch. Virol.* 144: 1249–1273.
Young, J.M., Saddler, G.S., Takikawa, Y., De Boer, S.H., Vauterin, L., Gardan, L., Gvozdyak, R.I., and Stead, D.E. 1996. Names of plant pathogenic bacteria, 1864–1995. *Rev. Plant Pathol.* 75: 721–763.

This book will be useful to students, teachers, and researchers in seed pathology and seed technology. Personnel working in seed health testing laboratories, plant quarantine, and agro-industries will find this book helpful in formulating strategies for testing, interception, and control of pathogens occurring as internal infections.

Dalbir Singh
S.B. Mathur

Acknowledgments

The authors are grateful to Dr. Carmen Nieves Mortensen, Associate Professor, and Mr. S.E. Albrechtsen, former Associate Professor at the Danish Government Institute of Seed Pathology for Developing Countries, Copenhagen, for perusing Chapters 6 (Seed Infection by Bacteria) and 7 (Seed Infection by Viruses), respectively, and making critical comments and suggestions. We thank Prof. Thomas W. Carroll, Department of Plant Pathology, Montana State University, Bozeman, U.S.A., and Dr. Andy J. Maule, Department of Virology, John Innes Centre, Norwich, U.K., for providing literature on viruses.

We wish to thank the publishers and executives of journals and books, and individuals for granting permission to reproduce figures from their publications. Due acknowledgment has been made for such figures. Special thanks are due to the following individuals for providing photographs from their files: Prof. Rolland R. Dute, Auburn University, Auburn, Alabama; Prof. S.V. Thomson, Utah State University, Logan; Prof. A.M. Alvarez, University of Hawaii, Manoa; Dr. A. Halfon-Meiri, The Volcani Centre, Bet Dagan, Israel; Dr. M.J. Christey, Christchurch, New Zealand; and Dr. Eigil de Neergaard, Royal Veterinary and Agricultural University, Copenhagen, Denmark.

We thank Ms. Anette Højbjerg Hansen for her patience during the computer typing of the manuscript and Mr. Magdi El-din Ragab for his skillful cooperation in arranging the figures. We thank Ms. Henriette Westh for processing the final manuscript for submission.

Dalbir Singh is grateful to the Danish Ministry of Foreign Affairs (Danida) for supporting his visits to the Institute of Seed Pathology in Denmark for planning and writing the book. He is grateful to his colleagues at the Department of Botany, University of Rajasthan, Jaipur, for their interest and cooperation and to all his research collaborators for their cooperation and for allowing him free use of their contributions. He is especially thankful to Prof. Tribhuwan Singh, University of Rajasthan, Jaipur, and Dr. Kailash Agarwal, Agarwal College, Jaipur, for useful discussions and for improving some of the figures used in the book. The gracious help of Dr. Dileep Kumar during the entire period of manuscript preparation is gratefully acknowledged. Thanks are also due to Shri Rajesh Benara for typing the manuscript, and to Shri Mehar Chand and Shri Ankur at Jaipur for help with illustrations.

Dalbir Singh expresses his deep gratitude to his wife, Prem Singh, and his children — Nidhi, Smita and Mayank — for their patience and cooperation while he was engaged in writing the book.

The Authors

 **Dalbir Singh, Ph.D.,** former Professor of Botany, University of Rajasthan, Jaipur, India, received his Master's degree in Botany in 1952 and his Ph.D. in Reproductive Biology and Developmental Morphology in 1959 from Agra University. For 40 years (1952 to 1992), he taught generations of graduate and postgraduate students, and conducted courses in reproductive biology, embryology, anatomy, seed pathology, and seed technology. Dr. Singh initiated the teaching of seed pathology and seed technology to postgraduate students in the Department of Botany at Jaipur in 1975.

For the past 50 years, Dr. Singh has been involved in research concerning the development and structure of seed in economically important families of angiosperms and the histopathology of seeds infected with fungal pathogens. Since 1973 he has been associated with the Danish Government Institute of Seed Pathology for Developing Countries (DGISP). He and his collaborators have made significant contributions to the histopathology of a large number of fungal pathogens in the seeds of cereals, oilseeds, legumes, and spices. His research also concerns histology of physiogenic disorders in pea and chickpea, nematode gall development and structure in wheat, and more recently (after 1985), the histopathology of seeds infected with bacteria. He has guided 40 successful Ph.D. candidates and has published 300 research papers. Dr. Singh was awarded the Birbal Sahni Medal of the Indian Botanical Society in 1992 for his outstanding research contributions.

Dr. Singh is an elected Fellow of the National Academy of Sciences. He visited the former U.S.S.R. in 1977 as a member of an Indian delegation of botanists under a bilateral exchange program. From 1986 to 1987, he was a Visiting Professor at ASEAN PLANT I, a Regional Plant Quarantine and Training Institute in Kuala Lumpur. Dr. Singh has been associated with several national and international botanical societies. He served as the Secretary of the Indian Botanical Society from 1986 to 1992 and as its President in 1995 and 1996. From 1993 to 1994, he was the President, Section of Botany, Indian Science Congress Association. He is currently the Additional Secretary of the International Society of Plant Morphologists.

S.B. Mathur, Ph.D., is the Director of the Danish Government Institute of Seed Pathology for Developing Countries (DGISP) in Copenhagen, Denmark, where he has spent most of his professional life as a pioneer in the field of seed pathology. For more than 35 years Dr. Mathur has been instrumental in realizing seed health as an important step toward fighting hunger in the third world. His primary objective has been to fight seed-borne diseases, not only to find cures, but more importantly to investigate ways to detect seed-borne infections in the laboratory and prevent outbreaks of diseases at an early stage, in both formal and informal seed sectors. Creating awareness of the importance of seed health, the relationship between seed health and food production and food security, and the impact of good quality seed on increase in yield has been a milestone in Dr. Mathur's life. His contributions to international agriculture have been recognized by the international community. In 1992 he was awarded the prestigious FIS World Seed Prize, presented by the International Seed Federation, Switzerland, and in 2002 he was awarded the Prof. K.M. Safeeulla Gold Medal by the University of Mysore, India.

The Institute of Seed Pathology in Denmark, the brain-child of Dr. Mathur, is financed and supported by Danida (Ministry of Foreign Affairs of Denmark). More than 550 scientists and technologists from 72 developing countries have been educated there and have conducted basic and applied research related to solving seed pathological problems. Dr. Mathur has trained more than 400 agricultural scientists from more than 70 countries in short courses, conducted in various countries of the developing world. He has been responsible for the introduction and monitoring of seed health at the Consultative Group on International Agricultural Research (CGIAR) Centers and routine checking of germplasm for seed health at national plant quarantine inspection laboratories.

Dr. Mathur is presently leading a group of renowned experts engaged in establishing two educational Seed Pathology Centers, one in India for Asia and the other in Tanzania for Africa. The major goal of these centers is to develop trained personnel who will be responsible for handling seed health issues and increase and improve food and seed production, especially for resource-poor farmers.

Contents

1 Introduction

The reports on the number of microorganisms associated with seeds have increased gradually during the latter half of the 20th century. This is obvious from the first and the most recent editions of *An Annotated List of Seed-Borne Diseases* by Noble, De Tempe, and Neergaard (1958), and Richardson (1990), respectively. The organisms occur with seed either as contaminants adhering to the seed surface, loosely mixed with seed, or as an infection present inside the seed tissues. This book describes penetration and location of microorganisms in seed tissues. Information on location of infection in seeds using histological techniques alone is considered. The presence of internal infection in seed detected in transmission studies is beyond the scope of this book.

In order to appreciate penetration, course of infection, location, and the effect of infection of microorganisms in floral and seed tissues, it is necessary to have a good understanding of the structure of flowers and changes in fertile appendages — stamen and carpel — leading to the formation of seed. Neergaard (1979) has included information on morphology and anatomy of seed in relation to transmission of pathogens. This account is fragmentary and does not include information on many critical steps in the formation of seed. Considerable new knowledge, including ultrastructure of reproductive components — embryo sac, endosperm, embryo, and their surrounding tissues — has been provided (Johri, 1984; Johri, Ambegaokar, and Srivastava, 1992). These data elucidate the contacts and barriers among the tissues of developing seed. The structure of mature seed, including one-seeded dry indehiscent fruits, is also highly variable in angiosperms (Netolitzky, 1926; Corner, 1976). The variations in size of the hilum, micropyle opening, nature and thickness of the cutile, and thickness of seed coat and pericarp have shown direct correlation with the penetration and location of fungal pathogens in certain host–parasite interfaces. Chapter 2 provides a concise up-to-date account of the structure and development of floral parts and the formation of seed. Chapter 3 deals with the structure of mature seed in selected families of angiosperms. Chapters 5 through 8 concern the histopathology of seed infections by fungi, bacteria, viruses, and nematodes. Chapter 9 describes physiogenic seed disorders. Chapter 10 includes a brief account of histopathological techniques and tips. Only a basic account is given in Chapter 10; several detailed books on plant microtechnique and transmission and scanning electron microscopy are available.

1.1 THE SEED

Biologically, seed is the ripened ovule. In angiosperms, to which a majority of the crop plants belong, the ovules are borne in the ovary, the basal part of the gynoecium (pistil). The seed formation takes place *in situ* through a series of integrated sequential steps in the life cycle of the flowering plant (Maheshwari, 1950, 1963; Johri, 1984). After pollination and fertilization, changes in different parts of the ovule, i.e., the embryo sac (zygote and primary endosperm nucleus), nucellus, chalaza, and integument, lead to the formation of seed. The term seed, when used *sensu lato*, includes one-seeded dry indehiscent fruits that are the dispersal propagative units in plants of several families such as Poaceae, Asteraceae, Apiaceae, and Chenopodiaceae. In the present treatment, the term seed is used in a loose sense.

Seed is an autonomous living unit and links successive generations. Structurally, it consists of an embryo (new plantlet), a protective covering — seed coat, pericarp, or both — and reserve food material, which may be present in the endosperm, perisperm, or embryo. It has the capacity to withstand desiccation and retain viability under unfavorable environments or until it germinates. These properties of seed make it an important commodity for storage as well as transport to new areas or countries, for planting or for edible purposes. The structure of seed is fairly constant in a species, but varies in different plant taxa (Netolitzky, 1926; Corner, 1976).

Seeds, if infected by microorganisms, will act as carriers. If the organisms remain viable they will result in development of disease in the new crop. Infected seeds are often responsible for the spread of diseases to new areas. For this reason seed has become an object of plant quarantine internationally. Countries also use domestic seed certification, including seed health testing, as a method of quality control of seed.

1.2 MICROORGANISMS IN SEED

Fungi, bacteria, viruses, and nematodes are known to be seed-borne (Neergaard, 1979; Maude, 1996; Agarwal and Sinclair, 1997). Fungi form a major group of pathogens that are seed-borne as well as seed-transmitted. In addition to saprophytes and parasites, fungi are known to form an inherent association with seeds of some members of Cistaceae, Ericaceae, and Orchidaceae. In the Orchidaceae, the seeds will not usually germinate without the presence of a mycorhizal fungus (Rayner, 1915). The seed coat in seeds of *Helianthemum chamaecistus* (Cistaceae) harbors a fungus that seems essential for normal germination of seed. In the absence of the fungus, the plumule fails to emerge, and roots also are not formed at the time of germination (Boursnell, 1950).

The list of saprophytic and parasitic fungi associated with seeds of different plants is very large (Richardson, 1990), and they belong to all fungal classes. The fungi that are discussed in this book and for which histopathological information is available belong to the division Eumycota, subdivision Mastigomycotina, class Oomycetes, subdivisions Ascomycotina, Basidiomycotina, and Deuteromycotina. The members of Deuteromycotina dominate and these fungi belong to the classes Hyphomycetes and Coelomycetes. The endophytic fungi are discussed separately.

A large number of viruses, including cryptoviruses and viroids, are known to be seed-borne, but the information on seeds infected by viruses is limited and mostly inconclusive. The better-studied viruses are barley stripe mosaic virus (BSMV) and pea seed-borne mosaic virus (PSbMV) due to the studies of Carroll and co-workers

(Carroll, 1969, 1974; Carroll and Mayhew, 1976a,b; Mayhew and Carroll, 1974) and Wang and Maule (1992, 1994), respectively. Similarly, the histopathology of seeds affected by bacteria is poorly studied, and the investigations are usually confined to seeds infected by *Acidovorax, Burkholderia, Clavibacter, Curtobacterium, Pantoea, Pseudomonas, Rathayibacter,* and *Xanthomonas.*

Seed-borne nematodes also occur as seed infestation or seed infection. The latter causes either seed gall formation (*Anguina* spp.) or symptomatic or symptomless infections.

1.3 HISTOPATHOLOGY

Ever since Cobb (1892; see Royle, 1976) proposed the *mechanical theory* of rust resistance indicating that morphological features, such as thick cuticle, waxy covering, small stomata, abundant leaf hairs, and upright leaves, might be responsible for the resistance of wheat varieties to *Puccinia graminis*, numerous studies on histology of infected plant parts, particularly leaves and stems, have been carried out. This information has been summarized in excellent reviews on (1) histology of defense (Akai, 1959; Royle, 1976; Schonbeck and Schlouster, 1976); (2) abilities of pathogens to breach host barriers (Dickinson, 1960; Emmett and Parberry, 1975; Dodman, 1979); and (3) physiology and biochemistry of penetration and infection (Flentje, 1959; Alberschein, Jones, and English, 1969; Mount, 1978; Durbin, 1979; Kollattukudy, 1985).

Although the first observation on internal presence of the mycelium of *Colletotrichum lindemuthianum* in cotyledons of seeds of *Phaseolus vulgaris* was made as early as 1883, further progress until 1950 was rather slow. It is during the latter half of the 20th century and more particularly after 1970 that several comprehensive reports have appeared on the penetration and location of microorganisms in seeds. Early information has been summarized by Baker (1972), Neergaard (1979), and Agarwal and Sinclair (1997).

Various histological techniques, e.g., the embryo extraction method initially used by Skvortzov (1937), whole-mount preparations of seed components (Maden et al., 1975; Singh, Mathur, and Neergaard, 1977), free hand sections, and microtome sections have been used. Microtome sections of weakly, moderately, and heavily infected seeds alone provide information on exact expanse of mycelium in seed and also the effects of host–parasite interactions (Singh, 1983). Although used primarily in the study of virus infections and in a limited way in the study of fungal infections, transmission electron microscopy (TEM) and scanning electron microscopy (SEM) have yielded valuable information, which certainly surpasses the results of light microscopy.

REFERENCES

Agarwal, V.K. and Sinclair, J.B. 1997. *Principles of Seed Pathology*, 2nd ed. CRC Press, Boca Raton, FL.

Akai, S. 1959. Histology of defence in plants. In *Plant Pathology: An Advanced Treatise.* Horsfall, J.G. and Dimond, J.G., Eds. Academic Press, London. Vol. 1, 391–434.

Albershein, P., Jones, T.M., and English, P.D. 1969. Biochemistry of the cell wall in relation to infective processes. *Ann. Rev. Phytopathol.* 7: 171–194.

Baker, K.F. 1972. Seed Pathology. In *Seed Biology.* Kozlowski, T.T., Ed. Academic Press, New York. Vol. 2, 317–416.

Boursnell, J.G. 1950. The symbiotic seed-borne fungus in the Cistaceae. I. Distribution and function of the fungus in the seedling and in the tissues of the mature plant. *Ann. Bot.* (Lond.) N.S. 14: 217–243.

Carroll, T.W. 1969. Electron microscopic evidence for the presence of Barley stripe mosaic virus in cells of barley embryos. *Virology* 37: 649–657.

Carroll, T.W. 1974. Barley stripe mosaic virus in sperm and vegetative cells of barley pollen. *Virology* 60: 21–28.

Carroll, T.W. and Mayhew, D.E. 1976a. Anther and pollen infection in relation to the pollen and seed transmissibility of two strains of Barley stripe mosaic virus in barley. *Can. J. Bot.* 54: 1604–1621.

Carroll, T.W. and Mayhew, D.E. 1976b. Occurrence of virus in developing ovules and embryo sacs of barley in relation to the seed transmissibility of Barley stripe mosaic virus. *Can. J. Bot.* 54: 2497–2512.

Corner, E.J.H. 1976. *Seeds of Dicotyledons.* Vols. 1 and 2. Cambridge University Press, Cambridge, U.K.

Dickinson, S. 1960. The mechanical ability to breach the host-barriers. In *Plant Pathology: An Advanced Treatise.* Horsfall, J.G. and Dimond, J.G, Eds. Academic Press, London. Vol. 1, 203–232.

Dodman, R.L. 1979. How the defenses are breached. In *Plant Disease: An Advanced Treatise.* Horsfall, J.G. and Cowling, E.B., Eds. Academic Press, New York. Vol. 4, 135–151.

Durbin, R.D. 1979. How the breachhead is widened. In *Plant Disease: An Advanced Treatise.* Horsfall, J.G. and Cowling, E.B., Eds. Academic Press, New York. Vol. 4, 155–162.

Emmett, R.W. and Parberry, D.G. 1975. Appressoria. *Ann. Rev. Phytopathol.* 13: 147–167.

Flentje, N.T. 1959. The physiology of penetration and infection. In *Plant Pathology, Problems and Progress*, 1908–1958. University of Wisconsin, Madison, 76–87.

Johri, B.M., Ed. 1984. *Embryology of Angiosperms.* Springer-Verlag, Berlin.

Johri, B.M., Ambegaokar, K.B., and Srivastava, P.S. 1992. *Comparative Embryology of Angiosperms.* Vols. 1 and 2. Springer-Verlag, Berlin.

Kolattukudy, P.E. 1985. Enzymatic penetration of the plant cuticle by fungal pathogens. *Ann. Rev. Phytopathol.* 23: 223–250.

Maden, S., Singh, D., Mathur, S.B., and Neergaard, P. 1975. Detection and location of seed-borne inoculum of *Ascochyta rabiei* and its transmission in chickpea (*Cicer arietinum*). *Seed Sci. Technol.* 3: 667–681.

Maheshwari, P. 1950. *An Introduction to the Embryology of Angiosperms.* McGraw-Hill, New York.

Maheshwari, P., Ed. 1963. *Recent Advances in the Embryology of Angiosperms.* International Society of Plant Morphologists, University of Delhi, India.

Mayhew, D.E. and Carroll, T.W. 1974. Barley stripe mosaic virus in the egg cell and egg sac of infected barley. *Virology* 58: 561–567.

Maude, R.B. 1996. *Seed-Borne Diseases and Their Control.* CAB International, Wallingford, U.K.

Mount, M.S. 1978. Tissue is disintegrated. In *Plant Disease: An Advanced Treatise.* Horsfall, J.G. and Cowling, E.B., Eds. Academic Press, New York. Vol. 3, 279–293.

Neergaard, P. 1979. *Seed Pathology.* Vols. 1 and 2. Macmillan Press, London.

Netolitzky, F. 1926. *Antomie der Angiospermen-Samen*. In Handbuch der Pflanzenanatomie. Linsbauer, K., Ed. Abt. 2, Teil 2, Bd. 10. G. Borntraeger, Berlin.

Noble, M., de Tempe, J., and Neergaard, P. 1958. *An Annotated List of Seed-Borne Diseases*. Commonwealth Mycological Institute, Kew, Richmond, Surrey, U.K.

Rayner, M.C. 1915. Obligate symbiosis in *Calluna vulgaris. Ann. Bot.* (Lond.) N.S. 29: 96–131.

Richardson, M.J. 1990. *An Annotated List of Seed-Borne Diseases*, 4th ed. Proceedings of the International Seed Testing Association, Wageningen, the Netherlands.

Royle, D.J. 1976. Structural features to plant diseases. In *Biochemical Aspects of Plant Parasite Relationships*. Frank, J. and Thresefall, D.R., Eds. Academic Press, New York, pp. 162–193.

Schonbeck, F. and Schlouster, E. 1976. Preformed substances as potential protectants. In *Physiological Plant Pathology*. Heitefuss, R. and Williams, P.H., Eds. Springer-Verlag, Berlin, pp. 653–678.

Singh, D. 1983. Histopathology of some seed-borne infections: a review of recent investigations. *Seed Sci. Technol.* 11: 651–663.

Singh, D., Mathur, S.B., and Neergaard, P. 1977. Histopathology of sunflower seeds infected by *Alternaria tenuis. Seed Sci. Technol.* 5: 579–586.

Skvortzov, S.S. 1937. A simple method for detecting hyphae of loose smut on wheat grains. *Plant Prot. Leningrad* 15: 90–91.

Wang, D. and Maule, A.J. 1992. Early embryo invasion as a determinant in pea of the seed transmission of Pea seed-borne mosaic virus. *J. Gen. Virol.* 73: 1615–1620.

Wang, D. and Maule, A.J. 1994. A model for seed transmission of a plant virus: genetic and structural analyses of pea embryo invasion by Pea seed-borne mosaic virus. *Plant Cell* 6: 777–787.

2 Reproductive Structures and Seed Formation

Seed is the end product of sexual reproduction that takes place through sequential changes in the reproductive shoot, the flower in angiosperms. The ovules are borne in a closed structure, the ovary. The ovules and seeds may get infected at any stage during their development either directly through the mother plant or through the infected floral parts, including bracts and nectaries. The contacts between the plant and flower and in turn with the ovule and seed are important in understanding the course of such infections. The interface of pathogen and host may cause morphological and anatomical changes in tissues of the flower and may also affect the reproductive cycle, including seed formation. Green ear disease of pearl millet, caused by *Sclerospora graminicola,* results in the transformation of florets into leafy structures. Inflorescence or flower infection by *Protomyces macrosporus* in coriander and *Albugo candida* in Brassicaceae cause hypertrophy in the latter, and fruit gall formation in the former. *Claviceps purpurea* and other species, when infecting cereal florets, cause the ovary to form fungal sclerotia instead of producing a kernel. The contents in the ovary are replaced by smut chlamydospores due to the infection of *Sphacelotheca, Tilletia, Tolyposporium,* and *Ustilago* in various cereal crops.

The structure of flowers, the reproductive cycle, and seed formation are described in this chapter. The aspects that may be relevant to pathogenesis are highlighted. An outline of the reproductive cycle is given in Figure 2.1.

2.1 FLOWER

The flowers may occur singly (cotton, kenaf, cucumber, pumpkin, and melon) or aggregated in inflorescence — small (tomato, potato, and brinjal) to massive (rice, wheat, pearl millet, sunflower, and coriander). Stalked or sessile flowers are borne loosely (coriander, carrot, and tomato) or compactly (wheat, pearl millet, maize, and sunflower), have an orderly arrangement, and depending on the order of development of flowers, the inflorescence is classified into indeterminate (racemose) or determinate (cymose) type. In the former, the order of development of flowers is acropetal, with the youngest bud and flower near the growing tip (*Brassica, Raphanus,* and *Linum*), whereas in the latter, it is basipetal, with the growing point ending in a flower and subsequent ones produced in the axil(s) of bracts below the terminal flower (*Lycopersicon, Sesamum,* and *Solanum*). The presence of accessory structures, e.g., spathe in spadix (maize); involucre of bracts in capitulum (sunflower) and umbel (coriander and carrot); and glumes, palea, and lemma with florets in simple

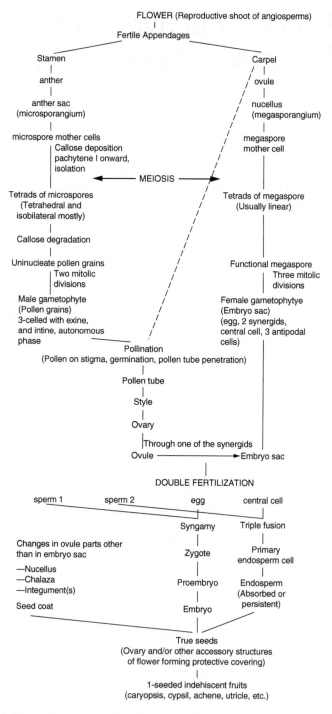

FIGURE 2.1 Schematic representation of reproductive cycle and seed formation, including one-seeded indehiscent fruits in angiosperms.

or compound spikes (wheat, barley, pearl millet, and rice), provide additional protection to flowers, but cause increased humidity around them. They may also provide a conducive environment for the development of pathogens that succeed in invading them.

The flower is a shoot of determinate growth. The distal end of the axis is swollen or cup-shaped, forming a thalamus or receptacle, which bears laterally floral organs. A typical flower of dicotyledons has four types of organs comprising two whorls of sterile appendages, the sepals and petals collectively called the calyx and corolla, respectively, and two whorls of fertile appendages, the stamens and carpels, collectively termed the androecium and gynoecium. The appendages are arranged on the receptacle in succession spirally or in whorls. Many variations occur in size, shape, color, and number of members in each whorl and in their organization. Usually in monocotyledons, there is only one whorl of sterile appendages, termed the perianth (onion and garlic). Flowers may be bisexual or unisexual. In the unisexual condition staminate and pistillate flowers are monoecious, or borne on one plant, as in maize, cucumber, pumpkin, and squash. If the staminate and pistillate flowers occur on different plants, as in papaya, they are dioecious.

In a flower, if the floral appendages are inserted successively one above the other below the gynoecium (carpel), the flower is hypogynous and the ovary is superior (Figure 2.2A). When the receptacle becomes concave and surrounds the ovary or is fused with the ovary wall so that the sepals, petals, and stamens arise from the top of the ovary (Figure 2.2B), the flower is epigynous and the ovary is inferior as in coriander, cumin, carrot, cucumber, pumpkin, squash, sunflower, lettuce, apple, and pear. In an epigynous flower, the ovary is exposed to the environment from initiation until maturity, i.e., the formation of fruit.

The flower in the grass family, Poaceae, is subsessile and bracteate; the dry chaffy lemma, which is usually awned, is the bract. The dry membranous structure on the posterior side is known as the palea, which represents two fused posterolaterally situated bracteoles. The palea encloses the flower and itself is enclosed by the lemma. The flower consists of lodicules (tepals), androecium, and gynoecium. Usually there are two fleshy, often hairy, lodicules situated in the anterolateral position. In *Bambusa*, there are six lodicules. Above the lodicules, there are usually three stamens and rarely more, six or less in *Oryza* and *Bambusa*, etc., two in *Anthraxon*, and one in *Uniola*. The gynoecium is tricarpellary and syncarpous and the ovary is unilocular with a single large ovule. The vascular anatomy has shown that the placentation is parietal and that two of the three placentae are sterile. The ovary bears two laterally situated feathery styles and stigma.

2.1.1 VASCULARIZATION OF FLOWER

The anatomical structures of the peduncle (axis of the inflorescence) and the pedicel are similar to that of the stem. The vascular cylinder may be continuous or split. The vascular bundles of the pedicel at a higher level break up and give out traces to the floral appendages in succession in a hypogynous flower (Figure 2.2A). The vascular traces meant for the floral appendages enter each unit either unbranched or after undergoing further branching. Each sepal generally receives three traces, a

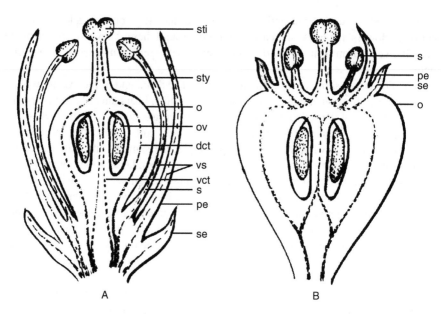

FIGURE 2.2 Structure and vascularization of angiospermous flower. A, Longitudinal section (Ls) showing vascular supply in hypogynous flower, traces of each whorl separate in the thalamus and diverge at corresponding places. B, Ls epigynous flower, traces for carpels are the first to diverge while those of calyx, corolla, and androecium diverge at levels above the ovary. (Abbreviations: dct, dorsal carpellary trace; o, ovary; ov, ovule; pe, petal; s, stamen; se, sepal; sti, stigma; sty, style; vct, ventral carpellary trace; vs, vascular supply.)

petal with a single median trace or a median trace with two or more laterals, and the stamen a single trace. The carpel is vascularized with a median (dorsal) and two lateral (ventral or marginal) traces. The ventral traces unite partially or completely into a single ventral trace. The ventral trace(s) are generally limited to the ovule-bearing part of the carpel. The vasculature in an epigynous flower differs only in the manner of resolution or departure of vascular traces from the main cylinder. The vascular traces of the whorls of appendages depart at levels higher than that of the gynoecium (Figure 2.2B). Thus, in a flower there is continuity in the vascular supply of different whorls with that of the mother plant through the peduncle and pedicel.

2.2 STERILE APPENDAGES

Sepals are more or less leaflike or bractlike in form, structure, and vasculature. They are usually green and rarely petaloid. Anatomically, the epidermis shows deposition of cutin and development of stomata and trichomes similar to those on foliage leaves. The mesophyll is undifferentiated or rarely differentiated into palisade and spongy tissues.

Petals vary in shape, size, and color. In most flowers, they form the most conspicuous whorl and help in insect pollination. The epidermis bears stomata, functional or undifferentiated and nonfunctioning, and may develop intercellular spaces over-arched by the cuticle. The mesophyll is only a few cells thick, except in flowers with fleshy petals. The cuticle is commonly striated (Martens, 1934). The fragrance of flowers is produced by volatile substances, mostly essential oils, occurring in the epidermal cells. In some plants, such as *Lupinus* and *Narcissus,* the fragrance originates in special glands called osmophors.

2.3 FERTILE APPENDAGES

The stamens and carpels, male and female appendages, respectively, are highly specialized and form the seat of the development of male and female gametophytes, pollen and embryo sac, in angiosperms. The development and structure of male and female gametophytes have been studied in numerous plants of dicotyledons and monocotyledons. Detailed information on this topic can be found in the publications of Schnarf (1929, 1931), Maheshwari (1950, 1963), Davis (1966), Johri (1984), and Johri, Ambegaokar, and Shrivastava (1992).

2.3.1 STAMEN

A typical stamen consists of the filament bearing a two-lobed and four-loculed anther (Figure 2.3A, B). The two lobes of the anther are separated by a sterile tissue, called the connective. The single vascular supply of stamen traverses the filament and may end at the base of the anther or may extend into the tissue of the connective. The vascular bundle is not connected by any vascular element with the sporogenous tissue. The ground tissue of the connective and filament consists of parenchymatous cells. The epidermis is cutinized and may have stomata.

The fertile region of the anther comprises four microsporangia, locules or pollen sacs, two per anther lobe. During development and initial organization, the pollen sacs of an anther lobe are distinct (Figure 2.3B), but at maturity due to confluence, these become one.

2.3.1.1 Development of Microsporangium and Microsporogenesis

The development of microsporangium is fairly uniform in angiosperms. The male archesporium differentiates as a row of cells in the hypodermal region in four corners. The archesporial cells divide periclinally. The outer derivatives divide anticlinally and periclinally forming anther wall layers other than the epidermis. Depending on the pattern of divisions in the primary parietal layer, four types of anther walls are recognized (Davis, 1966). The anther wall usually consists of an epidermis, an endothecium, one or two middle layers, and a tapetum (Figure 2.3C, D). Tapetum, the innermost wall layer that forms a jacket around the sporogenous tissue, is of dual origin, partly formed by the innermost derivatives of parietal layer and partly by the parenchymatous cells, adjacent to the sporogenous tissue of the connective (Periasamy and Swamy, 1966). Its cells become rich in cell contents and function to nourish the developing sporogenous cells. Depending on its mode of function, the tapetum is secretory or amoeboid in nature. In the latter, the cells lose cell walls, and contents migrate in between the sporogenous cells (Figure 2.3G, H).

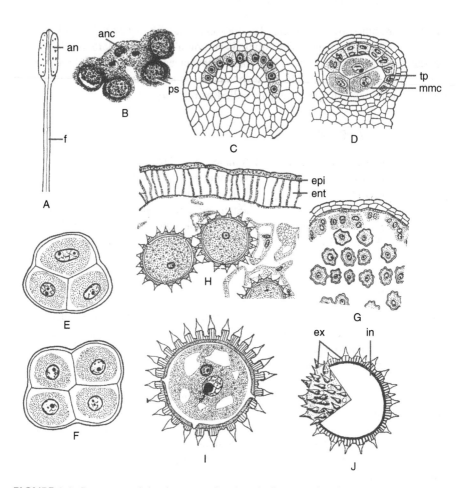

FIGURE 2.3 Structure and development of anther. A, Stamen with bilobed anther. B, Transverse section (Ts) bilobed anther showing tetrasporangiate condition. C, D, Sections of young anthers, note the organization of wall layers and microspore mother cells. E, F, Tetrahedral and isobilateral tetrads of microspores. G, H, Ts anther at microspore and uninucleate pollen grain stages. I, Two-nucleate pollen grain. J, Palynogram of mature pollen grain. (Abbreviations: an, anther; anc, anther-connective; ent, endothecium; epi, epidermis; ex, exine; f, filament; in, intine; mmc, micropore mother cell; ps, pollen sac; tp, tapetum.) (C to J, From Ramchandani, S., Joshi, P.C., and Pundir, N.S. 1966. *Indian Cotton J.* 20: 97–106.)

The middle layers are usually ephemeral. The outermost parietal layer is persistent, develops fibrous thickenings, and is called the endothecium (Figure 2.3H). It helps in the dehiscence of anther lobes and the release of pollen grains.

The inner derivatives of the male archesporium divide mitotically to form the sporogenous tissue (Figure 2.3C, D). The microspore mother cells undergo meiosis forming tetrads of microspores (Figure 2.3E, F). The cytokinesis during meiotic division is either successive, heterotypic as well as homotypic divisions followed by wall formation, or simultaneous, with wall formation taking place at the end of homotypic division. The simultaneous type of wall formation is common in dicotyledons and the successive type is common in monocotyledons.

Recent studies using TEM and fluorescence microscopy have shown that during early stages of prophase of heterotypic division of meiosis, plasmodesmatal connections occur between the tapetum and sporogenous cells as well as among the sporogenous cells, forming a syncytium. Subsequently, from pachytene I to telophase I, the sporogenous cells gradually develop refractive walls of callose. The synthesis of callose is retarded during homotypic division. The callose isolates the microsporocytes and microspore tetrads from the tapetum and also from one another (Westerkeyn, 1962). The autoradiographic study of Heslop-Harrison and Mackenzie (1967) has shown that the labeled thymidine derivatives do not penetrate microspore mother cells or microspores when they are surrounded by callose. Any irregularity in callose formation results in male sterility (Frankel, Izhar, and Nitsen, 1969).

2.3.1.2 Development and Structure of Male Gametophyte

With the formation of microspore tetrads, the callose starts to dissolve due to the action of β-1,3-glucanase (Stiegelitz and Stern, 1973). The microspores of a tetrad usually separate, and the uninucleate microspores become more or less spherical (Figure 2.3G, H). Further development of uninucleate microspores follows a uniform pattern in angiosperms (Figure 2.4A to F). The microspore nucleus divides mitotically forming a generative cell close to the wall and the vegetative cell. Electron microscope studies have revealed that the formation of a wall around the generative cell corresponds to that in other cells (Karas and Cass, 1976). The generative cell acquires a spindle shape in the pollen grain. The generative cell divides *in situ* (Figure 2.4G, H) or after pollination in the pollen tube to form two sperms (Figure 2.4I, J).

Mature pollen grains have a double wall, the exine and the intine (Figure 2.3I, J). The intine is thin and made up of pectocellulose while the principal component of the exine is sporopollenin, derived from tapetum. The exine is thick and remarkably durable. The intine and the exine also contain enzymes (proteins) that are released at the time of fertilization and degrade the cuticle of stigma papillae (Tsinger and Petrovskaya-Baranova, 1961). The intine proteins are the secretory products of pollen protoplast whereas the proteins of the exine originate from the tapetum. The exine proteins are involved in the penetration of the pollen tube into the stigma and also in pollen–stigma interaction that determines the incompatibility relationship (Heslop-Harrison, 1975).

2.3.2 CARPEL

One of the distinctive features of the angiosperms is the carpel, which bears the ovules. A typical carpel consists of three parts, the basal fertile region — the ovary — and two sterile parts, the style and the stigma. The gynoecium is termed apocarpous when carpels in a flower are free, or syncarpous when carpels become fused. In some angiosperms, the carpels are not completely closed, e.g., *Degenaria, Drimys,* and *Reseda.* Butomaceae, Hydrocharitaceae, and intraovarian pollen grains have been reported in *Butomopsis* and *Reseda.*

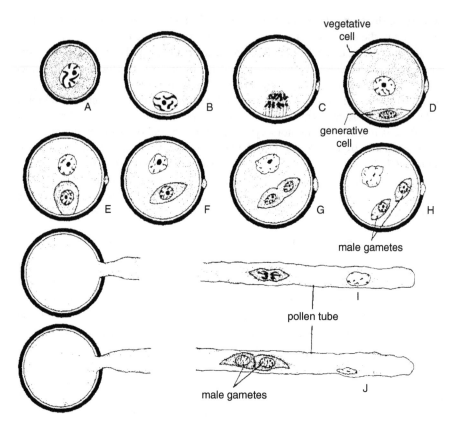

FIGURE 2.4 Development of male gametophyte. A, Newly formed microspore. B, Microspore showing vacuolation and wallward position of nucleus. C, Microspore nucleus dividing. D, Two-celled microspore. E, Generative cell losing contact with wall. F, Generative cell lying free in the cytoplasm of vegetative cell. G, H, Division of generative cell in pollen grain to form male gametes. I, J, Division of generative cell in pollen tube to form male gametes. (From Maheshwari, P. 1950. *An Introduction to the Embryology of Angiosperms.* McGraw-Hill, New York.)

2.3.2.1 Ovary

The ovary bears ovules internally at distinctive sites called placentae. The ovary in syncarpous gynoecia is multilocular with axile placentation or unilocular having parietal, superficial, and free central placentae. In the unilocular ovary, with one or several ovules, the placentation may be basal or pendulous. In the monocarpellary ovary, the placentation is marginal.

As mentioned previously, the carpel is vascularized with a median (dorsal) and two lateral (ventral or marginal) traces. The ventral traces unite partially or completely and give out the ovular supply. Before fertilization, the ovary wall comprises parenchyma and vascular bundles. It may have secretory canals as in Apiaceae (coriander and cumin) and calcium oxalate crystals as in Asteraceae. The epidermis is cuticularized, and the stomata and hairs may occur (Figure 2.5A, B, D, E).

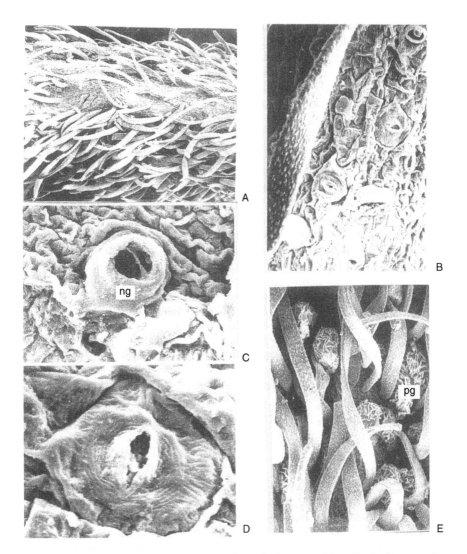

FIGURE 2.5 SEM photographs of ovary surface of *Cassia occidentalis*. A, Ovary surface showing trichomes. B, Same showing stomata and trichomes. C, Cup-shaped glands on ovary surface having thick cuticle around globular head. D, Stomata magnified, guard cells with arching cuticular rim. E, Trichomes with pollen grains having reticulate exine. (Abbreviations: ng, nectary gland; pg, pollen grain; st, stomata.) (From Sharangpani, P.R. and Shirke, D.R. 1996. *Phytomorphology* 46: 277–281.)

2.3.2.2 Style

The sterile part of the carpel between the stigma and ovary constitutes the style. Single and free carpels usually have one style, while in syncarpous gynoecia, the style may be united or partly or completely free. The style forms the pathway for pollen tubes to reach placentae and ovules. It has a specialized tissue that provides nutrients to the pollen tube and permits it to reach the destination. Arber (1937) called this tissue the transmitting tissue. Based on its distribution, the styles are classified as hollow and solid. The hollow style, which is common in monocotyledons, is characterized by the presence of a stylar canal, lined entirely or in longitudinal strips by glandular transmitting tissue. In *Lilium longiflorum*, the cells of the transmitting tissue are rich in organelles and contain abundant multivesicular bodies (Rosen and Thomas, 1970; Dashek, Thomas, and Rosen, 1971).

The solid styles lack stylar canal and show one or more strands of transmitting tissue, leading to the placentae of ovary. The cells of the transmitting tissue are elongated, rich in cytoplasm, and possess intercellular spaces. The pollen tubes pass through the intercellular spaces, which, as reported in *Lycopersicon*, possess a viscous fluid (Cresti et al., 1976). A solid style is common in dicotyledons.

A third type of style, *half-closed,* has been reported in some members of *Cactaceae* (Hanf, 1935) and *Artabotrys* (Rao and Gupta, 1951). In these plants the style is hollow, and the transmitting tissue develops only on one side of the stylar canal.

In addition to the transmitting tissue, the style consists of parenchyma with vascular supply. The epidermis may have stomata and is covered by the cuticle.

2.3.2.3 Stigma

The distal part of the carpel, having special features to facilitate pollen reception, germination, and penetration of the pollen tube into the closed carpel, is called the stigma. The stigmatic surface is generally papillate or hairy and rarely smooth. In *Poaceae* and other wind-pollinated plants, the stigmatic surface develops into long branched hairs. According to Konar and Linskens (1966a), the stigma in *Petunia* comprises two zones — an upper secretory zone of epidermis and the lower storage zone. The epidermal cells are covered with the cuticle. Mattson et al. (1974) observed the presence of an extracuticular proteinaceous layer (pellicle) on the stigma papillae of many angiosperms. The pellicle remains intact in fresh stigma, but forms cracks and fissures on older stigma through which the cuticle can be seen after staining with benzpyrene and observing with a fluorescent microscope.

The stigma secretes the stigmatic fluid, which in *Petunia* consists of oil, sugars, and amino acids (Konar and Linskens, 1966b). The protein contents of the stigmatic exudate are specific recognition factors that interact with proteins released from the pollen. In case of acceptance or compatible reaction, the pollen germinates, forming a tube, which penetrates the cuticle of the stigma, whereas rejection is indicated by the failure of pollen to germinate. Heslop-Harrison et al. (1975) using fluorochromatic and immunofluorescence methods have concluded that the stigma pellicle is the site of recognition responses. In compatible reactions, the ejected pollen proteins bind very quickly to the pellicle, and it is in the binding zone that erosion of the cuticle takes place to facilitate pollen tube entry. In *Raphanus* (Dickinson and Lewis, 1973) and *Iberis* (Heslop-Harrison, Knox, and Heslop-Harrison, 1974), the rejection reaction is signified by the deposition of callose in both the pollen tube and the stigma papillae within 4 to 6 hours of pollination. Dickinson and Lewis (1975) observed that in *Raphanus* the incompatible pollen tubes may penetrate the papillar wall through enzymatic action, but soon after penetration a lenticular callose reaction body is formed in the region of entrance in response to penetration by the incompatible tube. Pollen tube growth ceases after formation of the reaction body.

Recently, Atkinson et al. (1993) and Anderson et al. (1996) detected proteinase inhibitors in *Nicotiana alata* stigmas, and they believe that these may be involved in protecting the sexual tissues against potential predators and pathogens. Anderson et al. (1996) have found that the most abundant defense-related molecules in the stigma of *N. alata* are a series of serine proteinase inhibitors.

2.4 NECTARIES

The nectaries occur on flowers (floral nectaries) and on vegetative parts (extrafloral nectaries). The floral nectaries are found in most insect- and bird-pollinated flowers. Fahn (1953) has classified nectaries on the basis of their distribution on floral parts into perigonial (perianth), toral (torus or receptacle), staminal, ovarial (Figure 2.5 C), and stylar nectaries. A nectary is usually composed of small cells with thin walls, relatively large nuclei, dense cytoplasm, and small vacuoles. The tissue contains branches of vascular bundles with a high proportion of phloem elements (Frei, 1955; Frey-Wyssling, 1955). According to Agthe (1951), the type of vascular tissue present in the nectariferous tissue is correlated with the sugar concentration of the nectar. If the sugar concentration in the nectar is high, the ends of vascular branches consist only of phloem. When nectaries produce nectar with low sugar concentration, the branch endings comprise equal amounts of xylem and phloem.

The exudation of nectar from the nectary depends on the structure of the tissue secreting nectar. If the nectar is secreted from the epidermal cells, which lack an observable cuticle, nectar diffuses through the wall. In case the secretory epidermal cells are covered by a cuticle, exudation takes place through the pores in the cuticle, by its rupture or by the cuticle being permeable (Frey-Wyssling, 1933). When the secretion comes from parenchymatous cells, the nectar is collected in the intercellular spaces from where it is exuded through stomata, which remain open as the guard cells are not able to close the opening.

2.5 OVULE

The ovule developing from the placenta in the ovary is the site for the formation of an embryo sac (female gametophyte) and is the forerunner of seed (Figure 2.6A). The primordia for ovules are two- or three-zonate. The three-zonate primordium generally gives rise to a large-size ovule as found in Brassicaceae, Cucurbitaceae, Malvaceae, Euphorbiaceae, Apiaceae, Solanaceae, Poaceae, etc., and the two-zonate primordium forms primarily small-size ovules (Kordyum, 1968; Bouman, 1984).

2.5.1 STRUCTURE AND TYPES OF OVULES

A normal ovule has the funiculus, chalaza, nucellus, and integument (Figure 2.7A). The size of the funicle is variable and when absent, the ovule is sessile. The part of the funicle that gets adnate to the body of the ovule is called the raphe. The opening in the integument forming a passage above the nucellus is called the micropyle. The

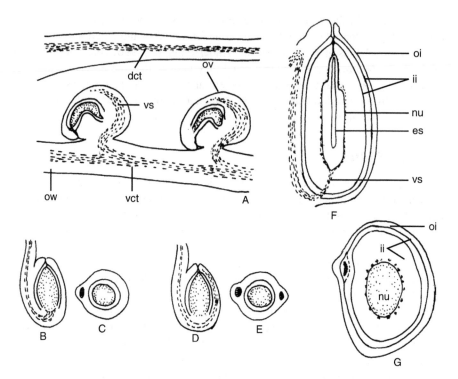

FIGURE 2.6 Vascular supply of ovule. A, Ls of portion of carpel of pea showing vascular supply of developing ovules from the ventral carpellary trace. B, C, Ls and Ts ovule showing raphae vascular bundle terminating at chalaza. D, E, Ls and Ts ovule to show postchalazal development of vascular supply. F, G, Ls and Ts ovule of *Ricinus* showing pachychalaza and pachychalazal bundle. (Abbreviations: dct, dorsal carpellary trace; es, embryo sac; ii, inner integument; nu, nucellus; oi, outer integument; ov, ovule; ow, ovary wall; vct, ventral carpellary trace; vs, vascular supply.) (A, Adapted and redrawn from Hayward, H.E. 1938. *The Structure of Economic Plants*. Macmillan, New York; F and G, From Singh, R.P. 1954. *Phytomorphology* 4: 118–121.)

scar left on the seed after it has separated from the funicle is the hilum. The funicle, raphe (when present), chalaza, and nucellus form a continuous tissue — the central axis of the ovule — and sharp demarcation lines cannot be drawn between them (Figure 2.7A). Generally five types of ovules based on the turning of ovule on stalk and funicle and curvature in its body are recognized. These are atropous (orthotropous), hemitropous (hemianatropous), anatropous, campylotropous, and amphitropous. The anatropous ovule is common in the angiosperms. Bocquet (1959) regards orthotropous and anatropous ovules as basic types, each forming a series of derived campylotropous and amphitropous conditions. The ontogeny and the form of vascular supply in fully developed ovule, straight or curved, permit the identification of ovules in the two series, i.e., ortho-campylotropous and ortho-amphitropous, and ana-campylotropous and ana-amphitropous types. If the body of the ovule takes a complete turn on its stalk and funicle so that the micropyle faces upward and the

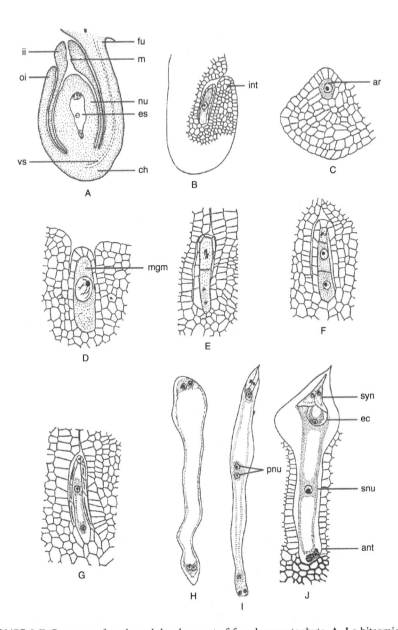

FIGURE 2.7 Structure of ovule and development of female gametophyte. A, Ls bitegmic and crassinucellar ovule. B to J, *Sesamum indicum*. B, Ls unitegmic and tenuinucellar ovule. C, Ls ovular primordium with female archesporium. D to F, Ls parts of ovules with megaspore mother cell, dyad with dividing nuclei and triad with nucleus of the micropylar dyad cell in dividing stage. G to J, Two-, four-, and eight-nucleate, and organized embryo sacs. (Abbreviations: ant, antipodal cells; ar, archesporium; ch, chalaza; ec, egg cell; es, embryo sac; fu, funiculus; ii, inner integument; int, integument; m, micropyle; mgm, megaspore mother cell; nu, nucellus; oi, outer integument; pnu, polar nuclei; snu, secondary nucleus; syn, synergid; vs, vascular supply.) (B to J, From Singh, S.P. 1960. *Phytomorphology* 10: 65–82.)

long funicle almost completely surrounds it, the ovule is called the circinotropous type.

The ovules are also classified as crassinucellar and tenuinucellar. In the former, parietal layers exist between the nucellar epidermis and the female gametophyte (Figure 2.7A), whereas in the latter the nucellar epidermis alone covers the female gametophyte (Figure 2.7B, E). The tenuinucellar ovules are characteristic of sympetalae (*Lycopersicon, Solanum, Sesamum, Helianthus,* and *Lactuca*) while other dicotyledons (*Gossypium, Hibiscus, Glycine, Vigna, Phaseolus, Pisum, Cucurbita,* and *Cucumis*) and monocotyledons (*Triticum, Hordeum, Zea, Oryza,* and *Allium*) have crassinucellar ovules.

The ovules have one (Figure 2.7B) or two integuments (Figure 2.7A). In bitegmic ovules, the inner integument is largely of dermal origin except in Euphorbiaceae (Bor and Bouman, 1974) and Malvaceae (Joshi, Wadhwani, and Johri, 1967; Kumar and Singh, 1990), but the outer integument or the single integument is of subdermal or dermal origin. The integument arises as a rimlike outgrowth and grows to enclose the nucellus to various extents, having an opening at the apex, the micropyle. The micropyle may be formed by both the integuments or by the outer or inner integument alone. Electron microscopy has shown that the micropyle in *Beta vulgaris* contains a fibrillar periodic acid-Schiff positive substance and is often covered by a thin sheet or hymen (Olesen and Bruun, 1990).

2.5.2 Vascular Supply of Ovule

The ovule in angiosperms commonly receives a single vascular bundle from the ventral carpellary vein (Figure 2.6A), rarely two or more in *Sechium and Sicyos* (Puri, 1954; Singh, 1965). The vascular supply ends as such or after fanning out in the chalaza (Figure 2.6B, C). The ovular supply may extend in the outer (Cucurbitaceae and Fabaceae) (Figure 2.6D, E) or the single integument (Asteraceae and Convolvulaceae) on antiraphe side and rarely in the inner integument as in Euphorbiaceae (Figure 2.6F, G). Integumentary vascularization is rare in monocotyledons.

Large ovules usually have a developed vascular supply, whereas small ones show reductions. Very small ovules as in *Orchidaceae* have no trace of vascular supply. The integumentary vascular supply may be unbranched, or branched in some cases, forming a network of bundles (Kuhn, 1928). Vascular elements, xylem and phloem, may become differentiated in later developmental stages, or it may not take place, and the vascular supply consists of procambial strands only.

Nucellar tracheids are known in the Asclepiadaceae, Capparidaceae, Casuarinaceae, Amantiferae, Liliaceae, and Olacaceae. The tracheids are annular or spiral, very small, slender, and usually isolated or in small clusters.

2.5.3 Cuticles in Ovule

Cuticles are reported to be present in ovules from early stages of development. The ovule primordium bears a cuticle. After the development of the integuments in a bitegmic and crassinucellar ovule, as many as five cuticular layers may be identified: (1) on the outside of the outer integument and the funiculus; (2) on the inside of the outer integument; (3) on the outer side of the inner integument; (4) on the inside of the inner integument; and (5) on the nucellus. In unitegmic crassinucellar ovules, the cuticles for the single integument and the nucellus are present. In unitegmic and tenuinucellar ovules, however, since cells of the nucellus are disorganized early, the cuticle may also be lost.

2.5.4 SPECIAL STRUCTURES IN OVULES

The ovule may have the development of some structures that continue differentiation in the developing seed. Epistase is organized at the micropylar end of the nucellus. It is a caplike structure formed from the nucellar epidermis or its derivatives. Its cells become cutinized.

The nucellar cells on the chalazal side form a *hypostase* (van Tiegham, 1901). In literature the term hypostase has been used in a rather loose sense. Its features are diverse in different taxa. The cells are rich in cytoplasm, or accumulate tannin-like substances, or become cutinized, callosic, lignified, or suberised. More than one type of cell may differentiate to form a hypostase as in *Carica* (Dathan and Singh, 1970) and *Passiflora* (Dathan and Singh, 1973). Multiple functions have been attributed to the hypostase. It acts as connecting tissue between the vascular supply and the embryo sac, facilitating the transport of nutrition (Johansen, 1928; Venkata Rao, 1953; Tilton, 1980). When the cells are thick-walled, the hypostase forms a barrier or protective tissue outside the embryo sac.

Another important feature reported in ovules of many families of dicotyledons and monocotyledons is the differentiation of an *integumentary tapetum* or *endothelium*. In tenuinucellar ovules, the nucellus disorganizes early and the megaspore mother cell or the female gametophyte is surrounded by the inner epidermis of the integument. Its cells enlarge radially and become densely cytoplasmic (Figure 2.7G, J). The endothelium is common in unitegmic ovules and rarely formed in bitegmic ovules (*Linaceae*). In the latter, the cells of the inner epidermis of the inner integument form the endothelium. The endothelial cells are separated from the embryo sac by a cuticle whose thickness varies at different sites and alters during seed development (Erdelska, 1975). It is suggested that the endothelium translocates nutrients, derived from the integumentary tissue to the embryo sac (Esser, 1963; Masand and Kapil, 1966).

2.6 DEVELOPMENT AND STRUCTURE OF FEMALE GAMETOPHYTE

Usually one, rarely more, archesporial cells differentiate in the hypodermis of the nucellus due to their large size, prominent nucleus, and dense cytoplasm (Figure 2.7C). With or without cutting of a parietal cell, the female archesporium functions as a megaspore mother cell (Figure 2.7D). The megaspore mother cell undergoes meiosis, forming four megaspores (Figure 2.7E, F). The three micropylar megaspores degenerate and the chalazal megaspore functions (Figure 2.7G). The nucleus undergoes three mitotic divisions to form eight nuclei (Figure 2.7G to I). Three of the nuclei at the micropylar end form the egg apparatus, an egg and two synergids, while three nuclei at the chalazal end form the antipodal cells. One nucleus from each end moves to the center to form the polar nuclei. Thus an eight-nucleate, seven-celled female gametophyte or embryo sac is formed (Figure 2.7I, J). This type of embryo sac development in which only one megaspore forms the embryo sac is known as the monosporic type. When two or all four megaspore nuclei form the embryo sacs, they are known as bisporic and tetrasporic types, respectively. Each of these categories has a number of subtypes (Maheshwari, 1950, 1963).

The female archesporium and megaspore mother cell have plasmodesmatal connections with the nucellar cells. Callose is deposited during megasporogenesis. It develops in species with monosporic and bisporic embryo sacs, but is absent in tetrasporic forms (Rodkiewicz, 1970). In the monosporic polygonum type of embryo sac, the callose disappears completely from the chalazal functional megaspore. Plasmodesmata are usually absent during callose development between megaspores and the nucellus. Cytoplasmic connections are also absent in the transverse walls of cells in dyads and tetrads. The functional megaspore and two- and four-nucleate embryo sacs have plasmodesmatal connections to the nucellus (Sehulz and Jensen, 1986; Folsom and Cass, 1988, 1989).

The ultrastructural studies of the embryo sac have shown that the egg and synergids are surrounded by a wall at the micropylar end. The lower one third part of these cells is surrounded by the plasma membrane only (Jensen, 1965a, b; Folsom and Peterson, 1984; Maeda and Maeda, 1990; Willemse and van Went, 1984). Yan, Yang, and Jensen (1991), who have carried out detailed investigations on the developing embryo sac of *Helianthus annuus*, have observed that in the young embryo sac, two days before anthesis, egg, synergids and central cell were completely surrounded by walls. The chalazal portion of the walls of the egg, synergids, and the micropylar part of the central cell disappeared one day before anthesis. The chalazal and lateral wall of the central cell remained intact and became thick.

The synergids are characterized by the presence of a filiform apparatus at the micropylar end. It consists of a mass of wall projections extending deep into the cytoplasm. The presence of plasmodesmata between the synergid and the central cell are reported in a few species (Morgensen and Suthar, 1979; Wilms, 1981; Vijayraghavan and Bhat, 1983; Willemse and van Went, 1984; Folsom and Peterson, 1984). In addition to plasmodesmata, Wang and Wang (1991) have observed some vesicles between the synergids and central cell and believe that these may be involved in the transport of metabolites between synergids and the central cell.

The plasma membrane and the cell wall of the egg are usually intact. Plasmodesmata are rarely observed between egg and the central cell cytoplasm in the micropylar region in maize and rice (van Lammereen, 1986; Maeda and Maeda, 1990). Maeda and Maeda (1990) consider that the egg cell has transmembranal and symplastic connections with the central cell, but apoplastic connections with nucellar cells.

The antipodal cells are usually small and ephemeral but in Poaceae they multiply, become many-celled and persist during the early stages of grain formation. These cells develop wall invaginations on the side adjacent to nucellar cells (Newcomb, 1973; Maze and Lin, 1975; Wilms, 1981; Engell, 1994). The presence of plasmodesmatal connections between antipodal cells and the surrounding nucellar cells has been observed in *Capsella* (Schulz and Jensen, 1972) and *Helianthus* (Newcomb, 1973).

The embryo sac wall lacks plasmodesmatal connections with the surrounding tissue (Orel and Shmaraev, 1987; Folsom and Cass, 1988; Johansson and Walles, 1993). Its wall shows the formation of protuberances on the internal surface and microinvagination of plasmalemma (Figure 2.12A; see p. 30). These formations are similar to those of the transfer cells involved in short-distance transport of metabolites.

2.7 FERTILIZATION

Double fertilization is characteristic of angiosperms. The pollen tube is porogamous — entering the ovule through the micropyle (Figure 2.8A); chalazogamous — entering through the chalaza (Figure 2.8B); and, very rarely, mesogamous — penetrating laterally through integuments between the micropyle and chalaza. In almost all the economically important plants, the pollen tube enters the ovule through the micropyle and is porogamous. Under all the above conditions, the entry of the pollen tube into the embryo sac is at the micropylar end. The pollen tube enters through one of the synergids (Figure 2.8C), and the sperms are released in the cytoplasm of the synergid through a subterminal or terminal pore (Cass and Jensen, 1970; Jensen, 1973). One sperm comes in contact with the egg plasma membrane and the other sperm come in contact with the plasma membrane of the central cell (Figure 2.8D, E). An opening in the fused plasma membranes of egg and sperm allows the entry of the sperm nucleus into the egg. Similarly, the other sperm nucleus enters the central cell. There is no evidence of the entry of the pollen tube or sperm cytoplasm in the egg or in the central cell.

The sperm nucleus fuses with the egg nucleus. Three types of karyogamy — premitotic, postmitotic, and intermediate — are involved in their fusion (Gerassimova-Navashina, 1960). The phenomenon is known as syngamy. The other male nucleus, which comes to lie close to the polar nuclei, or their fusion product, the secondary nucleus, fuses with them to achieve triple fusion, forming the triploid primary endosperm nucleus.

2.8 SEED DEVELOPMENT

Pollination and fertilization provide the stimulus for the development of fruit and seed. The fertilized ovule undergoes changes in all parts during seed formation. The ovule enlarges, the zygote forms the embryo, and the primary endosperm cell with the triploid nucleus forms the endosperm, which is required to provide nutrition to the developing embryo. The nucellus, chalaza, and the integuments undergo varied degrees of structural change. The development of seed is relatively constant in a group. A general account of the development of the endosperm, embryo, and other parts of the ovule is given here.

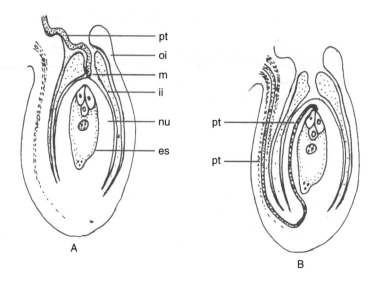

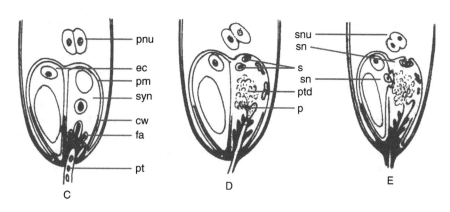

FIGURE 2.8 Course of pollen tube in ovule and embryo sac and the transfer of sperms. A, Ls ovule showing pollen tube entry through micropyle (porogamy). B, Ls ovule showing pollen tube traversing along the raphe bundle and entering nucellus through the chalazal end and the embryo sac at the micropylar end (chalazogamy). C to E, Entry of pollen tube into one of the synergids (C), release of its contents into the synergid (D), and subsequent movement of the two male gametes toward the egg and central cell. (Abbreviations: cw, cell wall; ec, egg cell; es, embryo sac; fa, filiform apparatus; ii, inner integument; m, micropyle; nu, nucellus; oi, outer integument; p, pore; pm, plasma membrane; pnu, polar nuclei; pt, pollen tube; ptd, pollen tube discharge; s, sperm; sn, sperm nuclei; snu, secondary nucleus; syn, synergid.) (C to E, From Jensen, W.A. 1973. *Bioscience* 23: 21–27. With permission.)

2.8.1 ENDOSPERM

The primary endosperm nucleus divides mitotically, and depending on the manner of wall formation during its development, three main types of endosperms are recognized. (1) If the first few divisions are not followed by wall formation and nuclei lie free in the cytoplasm, the endosperm is of the *nuclear type* (Figure 2.9A, B). (2) If the first and subsequent nuclear divisions are followed by cell wall formation, the endosperm is of the *cellular type* (Figure 2.9C, D). (3) It is in the intermediate type, in which the first nuclear division is followed by cell wall formation, that two unequal chambers are usually formed. The micropylar chamber is larger, and its nucleus divides repeatedly without wall formation. In the small chalazal chamber, the nucleus may undergo a few free nuclear divisions or it may remain undivided. This endosperm is of the *helobial type*. During further development, cell wall formation takes place in the micropylar chamber. Nuclear endosperms also ultimately become cellular. The endosperm development is of the nuclear type in most of the food plants (Brassicaceae, Malvaceae, Tiliaceae, Fabaceae, Cucurbitaceae, Apiaceae, Solanaceae, Asteraceae, and Poaceae) and cellular in *Sesamum* (Pedaliaceae).

The female gametophyte remains sac-like, and food material is translocated into it from the surrounding tissue. In several families, portions of endosperm in the chalazal and/or micropylar regions undergo tubular elongation to form endosperm haustoria, which function to draw nutrition. Endosperm haustoria have been reported in the nuclear as well as the cellular types of endosperms. Among the families of food plants, some genera of Fabaceae, Cucurbitaceae, and Pedaliaceae (sesame) develop haustoria. In Cucurbitaceae and Fabaceae, a chalazal, tubular, and coenocytic or cellular haustorium with dense cytoplasm, has been reported. The coenocytic haustoria in Cucurbitaceae (Chopra, 1955; Singh, 1957) and Fabaceae (Rau, 1953), when studied in living materials, show protoplasmic streaming. Dute and Peterson (1992), who studied the endosperm development in soybean, have reported the development of wall ingrowths in the chalazal endosperm haustorium, providing evidence that the haustorium functions to absorb nutrients from the surrounding nucellar tissue that shows lysis. Johansson and Walles (1993) report wall ingrowths along the whole embryo sac boundary in faba bean (Figure 2.12A).

The endosperm tissue is unique to angiosperms. It is rich in food materials, and in many plants almost the entire endosperm is absorbed by the developing embryo (Fabaceae and Cucurbitaceae), while in others it is persistent (Poaceae, Solanaceae, Euphorbiaceae, *Sesamum, Linum*, Malvaceae). However, little is known regarding the time when the embryo begins to utilize the endosperm for its nutrition. Recent studies have shown that the endosperm, during the early stages of development, needs adequate nutrients for its growth, and only in late embryogeny does it have a pool of reserve materials, which are utilized for the growth of embryo (Newcomb, 1973; Yeung and Clutter, 1978).

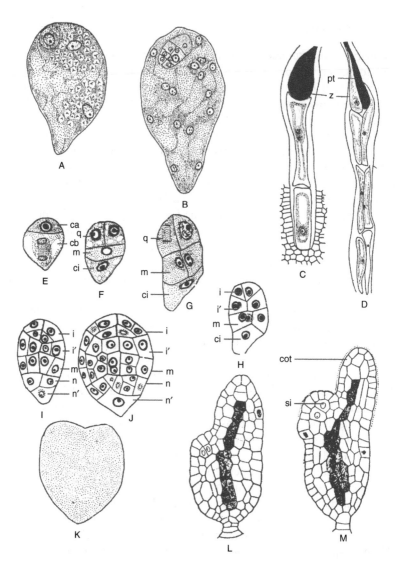

FIGURE 2.9 Development of endosperm and embryo. A, B, *Arachis hypogaea*. Embryo sacs with two and several endosperm nuclei (nuclear type). C, D, *Sesamum indicum*. Ls embryo sac to show initial stages in cellular type of endosperm development. E to K, Development of embryo in *Viola tricolor*. Note the bilateral symmetry in cordate embryo (dicotyledon type). L, M, Late stages from the development of embryo in *Najas lacerata* (monocot) showing differentiation of shoot initials on one side and single cotyledon in terminal position. (Abbreviations: cot, cotyledon; pt, pollen tube; si, shoot initials; z, zygote; the rest are the usual abbreviations used for terms that trace embryo development.) (A, B, From Prakash, S. 1960. *Phytomorphology* 10: 60–64; C, D, From Singh, S.P. 1960. *Phytomorphology* 10: 65–82; E to K, Singh, D. 1963. *J. Indian Bot. Soc.* 42: 448–462; L, M, Redrawn from Swamy, B.G.L. and Krishnanurthy, K.V. 1980. *From Flower to Fruit — Embryology of Flowering Plants.* Tata McGraw-Hill, New Delhi.)

2.8.2 EMBRYO

The zygote cytoplasm becomes homogeneous, the vacuole disappears, and the cell organelles show distinct polarization. It develops the cell wall all around. It usually divides after the primary endosperm nucleus has undergone several divisions. The divisions in the zygote follow a schematic order, and the pattern is characteristic of a species. During the development of the embryo, two main stages are identified: the formation of the proembryo, and the formation of the embryo proper. The mode of development in dicotyledons and monocotyledons has been shown in *Viola tricolor* and *Najas lacerata*, respectively (Figure 2.9E to M). The development of the proembryo is similar, maintaining bilateral symmetry in two groups. Main embryo types are recognized depending on the plane of divisions in cells of the two-celled proembryo and the contribution made by cells of the four-celled proembryo in the formation of parts of the embryo proper (Souéges, 1932; Johansen, 1950; Maheshwari, 1950; Crété, 1963). The proembryo may have a conspicuous or an inconspicuous suspensor. For detailed information on embryo development in angiosperms, the reader should refer to *Plant Embryology* by Johansen (1950).

During the development of the embryo proper and the differentiation of organs, bilateral symmetry is maintained throughout in dicotyledons (Figure 2.9E to K), while in monocotyledons, the globular proembryo acquires unilaterality. The differentiation of structures takes place along one face of the developing embryo; the other remains smooth and barren (Figure 2.9L, M). The mature embryo comprises an embryonal axis made up of the plumule (epicotyl, shoot apex), the hypocotyl — radicle axis — and one or two cotyledons. The shape and size of the embryo and the cotyledons show considerable variation in dicots and monocots. The cotyledons are thick or thin, straight or curved, with margins smooth or curled. The shoot apex lies between the two cotyledons in dicots (Figure 2.10A) and by the side of the single cotyledon in monocots (Figure 2.10B).

Generally, the embryo in seed is fully differentiated into the radicle, the plumule (epicotyl), and the cotyledons, but in some plants, it is reduced and lacks differentiation. The coiled embryo of *Cuscuta* is devoid of cotyledons and radicle. The embryo in Orobanchaceae (Tiagi, 1951, 1963), Orchidaceae (Poddubnaja-Arnoldi, 1967), and several other families, mostly of parasitic and saprophytic flowering plants, is without the differentiation of organs or even tissue.

The structure of the embryo is unique in *Poaceae*. The single cotyledon is shield-shaped and called the scutellum. The embryo has a few additional organs, viz., the coleoptile, the coleorhiza, and the epiblast (Figure 2.10C). The epiblast is absent in *Zea* and *Sorghum* embryos.

The basal part of the embryo that does not participate in the formation of embryo proper is known as the suspensor. It is inconspicuous in most angiosperms, but in Fabaceae (Maheshwari, 1950; Johri, 1984), Brassicaceae (Maheshwari, 1950), Orchidaceae (Poddubnaja-Arnoldi, 1967), Crassulaceae, and Rubiaceae (Bhatnagar and Johri, 1972), the suspensors are well developed and probably haustorial. The ultrastructure of the developing embryo of soybean shows wall invaginations (Figure 2.11A to C) in cells of the suspensor (Dute et al., 1989). Wall ingrowths in suspensor cells have been observed in several species (Schulz and

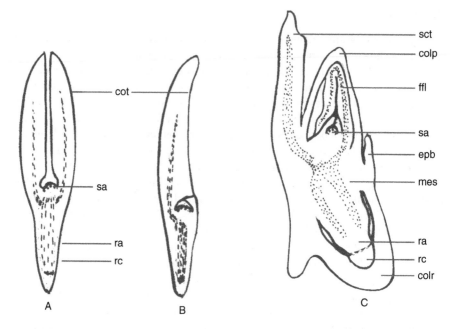

FIGURE 2.10 Diagrams of dicot A, monocot B, and grass C, embryos to show the relative position of cotyledons and plumule, and additional structures, coleoptile, coleorhiza, and epiblast in grass embryo. (Abbreviations: colp, coleoptile; colr, coleorhiza; cot, cotyledon; epb, epiblast; ffl, first foliage leaf; mes, mesocotyl; ra, radicle; rc, root cap; sa, shoot apex; sct, scutellum.)

Jensen, 1969; Newcomb and Fowke, 1974; Yeung and Clutter, 1979). The suspensor in *Phaseolus coccineus* comprises nearly 200 cells. Experimental studies in *P. coccineus* after administration of ^{14}C-sucrose to excised pods and seeds have provided direct evidence that the suspensor is the preferred major site of uptake of metabolites for the embryo, particularly at the globular-heart shaped stages (Yeung, 1980).

Plasmodesmata have been reported in the cell walls between the suspensor and endosperm (Figure 2.12C). They are also abundant between individual suspensor cells (Figure 2.12B) and between suspensor cells and the embryo proper in faba bean (Johansson and Walles, 1993). The distribution of wall ingrowths and plasmodesmata shows the occurrence of apoplastic as well as symplastic pathways for transport of nutrients from the endosperm to the embryo through the suspensor.

2.8.3 Changes in Nucellus

The nucellus in tenuinucellar ovules is absorbed during the prefertilization stages, and the developing female gametophyte comes in direct contact with the inner epidermis of the inner integument or that of the single integument. In crassinucellar ovules, the nucellar cells are polygonal and rich in cytoplasmic contents in unfertilized ovules. After fertilization, as the endosperm and the embryo grow, and the

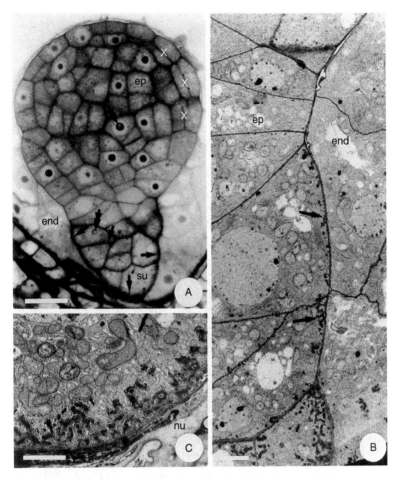

FIGURE 2.11 Young embryo of *Glycine max*. A, Light photomicrograph of embryo differentiated into embryo proper and suspensor with wall ingrowths in cells of suspensor (arrows). B, Transmission electron photomicrograph at the junction between the embryo proper, suspensor (arrows), and the endosperm cells. Note that wall ingrowths are associated with the suspensor and not with the embryo proper. Also the concentration of ingrowths increases toward the tip of the suspensor. C, Wall ingrowths in tip cell where suspensor abuts the crushed nucellar cells. (Abbreviations: end, endosperm; ep, embryo proper; nu, remnants of nucellus; su, suspensor.) (From Dute, R.R., Peterson, C.M., and Rushing, A.E. 1989. *Ann. Bot.* 64: 123–135. With permission.)

nucellar cells enlarge and are digested from inside to outside. In most species the nucellus is represented in seed by a few greatly compressed peripheral layers. The persistent nucellar remains are covered by a cuticle.

The nucellus in Amaranthaceae, Chenopodiaceae, Polygonaceae, Piperaceae, and Scitamineae increases in volume, stores reserve food materials, and persists. It is called the perisperm and serves as an accessory nutritive tissue, supplementing the endosperm.

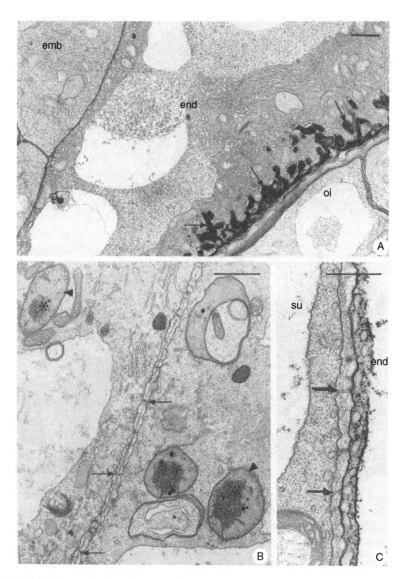

FIGURE 2.12 Transmission electron microscope photomicrographs showing contacts between embryo sac and integument, and embryo in *Vicia faba*. A, Part of the embryo sac adjacent to the outer integument with wall ingrowths in embryo sac boundary. B, Suspensor cells with plastids (arrowheads) and plasmodesmata (arrows) in the wall between the individual cell. C, Plasmodesmata (arrows) in the wall between the endosperm and a suspensor cell. (Abbreviations: emb, embryo; end, endosperm; oi, outer integument, su, suspensor.) (A to C, From Johansson, M. and Walles, B. 1993. *Int. J. Plant Sci.* 154: 535–549. With permission.)

2.8.4 CHANGES IN CHALAZA

The chalaza may or may not undergo significant changes in seed. In the former condition it is punctiform or unspecialized. However, in Euphorbiaceae, Asteraceae, Meliaceae, Annonaceae, and Myristicaceae, the chalaza undergoes general amplification contributing a significant portion of the mature seed. Such seeds are called *pachychalazal* (Periasamy, 1962). If the growth in chalaza is localized, forming a band or hoop around the nucellus, the chalaza is called the *perichalaza* (Corner, 1976).

The chalazal tissue may undergo differentiation similar to that taking place in the integuments (outer or single one), or it may have independent changes. In most seeds the epidermis in this region becomes thick walled and lignified.

2.8.5 CHANGES IN INTEGUMENTS

The ovules are bi- or unitegmic. In a bitegmic ovule, both integuments form the seed coat in Brassicaceae (Rathore and Singh, 1968; Prasad, 1974; Harris, 1991), Malvaceae (Joshi, Wadhwani, and Johri, 1967; Kumar and Singh, 1990), Euphorbiaceae (Singh, 1954), Amaranthaceae, Chenopdiaceae (Taneja, 1981), Linaceae (Boesewinkel, 1980), and Liliaceae (Sulbha, 1954), whereas in many families the inner integument degenerates and the outer integument alone forms the seed coat, e.g., Fabaceae (Corner, 1976; Pandey and Jha, 1988) and Cucurbitaceae (Singh, 1953; Singh and Dathan, 1972, 1990). In Poaceae, it is the inner integument that forms the seed coat, the outer one degenerates (Bradbury, Cull, and MacMarters, 1956; Bradbury, MacMasters, and Cull, 1956; Naryanaswami, 1953; Chandra, 1963, 1976). Seeds in some of the Poaceae are without a seed coat.

The cells of the integuments divide anticlinally to keep pace with the growing seed. This is also attained through the enlargement of cells. The integuments may or may not undergo periclinal divisions, and they are called multiplicative and nonmultiplicative, respectively. Both outer and inner integuments are nonmultiplicative in Amaranthaceae, Chenopodiaceae, and Liliaceae; the outer integument is multiplicative in Fabaceae and Cucurbitaceae; and the inner integument is multiplicative in Malvaceae, Euphorbiaceae, and Linaceae. Multiplication in cell layers may be diffuse, spread all over, or be localized, confined to certain zones. The most striking example of localized multiplication is Cucurbitaceae (Singh, 1953; Singh and Dathan, 1972, 1990) in which the outer epidermis of the outer integument and its derivatives undergo periclinal divisions (Figure 2.15B, G).

During development the integument undergoes differentiation, absorption of cell layers, and thickening of cell walls. In some cases, the deposition of pigmented material, mucilage, tannin, and crystals takes place. One or more layers of seed coat develop characteristic thickenings, forming the main mechanical tissue. The place of origin in the integument and the structure of cells of the main mechanical layer are characteristic in the seeds of a family. Depending on whether the main mechanical layer differentiates in the outer or inner integument, Corner (1976) has classified dicotyledonous seeds formed from bitegmic ovules into *testal* or *tegmic* types. Each of these types is further subdivided into exotestal, mesotestal, and endotestal, and exotegmic, mesotegmic, and endotegmic on the basis of the position of the main mechanical layer in the outer epidermis, middle layers, or inner epidermis of the outer or the inner integument, respectively.

2.9 SEED COAT DEVELOPMENT IN SELECTED GENERA

The seed coat development for some economically important plants, e.g., for true seeds of *Brassica* (Brassicaceae), *Hibiscus, Gossypium* (Malvaceae), *Crotalaria* (Fabaceae), *Lycopersicon* (Solanaceae), *Cucurbita, Sechium* (Cucurbitaceae), and one-seeded indehiscent fruits of *Lactuca* (Asteraceae) and *Triticum (Poaceae)*, is described.

2.9.1 *BRASSICA*

The ovule is bitegmic; initially each integument is two or three-layered. Both the integuments show slight multiplication. The outer integument becomes four-layered (Figure 2.13A). The cells in the inner epidermis of the outer integument elongate radially, and those of the other layers stretch tangentially and lose contents. The cells of the inner epidermis acquire U-shaped thickenings, forming the main mechanical layer in the mature seed coat (Figure 2.13C, D).

The inner integument becomes many-layered (Figure 2.13B), and its cells enlarge and undergo absorption during seed development (Figure 2.13C). The inner epidermis persists and forms the endothelium. Its cells develop pigmented material and some proteinaceous bodies (Vaughan, 1956, 1959; Rathore and Singh, 1968; Prasad, 1974). The mature seed coat consists of layers of the outer integument and the endothelium (Figure 2.13D).

2.9.2 *CROTALARIA*

The ovules are bitegmic (Figure 2.13E, F) but the inner integument disappears during development (Figure 2.13G, H) and the outer integument alone forms the seed coat. The outer integument is multiplicative. The cells of the outer epidermis elongate radially and form the palisade layer of sclereids or macrosclereids (Figure 2.13G to I). This is the main mechanical layer in fabaceous seed.

The cells of the subepidermal layer become columnar, develop unequal thickenings, and form the hourglass cells (Figure 2.13H). The remaining layers remain thin-walled and stretch tangentially, and are partly digested during development (Corner, 1976; Pandey and Jha, 1988).

The differentiation in the hilar region in developing fabaceous seed and its structure at maturity are quite characteristic (Baker and Mebrahtu, 1990), The region consists of rim-aril (when present), counter palisade (differentiated in the funiculus), hilar fissure, palisade layer (macrosclereids), tracheid bar, stellate parenchyma, and aerenchyma (Figure 2.13I). The development of hilar fissure, counter palisade, tracheid bar, and stellate parenchyma begins quite early and is nearly completed by the time 35% maximum seed size is attained in soybean (Baker and Mebrahtu, 1990).

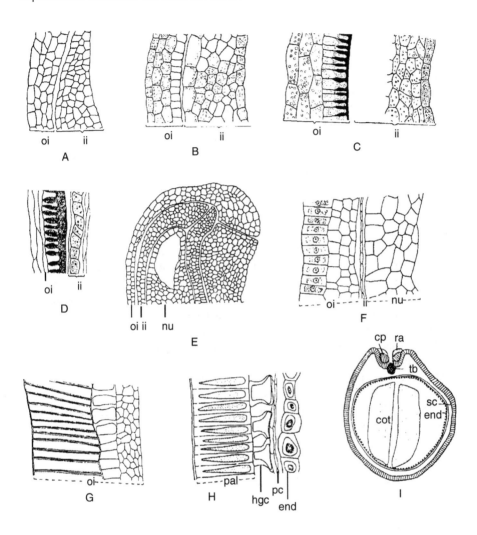

FIGURE 2.13 Development and structure of seed coat in Brassicaceae and Fabaceae. A to D, *Brassica campestris* var. yellow sarson. A to C, Ls portions of seed coat at various stages of ovule and seed development showing disintegration of layers of the inner integument and the development of characteristic thickenings in inner epidermis of outer integument in C. D, Ls part of mature seed coat. E to I, *Crotalaria varrucosa*. E, Ls micropylar part of ovule at organized embryo sac stage. F, Ls part of developing seed coat. G, H, Ts developing and mature seed coat showing palisade cells and hourglass cells. I, Ts mature seed with rim aril, counter palisade, and tracheid bar in the hilar region. (Abbreviations: cot, cotyledon; cp, counter palisade layer; end, endosperm; es, embryo sac; hgc, hourglass cells; ii, inner integument; oi, outer integument; pal, palisade layer; pc parenchyma cells; ra, rim aril; sc, seed coat; tb, tracheid bar.) (A to D, From Rathore, R.K.S. and Singh, R.P. 1968. *J. Indian Bot. Soc.* 47: 341–349; E to I, From Pandey, A.K. and Jha, S.S. 1988. *Flora* 181: 417–426.)

2.9.3 HIBISCUS AND GOSSYPIUM

The outer integument is three-layered in *H. ficulneus* and four- to six-layered in *Gossypium* (Figure 2.14A, B, H). It is nonmultiplicative. The inner integument is initially three-layered and multiplicative, becoming seven- to nine-layered in *Hibiscus* (Figure 2.14C, D) and eight- to fifteen-layered in *Gossypium* (Figure 2.14H). Both the integuments form the seed coat. The cells of the outer epidermis of the outer integument elongate tangentially and lose their contents, but hair initials remain squarish and finally undergo radial enlargement into hairs (Figure 2.14H). Stomata (Figure 2.14F, G) are found in the epidermis of the entire organism, but they are aggregated more in the chalaza. The middle layers are almost squeezed in *Hibiscus*, but they are differentiated into an outer zone of pigmented cells and an inner zone of colorless cells in *Gossypium* (Figure 2.14I). The cells in the inner epidermis elongate radially, lose their contents, and develop thickenings on the inner tangential and radial walls in *H. ficuleuns* (Figure 2.14E), but they remain thin-walled in *Gossypium*.

The inner integument differentiates into four zones. The cells of the outer epidermis enlarge radially and become thick-walled, forming the main mechanical layer (Figure 2.14E, I). The cells of four or five subdermal layers enlarge, become thick-walled, and accumulate tanniferous contents. The remaining middle layers remain thin-walled, lose their contents, and are partially digested, forming the colorless zone (Figure 2.14C, D, and I). The cells of the inner epidermis stretch tangentially and accumulate tanniferous contents in *Hibiscus* (Kumar and Singh, 1990), but the inner epidermis develops characteristic thickenings in *Gossypium*, called the fringe layer (Joshi, Wadhwani, and Johri, 1967).

The development of the seed coat in *Gossypium* is similar to *Hibiscus*, but the lint and fuzz hairs are distributed all over the surface of the seed.

2.9.4 CUCURBITA AND SECHIUM

The ovules are bitegmic, and the two- to four-layered inner integument (Figure 2.15A, F) is absorbed during ontogeny. The seed coat develops from the outer integument, which shows localized multiplication. By two successive periclinal divisions the outer epidermis forms three layers, which are designated e, e″, and e′, from outside to inside (Figure 2.15B, G). The second division does not take place throughout the surface in *Sechium*. The subepidermal layer e is, therefore, discontinuous in *Sechium* (Figure 2.15H). The cells in e divide only anticlinally to keep pace with the growing seed and later on enlarge tangentially and become thick-walled (sclerotic), forming the main mechanical layer in *Cucurbita*. The cells of e and e″ divide periclinally and the outermost derivatives enlarge radially, forming seed epidermis. The remaining layers become thick-walled and lignified, forming the hypodermis (Figure 2.15C to E).

The cells of the ovular hypodermis divide once or twice periclinally, enlarge, become stellate, and develop prominent air spaces to form aerenchyma. They become thick-walled and lignified in *Cucurbita*. The cells of other layers enlarge and those on the outer side develop air spaces while those on the inner side remain compact, forming a chlorenchymatous or parenchymatous zone. This tissue at maturity is separated from the outer hard coat and forms the inner coat (Singh and Dathan, 1972).

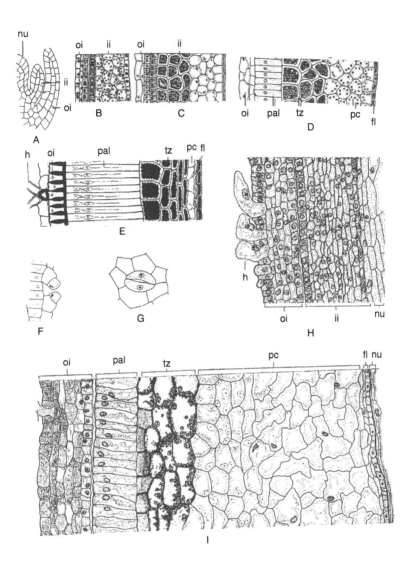

FIGURE 2.14 Development of seed coat in *Hibiscus* and *Gossypium*. A to F, *H. ficulineus*. A, Ls part of ovule showing outer and inner integuments. B to D, Ls integuments at different stages of development showing increase in number of layers in inner integument and radial elongation of outer epidermis. E, Ls part of mature seed coat showing epidermal hairs. F, G, Stomata in epidermis in section and surface view. H, I, *G. arboreum* var. Bhoj Ls part of developing seed showing development of hairs from epidermal cells of the outer integument and differentiation in the inner integument. (Abbreviations: fl, fringe layer; h, hair; ii, inner integument; oi, outer integument; nu, nucellus; pal, palisade cells; pc, parenchyma cells; tz, tannin zone.) (A to G, From Kumar, P. and Singh, D. 1990. *Phytomorphology* 40: 179–188; H, I, From Joshi, P.C., Wadhwani, A.M., and Johri, B.M. 1967. *Proc. Natl. Inst. Sci. India* 33B: 37–93.)

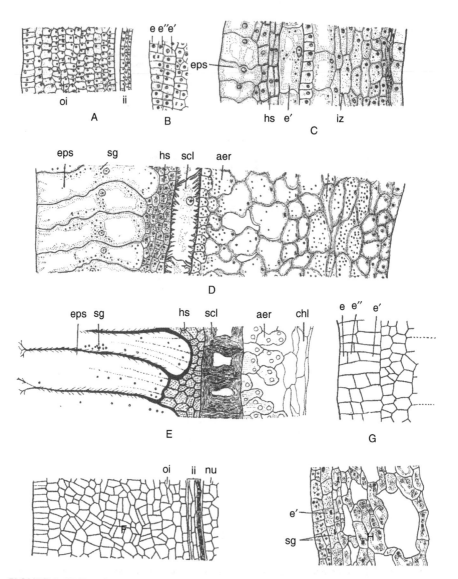

FIGURE 2.15 Development and structure of seed coat in *Cucurbita* and *Sechium*. A to C, *C. pepo*; D, E, *C. moschata*. A, Ls part of integuments. B, Outer region of outer integument showing periclinal divisions in the epidermis to form e, e″, and e′ layers. C, D, Ls part of developing seed coat showing radial elongation of epidermal cells and tangential elongation and initiation of thickenings in e, sclerenchymatous layer. E, Ts part of mature seed coat. F to H, *S. edule*. F, Ls part of integuments. G, Outer region of outer integument showing periclinal divisions in epidermis. H, Ts part of seed coat; cells possess starch grains but lack lignification. (Abbreviations: aer, aerenchyma; chl, chlorenchyma; e, e′, and e″, derivatives of ovular epidermis from outside to inside; eps, seed epidermis; hs, seed hypodermis; ii, inner integument; iz, inner zone of parenchyma cells; nu, nucellus; oi, outer integument; scl, sclerenchymatous layer; sg, starch grains.) (A to E, From Singh, D. and Dathan, A.S.R. 1972. *Phytomorphology* 46: 277–281; F to H, From Singh, D. 1965. *Curr. Sci.* 34: 696–697.)

In *Sechium edule*, the cells of e, e″ (wherever present), and e′ get packed with starch grains, but remain unlignified (Figure 2.15H). The remaining layers become aerenchymatous and acquire starch grains (Singh, 1965). The fruit in *Sechium* is one-seeded, and the seed shows many uncommon features, i.e., leathery testa, vivipary, and hypogeal germination.

2.9.5 LYCOPERSICON

The single integument is multiplicative and differentiates into three zones — the outer epidermis, the mesophyll, and the inner epidermis, which forms the endothelium (Figure 2.16B to D). During the advanced stages of seed development, the cells of the outer epidermis undergo enormous radial elongation. Their inner tangential and radial walls become thick, thickening and tapering from inside to outside on radial walls. The outer wall remains thin and readily separates at maturity (Figure 2.16E). The middle layers develop lysigenous cavities and are gradually absorbed. Only a few hypodermal layers persist. The cells of endothelium (Figure 2.16D, E) are flattened and accumulate pigmented contents (Saxena, 1970).

A succulent envelope (arillode) surrounds the mature seed. The primordium for the envelope originates in the placenta around the funiculus within 72 to 96 hours of pollination (Figure 2.16A). It grows rapidly and gradually surrounds the seed. Finally, the seed is completely surrounded (Figure 2.16F, H). The cells of the arillode are initially polygonal with prominent nuclei, but subsequently enlarge and acquire chloroplasts and starch grains (Figure 2.16 G). At maturity they lose their contents and their cell walls gelatinize.

2.9.6 LACTUCA

The cypsil, dry, one-seeded, indehiscent fruit, forms seed. The pericarp constitutes the protective covering, and the seed coat is thin when present, or it may be completely obliterated. Thus the integument and ovary wall both participate in forming the seed cover.

In *Lactuca*, the integument is six- or seven-layered, multiplicative, and differentiates into three and finally four zones (Figure 2.17A to C). The inner epidermis forms the endothelium, and three or four layers adjacent to it constitute the periendothelium (Figure 2.17C). All the layers except the epidermis and the hypodermis are absorbed. The cells of the periendothelium undergo gelatinization before absorption. The epidermal cells develop cellulose thickenings on the radial walls (Figure 2.17E, F).

The ovary wall comprises eight or nine layers, and its cells stretch tangentially during development. The epidermal cells show pronounced gliding growth, forming spinescent structures. At maturity the pericarp is sinuate and in carinal (ridge) position bears a fibrovascular bundle (Figure 2.17D, E). Other components of the pericarp are the epidermis and one or two layers of narrow tangentially elongated sclereids (Figure 2.17F, H). The size of the epidermal cells, the amount of tannin, and the ratio of their body and spinescent region (Figure 2.17G) vary in different *Lactuca* species (Kaul and Singh, 1982).

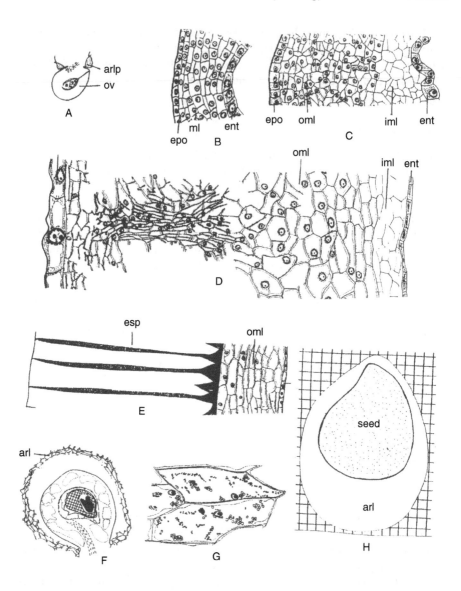

FIGURE 2.16 Development and structure of seed of *Lycopersicon esculentum*. A, Ls ovule at four-celled endosperm stage with arillode primordium (arlp). B to D, Ls part of integument and developing seed coat showing absorption of cells in outer middle layers. E, Ls part of immature seed coat comprising epidermal cells with inner tangential and radial walls lignified; reduced number of middle layers and the persistent endothelium. F, Ls developing seed with the arillode. G, Arillode cells from whole-mount preparation at green fruit stage. H, Diagram of fresh seed on graph to show the relative proportion of seed and arillode. (Abbreviations: arl, arillode; arlp, arillode primordium; ent, endothelium; epo, outer epidermis; eps, seed epidermis; iml, inner zone of middle layers; ml, middle layers; oml, outer zone of middle layers; ov, ovule.) (From Saxena, T. 1970. Ph.D. thesis, University of Rajasthan, Jaipur, India.)

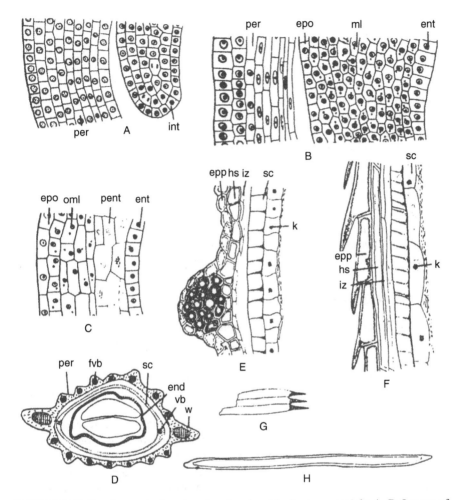

FIGURE 2.17 Development of seed coat and pericarp in *Lactuca scariola*. A, B, Ls part of ovary wall and integument at megaspore mother cell and four-nucleate embryo sac stages, respectively. C, Ls part of integument at organized embryo sac stage showing periendothelium. D, Ts mature cypsil having fibrovascular bundles in ridges. E, F, Ts and Ls portions of pericarp and seed coat, respectively. G, H, Epidermal cells and sclereid fiber from maceration. Note the proportion of the body and spine portions in epidermal cells. (Abbreviations: cot, cotyledon; end, endosperm; ent, endothelium; epo, outer epidermis of integument; epp, epidermis of pericarp; fvb, fibrovascular bundle; hs, hypodermis; int, integument; iz, inner zone of thin-walled cells; k, crystals; ml, middle layers; oml, outer middle layers; pent, periendothelium; per, pericarp; sc, seed coat; vb, vascular bundle; w, wing.) (From Kaul, V. and Singh, D. 1982. *Acta Biologica Cracoviensia* 24: 19–30.)

2.9.7 *TRITICUM*

The caryopsis development in Poaceae has been studied for most of the important genera. The ovule is bitegmic, the outer multilayered integument disintegrates, and the inner integument, which is usually two-layered, is either completely lost or the inner epidermis persists to form the seed coat. In *Triticum* the compressed inner integument contains pigment and is covered by a cuticle.

The ovary wall consists of thin-walled cells and is differentiated into outer epidermis, mesophyll, and inner epidermis. The epidermal cells elongate tangentially, lose their contents, and become thick-walled. The hypodermis also becomes weakly thickened but other mesophyll layers, except the inner epidermis and hypodermis, remain thin-walled and are partially digested. The cells of the inner epidermis enlarge tangentially or along the long axis of the seed and become thick-walled to form the *tube cells*, whereas those of the subepidermal layer elongate at right angles to the tube cells. These cells also become thick-walled and form the *cross cells*. Epicarp, cross cells, and tube cells form the characteristic layers of pericarp in caryopsis (Bradbury, Cull, and MacMasters, 1956; Bradbury, MacMasters, and Cull, 1956).

The pericarp is membranous in *Eleusine*, and both outer and inner epidermis of the inner integument form the protective seed coat. In rice (*Oryza sativa*), the hard siliceous husk is formed by the lemma and palea.

2.10 CONCLUDING REMARKS

A succession of events in the fertile appendages, stamen and carpel, of the flower leads to seed formation. The sterile as well as fertile appendages of the flower have direct connection with the mother plant through vascular traces. The ovular supply is a continuum of the vascular supply of the carpel through the placenta and funiculus. However, neither the pollen sac nor the embryo sac or embryo has any vascular connections with the surrounding sporophytic tissues.

Recent histochemical and electron microscopic studies have revealed several interesting aspects of microsporogenesis and microgametogenesis. Meiocytes have plasmodesmatal connections with the tapetum and among themselves during the early stages of heterotypic prophase. Subsequently, callose deposition isolates the microsporocytes and microspore tetrads from the tapetum and also from one another. The cell wall shows reduced permeability during this period. Callose is dissolved in tetrads, and uninucleate microspores are devoid of it. The male gametes (sperms) are cells, but only their nuclei enter the egg and central cell. The pollen grain walls, exine and intine, contain enzymes (proteins) that play an important role in the penetration of the pollen tube.

The megaspore mother cell has plasmodesmatal connections with the adjoining nucellar cells. Callose deposition takes place during megasporogenesis in monosporic and bisporic forms, but not in tetrasporic forms. Callose causes complete or partial isolation of cells, and it disappears completely from the functional megaspore. Two- and four-nucleate embryo sacs have plasmodesmatal connections with the surrounding nucellar cells, but these are severed in the organized embryo sac. In the organized embryo sac, egg, synergids, and the central cell are only partially surrounded by cell walls and are in contact through the plasma membrane.

Many interesting features, e.g., the presence of the extracuticular proteinaceous layer on stigma, pollen recognition responses between compatible proteins of stigmatic exudate and pollen proteins, penetration of pollen tube into the embryo sac through the synergid, and lack of evidence that the sperm cytoplasm enters the egg or the central cell, are recorded during pollination and fertilization.

The development of the endosperm may be nuclear, cellular, or helobial. Endosperm haustoria are found in several taxa of angiosperms. Available information shows absence of plasmodesmatal connections in the embryo sac wall and the surrounding tissue. Development of protuberances on the internal surface and the microinvagination of plasmalemma, similar to those of transfer cells, are observed. This indicates that the transfer of nutrients from the maternal tissues to the new sporophyte including the endosperm takes place apoplastically.

The development of the proembryo is similar in dicotyledons and monocotyledons. Differences develop during the organization of various parts of the embryo. The proembryo gets its nutrition through its basal suspensor cells, which develop wall ingrowths. During the later stages nutrition is obtained from the endosperm.

Changes take place in the nucellus, chalaza, and integument during seed development. Integuments undergo differentiation, absorption of cell layers, thickening of cell walls, and deposition of pigmented material and crystals etc. One or more layers of the seed coat develop characteristic thickenings, forming the main mechanical layer. The seed coat is testal if the main mechanical layer is formed in the outer integument and tegmic if it differentiates in the inner integument of a bitegmic ovule. Each type has several subtypes. The development and structure of the seed coat is fairly constant in a family.

REFERENCES

Agthe, C. 1951. Über die physiologische Herkunft des Pflanzen-nektars. *Ber. Schweiz. Bot. Ges.* 61: 240–277.

Anderson, M.A., Lee, M., Heath, R.L., Nielson, K.J., Craik, D.J., Guest, D.J., and Clarke, A.E. 1996. Defence-related molecules in the pistil of *Nicotiana alata*. *Plant Reproduction*. 14th International Congress of Sexual Plant Reproduction, Lorne (near Melbourne, Australia), Abstract, p. 1.

Arber, A. 1937. The interpretation of the flower: a study of some aspects of morphological thought. *Biol. Rev.* 12: 157–184.

Atkinson, A.H., Heath, R.L., Simpson, R.J., Clarke, A.E., and Anderson, M.A. 1993. Proteinase inhibitors in *Nicotiana alata* stigmas are derived from a precursor protein which is processed into five homologous inhibitors. *Plant Cell* 5: 203–213.

Baker, D.M. and Mebrahtu, T. 1990. Scanning electron microscopy examination of soybean hilum development. *Am. J. Bot.* 68: 544–550.

Bhatnagar, S.P. and Johri, B.M. 1972. Development of angiosperm seed. In *Seed Biology*. Kozlowski, T.T., Ed. Academic Press, New York, Vol. 1, pp. 77–149.

Bocquet, G. 1959. The campylotropous ovule. *Phytomorphology* 9: 222–227.

Boesewinkel, F.D. 1980. Development of ovule and testa of *Linum usitatissimum*. L. *Acta Bot. Neerl.* 29: 17–32.

Bor, J. and Bouman, F. 1974. Development of ovule and integuments in *Euphorbia milli* and *Codiaeum variegatum*. *Phytomorphology* 24: 280–296.

Bouman, F. 1984. *The Ovule. In Embryology of Angiosperms*. Johri, B.M., Ed. Springer-Verlag, Berlin, pp. 123–157.

Bradbury, D., Cull, I.M., and MacMasters, M.M. 1956. Structure of the mature wheat kernel. I. Gross anatomy and relationships of parts. *Cereal Chem.* 33: 329–342.

Bradbury, D., MacMasters, M.M., and Cull, J.M. 1956. Structure of mature wheat kernel. II. Microscopic structure of pericarp, seed coat and other coverings of the endosperm and germ of hard red winter wheat. *Cereal Chem.* 33: 342–360.

Bruun, L. 1987. The mature embryo sac of the sugar beet, *Beta vulgaris*: a structural investigation. *Nord. J. Bot.* 7: 543–551.

Cass, D.D. and Jensen, W.A. 1970. Fertilization in barley. *Am. J. Bot.* 57: 62–70.

Chandra, N. 1963. Morphological studies in the Graminae. IV. Embryology of *Eleusine indica* Gaertn. and *Dactyloctenium aegyptium* (Desf.) Beauv. *Proc. Indian Acad. Sci.* 58: 117–127.

Chandra, N. 1976. Embryology of some species of *Eragrostis*. *Acta Bot. Indica* 4: 36–43.

Chopra, R.N. 1955. Some observations on endosperm development in the Cucurbitaceae. *Phytomorphology* 5: 219–230.

Corner, E.J.H. 1976. *The Seeds of Dicotyledons*. Vols. 1 and 2. Cambridge University Press, Cambridge, U.K.

Cresti, M., Went, J.L., van Pacini, E., and Willemse, M.T.M. 1976. Ultrastructure of transmitting tissue of *Lycopersicon peruvianum* style. Development and histochemistry. *Planta* 132: 305–312.

Crété, P. 1963. Embryo. In *Recent Advances in the Embryology of Angiosperms*. Maheshwari, P., Ed. International Society of Plant Morphologists, University of Delhi, 171–220.

Dashek, W.V., Thomas, H.R., and Rosen, W.G. 1971. Secretory cells of lily pistils. II. Electron microscope cytochemistry of canal cells. *Am. J. Bot.* 38: 909–920.

Dathan, A.S.R. and Singh, D. 1970. Female gametophyte and seed of *Carica canda-marensis*. Hook. F. *Plant Sci.* 2: 52–60.

Dathan, A.S.R. and Singh, D. 1973. Development and structure of seed in *Tacsonia* Juss and *Passiflora* L. *Proc. Indian Acad. Sci.* 74B: 5–18.

Davis, G.L. 1966. *Systematic Embryology of the Angiosperms*. John Wiley & Sons, New York.

Dickinson, H.G. and Lewis, D. 1973. Cytochemical and ultrastructural differences between intraspecific compatible and incompatible pollination in *Raphanus*. *Proc. R. Soc. London B* 188: 327–344.

Dickinson, H.G. and Lewis, D. 1975. Interaction between pollen grain coating and the stigmatic surface during compatible and incompatible intraspecific pollination in *Raphanus*. In *The Biology of the Male Gamete*. Duckett, J.G. and Racey, P.A., Eds. Academic Press, London, pp. 165–175.

Dute, R.R. and Peterson, C.M. 1992. Early endosperm development in ovules of soybean, *Glycine max* (L.) Merr. (Fabaceae). *Ann. Bot.* 69: 263–271.

Dute, R.R., Peterson, C.M., and Rushing, A.E. 1989. Ultrastructural changes of the egg apparatus associated with fertilization and proembryo development of soybean, *Glycine max* (Fabaceae). *Ann. Bot.* 64: 123–135.

Engell, K. 1994. Embryology of barley. IV. Ultrastructure of the antipodal cells of *Hordeum vulgare*. L. cv. Boni before and after fertilization of the egg cell. *Sex Plant Reprod.* 7: 333–346.

Erdelska, O. 1975. Pre-fertilization development of ovule of *Jasione montana* L. *Phytomorphology* 25: 76–81.

Esser, K. 1963. Bildung und Abbau von Callose in den Samenanlagen der *Petunia hybrida*. *Z. Bot.* 51: 32–51.

Fahn, A. 1953. The topography of the nectary in the flower and its phylogenetical trend. *Phytomorphology* 3: 424–426.

Folsom, M.W. and Cass, D.D. 1988. The characteristics and fate of the soybean inner nucellus. *Acta Bot. Neerl.* 37: 387–394.

Folsom, M.W. and Cass, D.D. 1989. Embryo sac development in soybean: ultrastructure of megasporogenesis and early megagametogenesis. *Can. J. Bot.* 67: 2841–2849.

Folsom, M.W. and Peterson, C.M. 1984. Ultrastructural aspects of the mature embryo sac of soybean, *Glycine max* (L.) Merr. *Bot. Gaz.* 145:1–10.

Frankel, R., Izhar, S., and Nitsen, J. 1969. The timing of callose activity and cytoplasmic male sterility in *Petunia. Biochem. Genet.* 3: 451–455.

Frei, E. 1955. Die Innervierung der floralen Nektarien dikotyler Pflanzenfamilien. *Ber. Schweiz. Bot. Ges.* 65: 60–114.

Frey-Wyssling, A. 1933. Über die physiologische Bedeutung der extrafloralen Nektarien von *Hevea brasiliensis* Muell. *Ber. Schweiz. Bot. Ges.* 42: 109–122.

Frey-Wyssling, A. 1955. The phloem supply to the nectaries. *Acta Bot. Neerl.* 4: 358–369.

Gerassimova-Navashina, H. 1960. A contribution to the cytology of fertilization in flowering plants. *Nucleus* (Calcutta) 3: 111–120.

Hanf, M. 1935. Vergleichende und entwicklungsgeschichtliche Untersuchungen über Morphologie und Anatomie der Griffel und Griffelaste. *Beih. Bot. Ztbl.* 54A: 99–141.

Harris, W.H. 1991. Seed coat development in radish (*Raphanus sativus* L.). *Phytomorphology* 41: 341–349.

Hayward, H.E. 1938. *The Structure of Economic Plants.* Macmillan, New York.

Heslop-Harrison, J. 1975. Incompatibility and the pollen-stigma interaction. *Annu. Rev. Plant Physiol.* 26: 403–425.

Heslop-Harrison, J., Knox, R.B., Heslop-Harrison, Y., and Mattison, O. 1975. Pollen wall proteins: erosion and role in incompatibility responses. In *The Biology of the Male Gamete.* Duckett, J.C. and Racey, P.A., Eds. Academic Press, London, pp. 187–204.

Heslop-Harrison, J., Knox, R.B., and Heslop-Harrison, Y. 1974. Pollen wall proteins: exine held fractions associated with the incompatibility response in Cruciferae. *Theor. Appl. Genet.* 44: 133–137.

Heslop-Harrison, J. and Mackenzie, A. 1967. Autoradiography of (2-^{14}C) – thymidine derivative during meiosis and microsponogenesis in *Lilium* anthers. *J. Cell Sci.* 2: 387–400.

Jensen, W.A. 1965a. The ultrastructure and histochemistry of the synergids of cotton. *Am. J. Bot.* 52: 238–256.

Jensen, W.A. 1965b. The ultrastructure and composition of the egg and central cell of cotton. *Am. J. Bot.* 52: 781–797

Jensen, W.A. 1973. Fertilization in flowering plants. *Bioscience* 23:21–27.

Johansen, D.A. 1928. The hypostase: its presence and function in the ovule of Onagraceae. *Proc. Natl. Acad. Sci.* (Wash.). 14: 710–713.

Johansen, D.A. 1950. *Plant Embryology.* Chronica Botanica, Waltham, MA.

Johansson, M. and Walles, B. 1993. Functional anatomy of the ovule in broad bean, *Vicia faba* L. II. Ultrastructural development up to early embryogenesis. *Int. J. Plant Sci.* 154: 535–549.

Johri, B.M., Ed. 1984. *Embryology of Angiosperms.* Springer-Verlag, Berlin.

Johri, B.M., Ambegaokar, K.B., and Srivastava, P.S. 1992. *Comparative Embryology of Angiosperms.* Vols.1 and 2. Springer-Verlag, Berlin.

Joshi, P.C., Wadhwani, A.M. (nee Ramchandani, S.), and Johri, B.M. 1967. Morphological and embryological studies of *Gossypium. Proc. Natl. Inst. Sci. India* 33B: 37–93.

Karas, J. and Cass, D.D. 1976. Ultrastructural aspects of sperm cell formation in rye. Evidence for cell plate involvement in generative cell division. *Phytomorphology* 26: 36–45.

Kaul, V. and Singh, D. 1982. Embryology and development of fruit in Cichonieae. *Lactuca* Linn. *Acta Biologica Cracoviensia* 24:19–30.

Konar, R.N. and Linskens, H.F. 1966a. The morphology and anatomy of the stigma of *Petunia hybrida. Planta* 71: 356–371.

Konar, R.N. and Linskens, H.F. 1966b. Physiology and biochemistry of the stigma fluid of *Petunia hybrida. Planta* 71: 372–387.

Kordyum, E.L. 1968. Peculiarities of early ontogeny in ovules and different types of archesporium in some representatives in angiosperms. [In Russian.] *Tsitol. Genet.* 2: 415–424.

Kuhn, G. 1928. Beiträge zur Kenntnis der intraseminalen Leitbundel bei den Angiospermen. *Bot. Jahrb. Syst.* 61: 325–385.

Kumar, P. and Singh, D. 1990. Development and structure of seed coat in *Hibiscus. Phytomorphology* 40: 179–188.

Maeda, E. and Maeda, K. 1990. Ultrastructure of egg apparatus of rice (*Oryza sativa*) after anthesis. *Japan J. Crop Sci.* 59: 179–197.

Maheshwari, P. 1950. *An Introduction to the Embryology of Angiosperms.* McGraw-Hill, New York.

Maheshwari, P., Ed. 1963. *Recent Advances in the Embryology of Angiosperms.* International Society of Plant Morphologists, University of Delhi.

Martens, P. 1934. Recherches sur la cuticule: IV. Le relief cuticulaire et la differentiation epidermique des organes floraux. *Cellule* 43: 289–320.

Masand, P. and Kapil, R.N. 1966. Nutrition of the embryo sac and embryo — A morphological approach. *Phytomorphology* 16: 158–175.

Mattson, O., Knox, R.B., Heslop-Harrison, J., and Heslop-Harrison, Y. 1974. Protein pellicle of stigmatic papillae as a probable recognition site in incompatibility reactions. *Nature* (London) 247: 298–300.

Maze, J. and Lin, S.Ch. 1975. A study of mature gametophyte of *Stipa elmeri. Can. J. Bot.* 53: 2958–2977.

Morgensen, H.L. and Suthar, H.K. 1979. Ultrastructure of the egg apparatus of *Nicotiana tabacum* (Solanaceae) before and after fertilization. *Bot. Gaz.* 140: 168–179.

Naryanaswami, S. 1953. The structure and development of the caryopsis in some Indian millets — I. *Pennisetum typhoideum* Rich. *Phytomorphology* 3: 98–112.

Newcomb, W. 1973. The development of the embryo sac of sunflower *Helianthus annuus* before fertilization. *Can. J. Bot.* 51: 863–878.

Newcomb, W. and Fowke, L.C. 1974. *Stellaria media* embryogenesis. The development and structure of the suspensor. *Can. J. Bot.* 52: 607–614.

Olesen, P. and Bruun, L. 1990. A structural investigation of the ovule in sugar beet. *Beta vulgaris*: integuments and micropyle. *Nord. J. Bot.* 9: 499–506.

Orel, L.L. and Shmaraev, I.G. 1987. Ultrastructure of the *Triticum aestivum* embryo sac wall after fertilization. [In Russian.] *Botanicheskii Zhurmal* 72: 753–757.

Pandey, A.K. and Jha, S.S. 1988. Development and structure of seeds in some Genisteae (Papilionoideae-Leguminosae). *Flora* 181: 417–426.

Periasamy, K. 1962. The ruminate endosperm: development and types of rumination. In *Plant Embryology: A Symposium.* CSIR, New Delhi, pp. 62–74.

Periasamy, K. and Swamy, B.G.L. 1966. Morphology of anther tapetum in angiosperms. *Curr. Sci.* 35: 427–430.

Poddubnaya-Arnoldi, V.A. 1967. Comparative embryology of the Orchidaceae. *Phytomorphology* 17: 312–320.

Prasad, K. 1974. Studies in the Cruciferae gametophytes, structure and development of seed in *Eruca sativa* Mill. *J. Indian Bot. Soc.* 53: 24–33.

Prakash, S. 1960. The endosperm in *Arachis hypogaea* Linn. *Phytomorphology* 10: 60–64.

Puri, V. 1954. Studies in floral anatomy. VII. On placentation in the Cucurbitaceae. *Phytomorphology* 4: 278–299.

Ramchandani, S., Joshi, P.C., and Pundir, N.S. 1966. Seed development in *Gossypium* Linn. *Indian Cotton J.* 20: 97–106.

Rao, V.S. and Gupta, K. 1951. A few observations on the carpels of *Artohotrys. J. Univ. Bombay* 20: 62–65.

Rathore, R.K.S. and Singh, R.P. 1968. Embryological studies in *Brassica campestris*. L. var. yellow sarson Prain. *J. Indian Bot. Soc.* 47: 341–349.

Rau, M.A. 1953. Some observations on the endosperm in Papilionaceae. *Phytomorphology* 3: 209–222.

Rodkiewicz, B. 1970. Callose in cell wall during megasporogenesis in angiosperms. *Planta* 93: 39–47.

Rosen, W.G. and Thomas, H.R. 1970. Secretory cells of lily pistils. I. Fine structure and function. *Am. J. Bot.* 57: 1108–1114.

Saxena, T. 1970. Studies on the Development and Structure of Seed in Solanaceae. Ph.D. thesis, University of Rajasthan, Jaipur, India.

Schnarf, K. 1929. *Embryologie der Angiospermen.* G. Borntraeger, Berlin.

Schnarf, K. 1931. *Vergleichende Embryologie der Angiospermen.* G. Borntraeger, Berlin.

Schulz, P. and Jensen, W.A. 1969. *Capsella* embryogenesis: the suspensor and the basal cell. *Protoplasma* 67: 139–163.

Schulz, P. and Jensen, W.A. 1972. *Capsella* embryogenesis: the chalazal proliferating tissue. *J. Cell Sci.* 8: 201–207.

Schulz, P. and Jensen, W.A. 1986. Prefertilization ovule development in *Capsella*, the dyad, tetrad, developing megaspore and two-nucleate embryo sac. *Can. J. Bot.* 64: 875–884.

Sharangpani, P.R. and Shirke, D.R. 1996. Scanning electron microscopic studies on ovarian nectaries of *Cassia occidentalis* L. *Phytomorphology* 46: 277–281.

Singh, B. 1953. Studies on the structure and development of seeds of Cucurbitaceae. *Phytomorphology* 3: 224–239.

Singh, D. 1957. Endosperm and its chalazal haustorium in Cucurbitaceae. *Agra Univ. J. Resi (Sci.)* 6: 75–89.

Singh, D. 1963. Structure and development of ovule and seed of *Viola tricolor* and *Ionidium suffruticosum. J. Indian Bot. Soc.* 42: 448–462.

Singh, D. 1965. Ovule and seed development of *Sechium edule* Sw. A reinvestigation. *Curr. Sci.* 34: 696–697.

Singh, D. and Dathan, A.S.R. 1972. Structure and development of seed coat in Cucurbitaceae. VI. Seeds of *Cucurbita* L. *Phytomorphology* 22: 29–45.

Singh, D. and Dathan, A.S.R. 1990. Seed coat anatomy of the Cucurbitaceae. In *Biology and Utilization of the Cucurbitaceae.* Bates, D.M., Robinson, R.W., and Jeffrey, C., Eds. Comstock Publishing Associates, Cornell University Press, Ithaca, NY, pp. 225–238.

Singh, R.P. 1954. Structure and development of seeds in Euphorbiaceae *Ricinus communis* L. *Phytomorphology* 4: 118–121.

Singh, S.P. 1960. Morphological studies in some members of the family Pedaliaceae – I. *Sesamum indicum* D.C. *Phytomorphology* 10: 65–82.

Souéges, R. 1932. Les methodes de travail en embryologie vegetale. *Bull. Soc. Fr. Micr.* 1: 88–104.

Stiegelitz, H. and Stern, H. 1973. Regulation of β-1,3 glucanase activity in developing anthers of *Lilium* microsporocytes. *Dev. Biol.* 34: 169–173.

Sulbha, K. 1954. The embryology of *Iphigenia indica. Phytomorphology* 4: 180–191.

Swamy, B.G.L. and Krishnamurthy, K.V. 1980. *From Flower to Fruit — Embryology of Flowering Plants.* Tata McGraw-Hill, New Delhi.

Taneja, C.P. 1981. Structure and Development of Seed Coat in Some Centrospermae. Ph.D. thesis, University of Rajasthan, Jaipur, India.

Tiagi, B. 1951. Studies in the family Orobanchaceae — III. A contribution to the embryology of *Orobanche cernua* Loeff. and *O. aegyptiaca* Pers. *Phytomorphology* 1: 158–169.

Tiagi, B. 1963. Studies in the family Orobanchaceae – IV. Embryology of *Boschniackia himlaica* Hook. and *B. tuberosa* (Hook.) Jepson with remarks on the evolution of the family. *Bot. Notiser* 116: 81–93.

Tilton, V.R. 1980. Hypostase development in *Ornithogalum caudatum* (Liliaceae) and notes on other types of modifications in the chalaza of angiospermous ovules. *Can. J. Bot.* 58: 2059–2066.

Tsinger, V.N. and Petrovskaya-Baranova, T.P. 1961. The pollen grain-wall a living physiologically active structure. *Dokl. Acad. Nauk. SSSR* 138: 466–469.

Van Lammereen, A.A.M. 1986. A comparative ultrastructural study of the megagametophytes in two strains of *Zea mays* L. before and after fertilization. *Agric. Univ. Wageningen Papers* 86(1): 1–37.

Van Tieghem, P. 1901. L'hypotase, sa structure et son role constantes, sa position et sa forme variables. *Bulb. Mus. Hist. Nat.* 7: 412–418.

Vaughan, J.G. 1956. The seed coat structure of *Brassica integrifolia* (West) O.E. Schulz var. *Carinata* (A. Br.). *Phytomorphology* 6: 363–367.

Vaughan, J.G. 1959. The testa of some *Brassica* seeds of oriental origin. *Phytomorphology* 9: 107–110.

Venkata Rao, C. 1953. Contributions to the embryology of Sterculiaceae. V. *J. Indian Bot. Soc.* 32: 208–238.

Vijayaraghavan, M.R. and Bhat, U. 1983. Synergids before and after fertilization. *Phytomorphology* 33: 74–84.

Wang, V. and Wang, X. 1991. A new connection between the synergids and the central cell in *Vicia faba* L. *Phytomorphology* 41: 351–354.

Westerkeyn, L. 1962. Les parois microcytaires de nature callosique chez *Helleboris* et *Tradescantia*. *Cellule* 62: 225–255.

Willemse, M.T.M. and van Went, J.E.L. 1984. Female gametophyte. In *Embryology of Angiosperms*. Johri, B.M., Ed. Springer-Verlag, Berlin, pp. 159–196.

Wilms, H.J. 1981. Ultrastructure of the developing embryo sac of spinach. *Acta Bot. Neerl.* 30: 75–99.

Yan, H., Yang, H., and Jensen, W.A. 1991. Ultrastructure of the developing embryo sac of sunflower (*Helianthus annuus*) before and after fertilization. *Can. J. Bot.* 69: 191–202.

Yeung, E.C. 1980. Embryogeny of *Phaseolus*: the role of the suspensor. *Z. Pflanzenphysiol.* 96: 17–28.

Yeung, E.C. and Clutter, M.E. 1978. Embryogeny of *Phaseolus coccineus*: growth and microanatomy. *Protoplasma* 94: 19–40.

Yeung, E.C. and Clutter, M.E. 1979. Embryogeny of *Phaseolus coccineus*: the ultrastructure and development of the suspensor. *Can. J. Bot.* 57: 120–136.

3 Structure of Seeds

The seed habit is a significant advancement in the evolutionary history in the plant kingdom and has bestowed several advantages to seed-bearing plants. Biologically, the seed is the ripened ovule, as discussed in Chapter 2. However, the use of the term *seed* is not always restricted to this morphologically accurate definition. Seed is usually applied in the functional sense, i.e., as a unit of dissemination, a disseminule. In agriculture, in addition to true seeds, many one-seeded dry indehiscent fruits, namely caryopsis (Poaceae), cypsil (Asteraceae), cremocarp and mericarp (Apiaceae), and achene and utricle (Amaranthaceae and Chenopodiaceae), are termed seeds. The term seed is used *sensu lato* in this volume.

Seed structure is intimately concerned with the infection caused by plant microorganisms. Seed morphology and anatomy reveal the probable pathways of infection as well as the barriers to infection; therefore, we include a detailed general account of the structure of seed in angiosperms as well as brief specific accounts of seeds of 12 plant families that contribute to the majority of crop plants.

3.1 CONSTITUTION OF SEEDS

A viable seed consists of an embryo, a protective covering, and the reserve food material. The embryo comprises one or two (and rarely more) cotyledons, a plumule bud (epicotyl), a hypocotyl, and a radicle. The embryo varies in size and shape. The embryo is undifferentiated in many plants, particularly parasitic, saprophytic, and insectivorous plants. The cotyledons are absent in such cases. The embryo is narrow, elongated, and spirally coiled with plumular scales in *Cuscuta reflexa* (Johri and Tiagi, 1952).

The protective covering usually comprises a seed coat in true seeds, and a pericarp and a seed coat or a pericarp in one-seeded fruits (Apiaceae, Asteraceae, and Poaceae), bracts in *Oryza* and *Hordeum,* a calyx in *Brunnonia*, and a cuticle of endosperm in *Crinum*, Santalaceae, Loranthaceae, and Olacaceae. There is considerable variability in the anatomy of the seed coat and pericarp.

The reserve food material is stored in the endosperm (seeds called endospermic), the perisperm (persistent nucellar tissue), or the embryo (cotyledons), or in both the embryo and the endosperm. The major reserve foods are carbohydrates, fats and lipids, and proteins. Seeds also contain other minor reserves, some of which are nutritionally undesirable or even toxic, e.g., alkaloids, inhibitors, lectins, and phytins. The cuticula of different components of seed are distinct. The surface of each component is covered by a thick or thin cuticle.

3.2 EXOMORPHIC FEATURES

Variations in exomorphic seed characters are considerable and many of them are important with respect to the functional attributes of seeds. With the development of seed technology during and after the 1940s, several publications with characteristics and drawings of seeds have appeared (Korsmo, 1935; Murley, 1944, 1946, 1951; Isley, 1947; Martin, 1954; Brouwer and Stahlin, 1955; Heinisch, 1955; McClure, 1957; Dobrokhotov, 1961; Musil, 1963; Breggren, 1969). The U.S. Department of Agriculture (USDA) and some state agriculture departments have published seed handbooks (USDA, 1948, 1952, 1961; Bellue, 1952; Britton and Fuller, 1957). In the *Seed Identification Manual*, Martin and Barkley (1961) included 824 photographs of seeds of more than 600 plant species occurring in the United States. Gunn (1970a,b, 1971) provided seed characters and drawings of seeds of native and naturalized *Vicia* spp.

The common external features of seeds concern seed color, shape, size (length, width, and thickness), surface features, and the size, shape, and position of hilum, micropyle, and raphe. Other recognizable features, when present, are the appendages, e.g., wings, pappus, aril, caruncle, elaisome, spines, tubercles, and hairs. The main features of seed are fairly constant for a species; however, variations occur in seeds of different cultivars or clones, but such variations are not marked within the cultivar.

3.2.1 Color

Seeds may be monochrome or marked by points, mottles, and streaks. White, brown and brown derivatives, and black are by far the most common seed colors. Other colors, such as red, green, yellow, and double colors (e.g., the red and black of *Abrus precatarious*), are infrequent.

Seed pigment is reported to impart slight resistance to fungal pathogens. Singh and Singh (1979) have reported that the white seeds of sesame are most susceptible to *Macrophomina phaseolina*, the brown seeds showed weak incidence, and none of the black seed samples carried the infection. Glueck and Rooney (1978) found that kernel pigmentation in sorghum provides resistance to head mold caused by *Curvularia lunata* and *Phoma sorghina*. Stasz, Harmon, and Marx (1980) also found that fewer *Pythium ultimum* hyphae developed on the surface of colored seeds of pea.

3.2.2 Shape

Common seed shapes are spherical, subspherical, oblong, oval, ellipsoid, sublenticular, subpyramidal, cuboid, subcuboid, and reniform. Seeds may be turgid or compressed.

3.2.3 Size

Seeds vary immensely in size from dustlike particles to large coconuts and double coconuts. The feature is not very reliable, but seed length, width, thickness, and the ratio between length and width may be used in seed identification at the species and subspecies levels. It is difficult to determine the size of small seeds; Gunn's methods

(1971) of using an occular micrometer in a stereoscopic microscope or a Lufkin pocket slide caliper appear useful. Apparently, large seeds provide more exposed space for microorganisms as compared to small seeds, but there are no data to support such a conjecture.

3.2.4 SURFACE

The surface of the seed coat may be smooth or sculptured. Various types of ornamentation such as wrinkles, ridges and furrows, striations, reticulate, punctate, tuberculate, hairy, or spinescent, have been recorded. The surface of the seed coat is covered by the cuticle, a waxy, fatty hemicellulose or pectinaceous substance. It is thin or thick. The waxy coating may appear as irregular particles, crystalloid rodlets, filaments, flakes, or plates (Barthlott and Wollenweber, 1981). The gloss of the seed surface is due to the waxy coating, and the thin, uniform wax covering makes the seed surface shiny in fabaceous seeds.

The micromorphology of the seed coat surface described using SEM provides greater detail of sculpturing pattern, including the presence of minute hairs, micropores, cracks, and deposits (Figures 3.1 A through E). It has been studied extensively for soybean seeds (Wolf, Baker, and Bernard, 1981; Yaklich, Vigil, and Wergin, 1984). The surfaces of seed coats of different cultivars of soybean have been classified into three types: smooth, with pores (Figure 3.1A), and with a distinct type of waxy deposit called blooms. The pores of seed coat surfaces in the cultivars Williams, Guelph, Hoosier, and Jogan vary in number, size, and shape among the four cultivars. The seed coat surfaces in the cultivars Old Dominion, Laredo, Barchet, and Sooty are without pores (Figure 3.1B) (Kulik and Yaklich, 1991). The surfaces in Barchet and Sooty haves wax deposits called *bloom* (Figure 3.1C, D) believed to be derived from the endocarp layer of the pod (Newell and Hymovitz, 1978; Wolf, Baker, and Bernard, 1981).

Stomata on the seed surface have been reported in seeds of 19 families, most with endotestal or exotegmic seed coats. The presence of stomata is not known to be a constant feature of any family. Stomata are absent when the exotesta is made up of a compact palisade layer as in Fabaceae. Among the families with crop plants, stomata are found on the seed surface of Malvacae, Euphorbiaceae, and Papaveraceae. In cotton seed there is greater aggregation of stomata in the chalazal region. The pericarp is photosynthetic and commonly bears stomata. Cochrane and Duffus (1979) observed stomata in the pericarp epidermis on the ventral side at the apical end in some cultivars of wheat and barley. The lemma and palaea in *Oryza* (Azegami, Tabei, and Fukuda, 1988), wheat, and barley (Fukuda, Azegami, and Tabei, 1990), bear stomata in outer as well as inner epidermis.

Some of the above features of seed surface such as pores, stomata, and fissures are directly correlated with the functional attributes of seeds. Seed coats with pores are generally permeable (Calero, West, and Hinson, 1981; Wolf and Baker, 1980; Wolf, Baker, and Bernard, 1981). Kulik and Yaklich (1991) report that seeds of soybean cultivars that lack pores on the seed coat have a low rate of infection of *Phomopsis phaseoli* as compared to those with multiple pores.

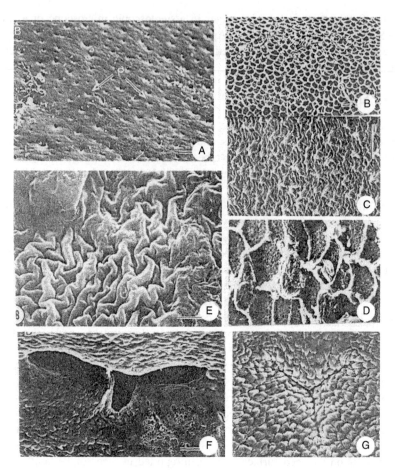

FIGURE 3.1 SEM photomicrographs of surface and micropyle of seeds of soybean cultivars. A, cultivar Williams seed coat surface with micropores. B to D, cultivar Sooty. B, Seed coat surface without pores. C, D, Seed coat surface showing areas without and with blooms, respectively. E, Cultivar Williams seed coat surface showing areas with and without deposits. F, G, Micropyle, open-type cultivar Williams and closed-type cultivar Laredo, respectively. (Abbreviation: p, pore.) (A, B, F, G, From Kulik, M.M. and Yaklich, R.W. 1991. *Crop Sci.* 31: 108–113; C to E, From Yaklich, R.W., Vigil, E.L., and Wergin, W.P. 1986. *Crop Sci.* 26: 616–624. With permission.)

3.2.5 MICROPYLE

The size and position of the micropyle are usually given in seed descriptions and, usually, in dry seeds the micropyle is described as occluded. SEM micrographs have indicated significant differences in measurements of the micropyle in seeds of different cultivars of soybean (Kulik and Yaklich, 1991). Kuklik and Yaklich divide the micropyle into two broad types: open (Figure 3.1F) and closed (Figure 3.1G). The average opening of the open micropyle is 0.44 mm^2 and the average opening of the closed micropyle is 0.18 mm^2. In soybean cultivars, the open type of micropyle is found in association with seed coats with pores, whereas the closed type is found with those without pores. Kulik and Yaklich (1991) found that hyphae of *Phomopsis phaseoli* were present on the seed coat and hilum, and their penetration via the *open type* of micropyle was far more prevalent than in seeds of cultivars with closed micropyles.

3.2.6 HILUM

The size, shape, and position of the hilum with respect to the micropyle and chalaza constitute important features. The hilum is generally considered to be a scar left after separation of the seed from the funiculus. Recent studies have shown that it evolves during seed development. Partial seed abscission is observed during hilum development (Pamplin, 1963; Baker and Mebrahtu, 1990). The scar is not the result of mechanical separation, but rather caused by the organization of an abscission layer. Hilum size varies from insignificant to quite prominent. In the latter, the cell organization in the hilar region may differ from that of the seed coat. The cells may be homogeneous or show tissue differentiation, as in fabaceous seed (Baker and Mebrahtu, 1990). The broad surface of hilum lacks the usual cuticle and may have fissures. Hilar fissure is characteristic in the seeds of Fabaceae.

3.2.7 RAPHE

In several types of ovules and seeds, the funiculus is adnate to the ovule surface. The abscission of seed occurs in the free part of the funiculus. The adnate part remains as a longitudinal ridge in seed and is called the raphe. Structurally, it usually resembles the seed coat, but sometimes it is distinctive.

3.2.8 SEED APPENDAGES

Seeds have various types of appendages, such as aril (Figure 3.2E, F), caruncle (Figure 3.2C, D), strophiole, wings (Figure 3.2A, B), and hairs. The aril is a soft, succulent local outgrowth of seeds of varied origin. The aril primordium may develop from any part of the ovule (e.g., funicle, raphe, and chalaza), the placenta, or the carpel. It may enclose the seed more or less completely as in *Passiflora* (Figure 3.2F) or it may form a localized outgrowth in *Turnera ulmifolia* (Figure 3.2E) and others. For detailed information on seed appendages in general and aril in particular, the reader should refer to Kapil, Bor, and Bouman (1980).

Caruncle (Figure 3.2C, D) is a small, disclike appendage, the attachment and growth of which are limited to the exostome rim. The micropyle may be seen in the center (*Ricinus* and *Euphorbia*). Strophioles are glandular or spongy, with proliferation limited only to the raphal region (*Chelidonium majus*).

Wings (Figure 3.2A, B) are flattened extensions of seed having optimal strength and a minimal biomass. Seed wings are often correlated with seed dispersal. Corner (1976) believes that the seed wings are local outgrowths of the testa. They may be peripheral or restricted to the raphe, chalaza, antiraphe, hilum, or funicle. Seed wings are rarely provided with vascular bundles, whereas fruit wings have a well-developed vascular supply.

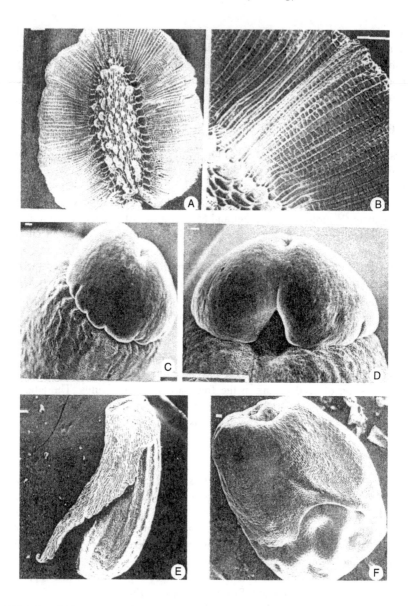

FIGURE 3.2 SEM photomicrographs of seed appendages. A, B, Winged seed of *Nemesia floribunda*, circular wing having reticulate patterns of elongated cells. C, D, Lobed caruncle of *Euphorbia lathyris*. E, Unilateral raphal aril of *Turnera ulmifolia*. F, Funicular aril covering the seed on all sides of *Passiflora suberosa*. (From Kapil, R.N., Bor, J., and Bouman, F. 1980. *Bot. Jahrb. Syst.* 101: 555–573. With permission.)

Seed hairs are common in many families, particularly Malvaceae, Cochlosperm-aceae, Asclepiadaceae, Convolvulaceae, Acanthaceae, and Polygalaceae. Usually the seed hairs are simple and unicellular with a thin or thick cuticle; however, multicell-ular hairs are known in some Rutaceae. The hairy structures on the seed surface in dry seed of *Lycopersicon* and some *Solanum* species are not true hairs. Saxena (1970) has shown that these are the radial walls of seed epidermal cells with thickening decreasing from inside to outside. These have been termed *spurious hairs* by Corner (1976) and *pseudohairs* by Rick (1978).

Seed appendages do not occur universally and they are rare in crop plants. They cause an increase in seed surface and this may promote association of microorganisms; however, no data exist to support this fact.

3.3 INTERNAL MORPHOLOGY

3.3.1 Gross Internal Morphology

The gross internal morphological features of the main components of seed, i.e., the seed coat, endosperm and perisperm, and embryo, give an idea of their spatial and topographic adjustments. The usefulness of these features as taxonomic pointers is widely recognized (Bouman, 1974). These features are also significant in the functional performance of seed.

Martin (1946) described the comparative morphology of seeds of 1287 genera of angiosperms and proposed a classification of seed types on the basis of the size of the embryo in relation to the endosperm, and on the differences in size, shape, and position of the embryo (Figure 3.3). He used embryo measurements in quarter units of a circle and designated five types: (1) small, with the embryo smaller than a quarter of internal space; (2) quarter, with a quarter or more but less than two quarters internal space; (3) half, with two quarters or more but less than three quarters internal space; (4) dominant, with more than three quarters internal space; and (5) total, with the embryo occupying the entire inner seed space.

On the basis of embryo position, Martin (1946) proposed three main categories of seed: *basal, peripheral,* and *axile*. Because of their overlapping features the basal and peripheral categories are merged into a single category called the peripheral. These categories are further subdivided on the basis of size and shape of the embryo. The peripheral type with the embryo at the micropylar end or oriented peripherally and with copious endosperm or perisperm has five subcategories: (1) rudimentary, where the embryo is small, globular to ovate oblong, and relatively undifferentiated (*Magnolia* and *Piper*); (2) broad, where the embryo is as wide as or wider than long, globular, or lenticular (*Nymphaea* and *Juncus*); (3) capitate, where the embryo is distally expanded (*Dioscorea, Tradescantia, Scirpus,* and *Carex*); and (4) lateral, where the embryo is lateral, inclined to expand along the periphery (Poaceae); and (5) peripheral, where the embryo is elongated, large, often curved, extending along the periphery, with cotyledons narrow or expanded (Amaranthaceae, Chenopodiaceae).

The *axile* type has small to large embryo, straight, variously curved or coiled, central, seeds small to large, and endospermic or nonendospermic. It has seven subtypes: (1) linear, where the embryo is much longer than broad, with the cotyledons straight, curved, or coiled (*Lilium* and *Allium*); (2 and 3) dwarf and micro, where the seeds are small, with the seed interior 0.2 to 0.3 mm long in (2) and less than 0.2 mm long in (3), the embryo is small or large (Orchidaceae, Burmanniaceae,

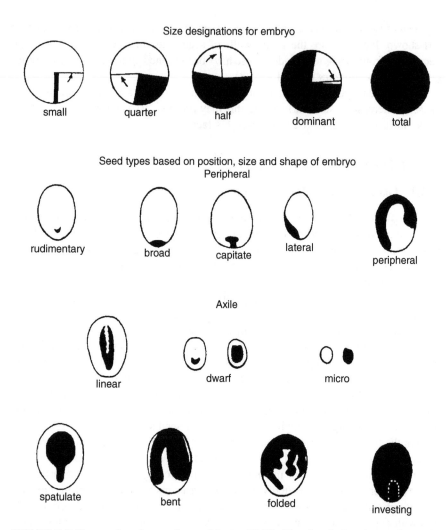

FIGURE 3.3 Types of seed according to Martin (1946). The size designations refer to the embryo–endosperm size ratio, represented volumetrically in quarter units of a circle. The two main groups, peripheral and axile seeds, are characterized based on the size, shape, and position of embryo in seed. (Adapted and redrawn from Martin, A.C. 1946. *Am. Midl. Nat.* 36: 513–660.)

and Orobanchaceae); (4) spatulate, where the embryo is erect, the hypocotyledonary axis is not enclosed by cotyledons or only slightly enclosed (*Corchorus, Sesamum,* and *Linum*); (5) bent, where the embryo is bent like a sac knife, and the cotyledons are expanded (Moraceae); (6) folded, the embryo is curved with folded cotyledons (*Gossypium, Hibiscus, Brassica, Eruca,* and *Ipomoea*); and (7) investing, where the embryo is erect and the cotyledons are large and enclose the hypocotyl axis (Cucurbitaceae and Mimosoideae).

Martin and Barkley (1961) used the above features in their *Seed Identification Manual.*

3.3.2 Seed Coat and Pericarp

The protective covering of seed is the seed coat in true seeds, pericarp, or bracts in one-seeded fruits. The seed has a main mechanical layer, which, as mentioned in Chapter 2, differentiates in the outer or the inner integument. Corner (1976) classifies dicotyledonous seeds into *testal* or *tegmic,* depending on whether the main mechanical layer has differentiated into the outer or the inner integument. Each category is subdivided on the basis of the place of differentiation of the main mechanical layer into exotestal, mesotestal, and endotestal, and exotegmic, mesotegmic, and endotegmic, respectively. Each subcategory is further divided into two or more types on the basis of the nature (shape and size) of cells in the main mechanical layer.

Corner (1976) did not consider unitegmic seeds (seeds formed from unitegmic ovules) in his classification, but he remarked that these can be called *exotestal* because the seed coat develops in an exostestal manner in these ovules. The seed coat in monocotyledons also has a main mechanical layer, and Corner's concept can be readily extended to the structure of the seed coat in this group as well (Maheshwari Devi et al., 1994).

The pericarp in caryopsis, cypsils, cremocarp, achene, and utricle is also characterized by the presence of mechanical (sclerenchymatous) layers and the presence of cuticle on the surface (Lavialle, 1912; Borthwick and Robbins, 1928; Bradbury et al., 1956a,b; Gupta, 1964; Bechtel and Pomeranz, 1978; Zeleznak and Verriano-Marston, 1982; Kaul and Singh, 1982).

For detailed information on seed coat structure the reader should refer to Netolitzky (1926), Singh, (1964), Vaughan (1970), and Corner (1976). *The Structure and Composition of Foods,* Vols. 1 to 4 (Winton and Winton, 1932–1939), give detailed information on the microscopic structure of various parts, namely the seed coat and pericarp, embryo, and endosperm and perisperm of grains and seeds used as food.

3.4 SEED STRUCTURE IN SELECTED FAMILIES

Seed characteristics of 12 families of angiosperms with common crop plants are given. The description includes external and internal characteristics. Special features and variations, when present, are indicated.

3.4.1 Brassicaceae (Cruciferae) (Figure 3.4A to E)

(Thompson, 1933; Sulbha, 1957; Rathore and Singh, 1968; Vaughan, 1970; Vaughan and Whitehouse, 1971; Prasad, 1974)

External: Seeds are small, globose, compressed or slightly flattened laterally. In the last type surface, contours usually have distinct radicle ridges and bent embryo, notched or cleft with a groove or line between the cotyledons and radicle; brown to black, yellow or white; surface is reticulate or pitted, with hilum and micropyle inconspicuous.

Internal: Seed coats two, endotestal, seed epidermis of cuboid or flattened cells with low (*Brassica nigra, B. juncea, B. rapa,* and *B. campestris*) or high content of

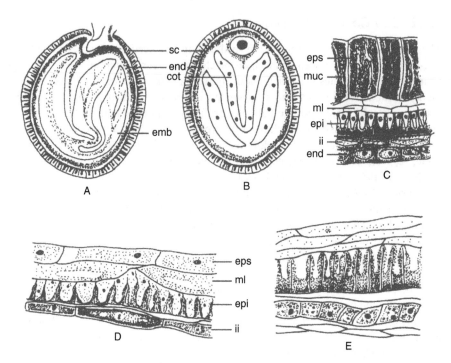

FIGURE 3.4 Structure of seed of Brassicaceae. A to C, *Eruca sativa*. A, B, Ls and Ts seed. C, Ts part of seed coat, cells of epidermis are full of mucilage and cells of inner epidermis (epi) of outer integument thick-walled. D, Ls part of seed coat of *Brassica juncea*. E, Ls part of seed coat of *B. campestris* var. yellow sarson. (Abbreviations: cot, cotyledon; emb, embryo; end, endosperm; epi, inner epidermis of outer integument consisting of characteristic thick-walled cells; eps, seed epidermis; ii, inner integument; ml, middle layers; muc, mucilage; sc, seed coat.) (A to C, From Prasad, K. 1974. *J. Indian Bot. Soc.* 53: 24–33; D, Sulbha, K. 1957. *J. Indian Bot. Soc.* 36: 292–301; E, Rathore, R.K.S. and Singh, R.P. 1968. *J. Indian Bot. Soc.* 47: 341–349.)

internal mucilage (*Eruca sativa*); mesophyll of outer seed coat thin-walled, the next layer formed by the inner epidermis of the outer integument is of cuboid or short radially elongated cells, thick-walled — inner and radial walls lignified forming the main mechanical layer (endotestal); the inner seed coat of thin-walled, compressed cells with protein bodies and pigmented material. A three-tiered basal body occurs in hilar region, subhilar tissue constitutes the solid core of the basal body and inner and outer integument after invagination forms the other parts. The basal body is two-tiered due to the obliteration of the lower (middle) tier in *Eruca sativa*.

Mucilage may diffuse or break through the outer wall and cuticle of the seed coat during seed maturity. When seeds are soaked in water, mucilage collects on the seed surface (*Eruca*).

Embryo: Bent, radicle dorsal or lateral, extend along the margin of seed; radicle lying along the edges of cotyledons or along the back of one cotyledon, cotyledons conduplicated or folded; reserve food material predominantly oil.

Endosperm: One- or two-layered, cells of outer layer contain aleurone grains.

3.4.2 MALVACEAE (FIGURE 3.5A TO H)

(Ramchandani, Joshi, and Pundir, 1966; Joshi, Wadhwani, and Johri, 1967; Kumar and Singh, 1990, 1991)

External: Seeds usually compressed, reniform with central notch or subglobose (*Abelmoschus esculentus*) or pyriform (*Gossypium*); usually brown to black; smooth, rough, rugose, or hairy (*Gossypium* and some *Hibiscus* spp.); hilum flush or depressed within the notch; micropyle inconspicuous; obliterated; stomata on seed surface, particularly at the chalazal end.

Internal: Seed coats two, exotegmic, differentiated into five or more zones: (1) seed epidermis, composed of thin-walled horizontal cells covered with thin cuticle, in *Gossypium* cells form lint as well as fuzz, sparse hairs in others; hairs are simple, one-celled or septate, thin-walled or lignified, hair bases are thick-walled; (2) outer mesophyll, one or more layers of outer integument, thin-walled, in *Gossypium* differentiated into an outer zone of pigmented cells and an inner of colorless cells; cells of inner epidermis inconspicuous, rarely with calcium oxalate crystals (*Hibiscus calycina*); (3) palisade or macroscleroid layer, thick-walled, lignified, light line area in the outer half of cells, the layer formed by outer epidermis of inner integument and is the main mechanical layer —seed coat exotegmic; (4) inner mesophyll usually multi-layered and differentiated into two zones: a pigmented zone of outer layers of small, thin-walled and tannin-pigmented cells and a colorless zone of thin-walled cells; (5) fringe layer, inner epidermis of inner integument, cells longitudinally elongated, thick-walled with pits on radial walls and pigmented contents. In *Gossypium* cells of inner mesophyll possess compound starch grains.

The vascular supply is prominent and terminates at the chalaza below a tanniferous pad.

Embryo: Peripheral, large, often curved, cotyledons two, foliaceous, folded or convoluted.

Endosperm: Scanty or one- or two-layered around the embryo, five- or six-layered at the micropylar and chalazal ends in *Gossypium*, also present between the folds of cotyledons. The cells of endosperm and embryo are rich in oil and proteins. Cotton seed contains the alkaloid *gossypol*.

3.4.3 LINACEAE (FIGURE 3.6A, B)

(Boesewinkel, 1980, 1984)

External: Seed small, flat or flattish, elliptic or elliptic-ovate, pointed at one end; smooth, lustrous; yellow to brown; hilum inconspicuous, micropyle obliterated.

Internal: Seed coats two, exotegmic; outer coat two-layered, epidermis of short, radially elongated cells with stratified mucilage deposits, cuticle thin, plicate, inner layer parenchymatous, inner coat, epidermis formed of tangentially elongated fibrous, thick-walled, lignified cells forming the main mechanical layer (exotegmic); mesophyll of crushed cells, inner epidermis — cells rectangular, thick-walled, pits on radial walls, tanniferous. Vascular supply terminates at the chalaza.

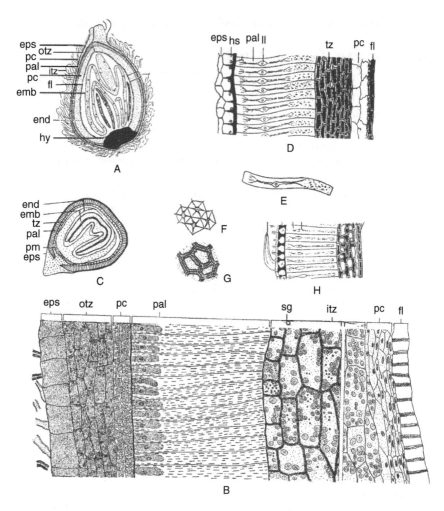

FIGURE 3.5 Structure of seed in Malvaceae. A, B, *Gossypium herbaceum*. A, Ls seed. B, Ls part of seed coat. C to G, *Abelmoschus esculentus*. C, Ts mature seed. D, Ts part of seed coat. E, F, Lateral view and view from top of palisade cells from maceration. G, Surface view of fringe layer (inner epidermis of inner integument). H, Ts part of mature seed coat of *Hibiscus cannabinus*. (Abbreviations: emb, embryo; end, endosperm; eps, seed epidermis; fl, fringe layer; hs, seed hypodermis; hy, hypostase; itz, inner tanniferous zone; ll, light line; otz, outer tanniferous zone; pal, palisade layer; pm, perisperm; sg, starch grains; tz, tanniferous zone.) (A, B, From Joshi, P.C., Wadhwani, A.M. [nee Ramchandani, S.], and Johri, B.M. 1967. *Proc. Natl. Inst. Sci. India,* 33B: 37–93; C to G, From Kumar, P. and Singh, D. 1991. *Acta Bot. India* 19: 62–67; H, From Kumar, P. and Singh, D. 1990. *Phytomorphology* 40: 179–188.)

Addition of water to seeds causes swelling of epidermal cells because of the swelling of mucilage. This causes ruptures in the cuticle, which remains attached with the peripheral wall at places; the free ends curl up.

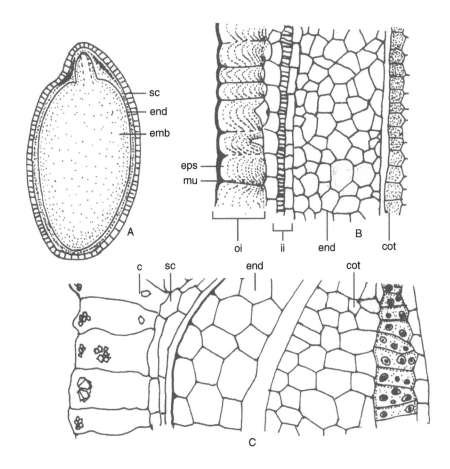

FIGURE 3.6 Structure of seed of *Linum usitatissimum* (A, B) and *Sesamum indicum* (C). A, Median Ls of seed. B, Ts of mature seed coat consisting of thick cuticle, radially elongated epidermal cells with mucilage, and thick-walled outer epidermis of inner integument. C, Ts part of seed, epidermal cells radially elongated with calcium oxalate crystals. (Abbreviations: c, calcium oxalate crystals; cot, cotyledon; emb, embryo; end, endosperm; eps, seed epidermis; ii, inner integument; mu, mucilage; oi, outer integument; sc, seed coat.) (A, B, Adapted from Boesewinkel, F.D. 1984. *Ber. Dtsch. Bot. Ges. Bd.* 97: 443–450; C, Adapted from Vaughan, J.A. 1970. *The Structure and Utilization of Oil Seeds.* Chapman & Hall, London.)

Embryo: Erect, axile, spatulate, cotyledons two, flat, cells with oil and protein as reserve food.

Endosperm: Three to six cells thick; cells are thin-walled and reserve food as oil and protein.

3.4.4 FABACEAE (LEGUMINOSAE) SUBFAMILY FABOIDEAE (PAPILIONATAE) (FIGURE 3.7A TO G)

(Vaughan, 1970; Lersten, 1981; Wolf, Baker, and Bernard, 1981; Yaklich, Vigil, and Wergin, 1984, 1986; Jha and Pandey, 1989; Baker and Mebrahtu, 1990; Kulik and Yaklich, 1991)

External: Seeds turgid, usually bean shaped (reniform, ovoid [*Cicer*] or nearly spherical [*Pisum*]), oblong, (rhomboid [*Trigonella*] or straight cylindrical [*Arachis*]), smooth or wrinkled; glossy or dull; monochrome (brown, black, green, red, cream, white, or their shades) or dichrome due to mottling or areas of two distinct colors; hilum conspicuous, nearly central or terminal, white or dull cream colored, without or with a colored margin, oval or elongate with a longitudinal split; micropyle inconspicuous to conspicuous, closed or open; arillate (rim-aril a collar-like outgrowth around the hilum, well to poorly developed) or nonarillate; lens (a raised area between the hilum and chalaza, area where water penetrates in the otherwise impermeable testa) distinct or inconspicuous, color as of testa or variable; raphe distinct or indistinct, discolored or with a different color *(Lablab)*.

Spermoderm patterns in SEM, smooth, reticulate, striate, tuberculate, rugose, and faveolate; seed coat with or without pores; covered with bloom or other deposits derived from endocarp of pod (Newell and Hymowitz, 1978; Wolf, Baker, and Bernard, 1981; Yaklich, Vigil, and Wergin, 1986).

Internal: Seed coat exotestal, seed epidermis — thick-walled palisade — like prismatic cells (Malpighian cells) with linea lucida, in transections outer and inner facets hexagonal, lumen linear and often substellate; weakly lignified or unlignified; hypodermis one-layered; rarely two-layered in hilar region (*Cajanus*), thick-walled, unlignified; hourglass cells with prominent air-spaces; remaining mesophyll including inner epidermis unspecialized, cells thin-walled, greatly compressed, a few outer layers distinct. In *Cicer*, the palisade layer is without thickening in kabuli seeds, but it is thick-walled and possesses pigmented contents in seeds of desi cultivars (Singh et al., 1984).

The chalaza is simple. The hilum is well differentiated, differentiation taking place in the young developing seed (Baker and Mebrahtu, 1990). The hilar layer is in continuation with the macrosclereid of the seed coat, radially elongated forming hilar palisade layer, funicle cells, opposite the hilar macrosclereids develop into an additional macrosclereid layer referred to as counter palisade layer (confined to hilar region only), palisade and counter palisade layers interrupted along the mid line forming suture or groove leading to the tracheid bar (hilar median groove), tracheid bar below hilar groove, cells elongate or isodiametric, lignified, thickenings reticulate; funicular remnants cover hilum-epihilum; rim aril present or absent; subhilar tissue of stellate parenchyma.

Seed vascular supply variable, usually extends into the antiraphe, postchalazal supply unbranched or branched; two recurrent bundles provided to the hilar region.

Embryo: Cotyledons two, thick, radicle exposed and embryonic axis inflexed; epicotyl with one or more buds, one or more seminal leaves, reserve food protein and starch in most pulses; protein and oil in oil seeds (*Arachis, Glycine*).

Endosperm: Scanty, copious in *Trigonella* and *Cyamopsis;* cells in *Trigonella* and *Cyamposis* thin-walled, mucilagenous.

Seed coat in *Arachis hypogaea* is atypical for Faboideae, characteristic malpighian cells and hourglass cells are lacking and so are the special features of hilum. Seed coat of unlignified cells with cellulose thickenings, epidermis of short palisade or squarish cells with thickenings on radial walls. Mesophyll and inner epidermis thin-walled enclosing vascular supply.

Seed of bambarra groundnut (*Vigna subterranean*; syn. *Vandezeia subterranea*) has features similar to those of other Faboideae members.

3.4.5 CUCURBITACEAE (FIGURE 3.8A TO F)

(Singh, 1953; Singh, 1965, 1968; Singh and Dathan, 1972, 1973, 1974, 1990)

External: Seeds medium to large, oval to ovate, ellipsoid or globose (*Trichosanthes dioica*); compressed or flattened, tumid or turgid, pointed or beaked at hilar end; white, pale or cream, brown to black, rarely red; smooth or sculptured with or without a distinct margin, winged in *Luffa cylindrica*; hilum inconspicuous; micropyle obliterated.

Fresh seeds in *Momordica* and *Trichosanthes* enclosed in an envelope of placental origin (placental aril), usually red.

Internal: Seed coat derived from outer integument only, generally consists of five identifiable zones: (1) seed epidermis, homocellular or heterocellular, cells large (epl) and small (eps), radially or horizontally enlarged; (2) hypodermis is a few to many layered, thin- or thick-walled; when multilayered, it may be distinguished into two zones — outer zone of thin-walled cells and inner zone of thick-walled cells; in *Luffa*, cells of the innermost hypodermal layer are radially elongated, thick-walled, lignified and pitted; (3) main sclerenchymatous layer (e′, the innermost derivative of ovular epidermis), cells thick-walled lignified-macrosclereids (*Luffa*), oesteosclereids (*Sicyos* and *Marah*), or astrosclereids (*Benincasa, Cucurbita, Cucumis, Citrullus, Trichosanthes, Lagenaria,* and *Momordica*), (4) aerenchyma one or many layered, cells stellate, thin-walled or weakly thick-walled, and (5) parenchyma or chlorenchyma, thin-walled cells with poor contents. The inner epidermis, which has sometimes been recognized as a distinct sixth zone, is indistinguishable from the adjacent cell layers.

In mature seed, the outer three zones usually remain together while the inner two zones — aerenchyma and parenchyma or chlorenchyma — detach from the main mechanical layer, forming two seed coats.

Solitary seed in *Sechium edule* (chou-chou) is large, viviparous, seed coat leathery — cells of epidermis, hypodermis and the so-called main mechanical layer remain thin-walled and acquire abundant starch grains.

Seed vascular supply traverses the inner layers in raphe and antiraphe, usually unbranched, branched and anastomising in *Momordica, Trichosanthes,* and *Cyclanthera,* three or more vascular bundles of ovary form ovular and seed supply in *Sechium* and *Sicyos.*

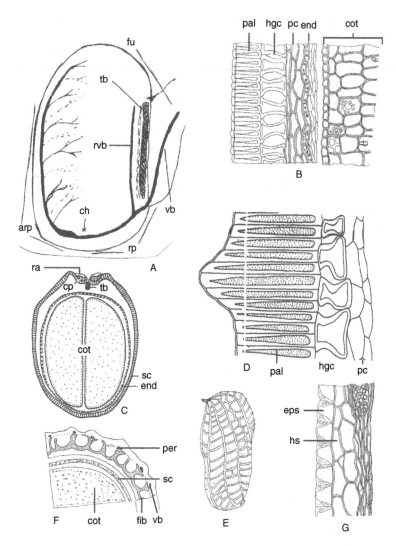

FIGURE 3.7 Seed structure in Fabaceae (Faboideae). A, Ls seed showing vascular supply in *Convalia*. B, Ts part of seed coat in *Glycine max*. Note the palisade (macrosclereid) layer, hourglass cells (osteosclereids), and parenchyma zone in seed coat. C, D, Ts seed and seed coat in *Melilotus alba*. Note the hilar organization consisting of palisade layer, counter palisade layer, funicular remnants on outer side, and tracheid bar and spongy parenchyma on inner side. E to G, *Arachis hypogaea*. E, Diagram of fruit. F, Ts part of fruit. G, Ts seed coat. Note simple epidermal cells and lack of hourglass cell layer. (Abbreviations: arp, anti-raphe; ch, chalaza; cot, cotyledon; cp, counter palisade; end, endosperm; eps, seed epidermis; fib, fiber zone; fu, funiculus; hgc, hourglass cells; hs, seed hypodermis; pal, palisade layer; pc, parenchyma cells; per, pericarp; ra, rim aril; rp, raphe; rvb, recurrent vascular bundle; sc, seed coat; tb, tracheid bar; vb, vascular bundle.) (B, E to G, Redrawn from Vaughan, J.A. 1970. *The Structure and Utilization of Oil Seeds.* Chapman & Hall, London; C, D, From Jha, S.S. and Pandey, A.K. 1989. *Phytomorphology* 39: 273–285.)

Chalaza simple; only outer one to four layers persist in nucellus, cells in outermost layer regular and covered by a thick cuticle.

Embryo: Cotyledons two, large, thick, fleshy, partially or completely investing the radicle, cotyledons multi-layered, hypodermis on adaxial surface palisade-like; fats and lipids and protein major reserve food material, procambial strands distinct in cotyledons.

Endosperm: One or two layers of greatly compressed cells.

3.4.6 APIACEAE (UMBELLIFERAE) (FIGURE 3.9 A TO H)

(Gupta, 1964; Sehgal, 1965; Arora, 1976)

Fruit dry, schizocarpic, splitting longitudinally into two parts, called mericarps (seed for agriculture). Most mericarps are flattened or concave on one side where they remain attached to the carpophore. In cross section, mericarps are asymmetrical with ridges and furrows. Pericarp forms the protective covering.

External: Seed (mericarp) elongate, narrowly oblong, planoconvex, flattened or plano or convexo-concave; usually five or more distinct or indistinct ridges, in *Daucus carota* and *Cuminum cyminum*, the primary ridges are intercalated by secondary ridges, which are in the form of multicellular emergences (vallecular ridges of Heywood and Dakshini, 1971); dark lines between ridges mark the position of oil ducts (vittae); surface with hairs, prickles, and warts; mericarp scar basal inconspicuous; stylopodium (lower swollen part of style) and calyx (sepals) usually persistent at the tip of mericarp.

Internal: Mericarp coat (pericarp) thick, differentiated into exocarp (outer epidermis or one or two subepidermal layers), mesocarp (middle layers) and endocarp (inner epidermis); epidermis covered with thick cuticle, stomata present, cells mostly rectangular, outer wall thick; mesocarp of two or three zones, outer zone mostly chlorenchymatous (*Daucus, Cumin,* and *Foeniculum*), sometimes parenchymatous; inner pericarp always parenchymatous; in some cases, e.g., *Coriandrum*, the cells of middle layers in mesocarp are tangentially elongated, thin-walled, lignified forming a fibrous zone; endocarp indistinct or distinct of tangentially or radially elongated thin or thick-walled cells.

Mericarps in mesocarp have five vascular bundles in ridges (one dorsal, two lateral, and two commisural bundles); alternate to vascular bundles, vittae occur in furrows. The vittae lie above the vascular bundles in *Coriandrum*. In *Daucus* vittae occur in both alternating vascular bundles and above the bundles.

The carpophore, an area common to the mericarps, is supplied by the ventral vascular bundle. It is flanked by two vittae. At maturity its cells become thick-walled and lignified. Finally it detaches from the tissue of the septum and forms the axis on which the two mericarps hang.

Seed Coat: Membranous, of more or less crushed cells or with persistent thin-walled outer epidermis.

Embryo: Small, basal, rudimentary or axile with two cotyledons.

Endosperm: Abundant, cell walls thick, reserve food as oil.

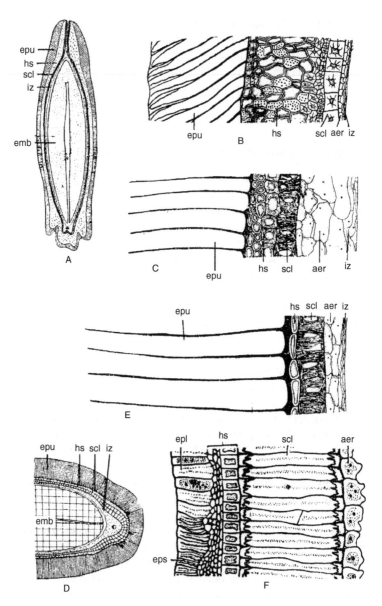

FIGURE 3.8 Structure of seed in Cucurbitaceae. A, B, *Benincasa hispida*. A, Ls seed. B, Ts part of seed coat. C, Ts part of seed coat of *Cucumis melo*. D, E, Ts seed and part of seed coat of *Cucumis sativus*. F, Ts part of seed coat of *Luffa acutangula*. (Abbreviations: aer, aerenchyma; emb, embryo; epl, large cells of seed epidermis; eps, small cells of seed epidermis; epu, uniform size of cells of seed epidermis; hs, seed hypodermis; iz, inner zone of parenchymatous cells; scl, sclerenchymatous cells.) (A, B, From Singh, D. and Dathan, A.S.R. 1978. In *Physiology of Sexual Reproduction in Flowering Plants*. Malik, C.P., Ed. Kalyani Publishers, New Delhi, India, pp. 292–299; C to E, From Singh, D. and Dathan, A.S.R. 1974. *New Bot.* 1: 8–22; F, From Singh, D. 1971. *J. Indian Bot. Soc.* 50A: 208–215.)

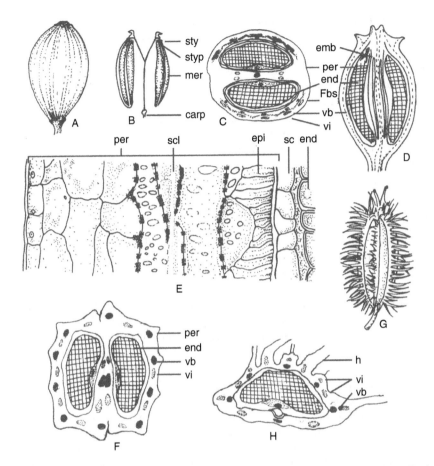

FIGURE 3.9 Structure of cremocarp in Apiaceae. A to D, *Coriandrum sativum*. A, B, Entire and splitted cremocarp showing parts. C, D, Ts and Ls cremocarp respectively. E, F, *Foeniculum vulgare*. E, Ls part of immature mericarp wall with sclerotic pitted cells in pericarp. F, Ts cremocarp showing vascular bundles in ridges and vittae in alternate positions. The carpophore has two vascular bundles and four vittae, two on each side of vascular bundles. G, H, *Daucus carota*, outline diagram of cremocarp and Ts of mericarp respectively. Note that the vittae occur above each vascular bundle and also in alternate positions. (Abbreviations: carp, carpophore; emb, embryo; end, endosperm; epi, inner epidermis of pericarp; fbs, fiber strands; h, hair; mer, mericarp; per, pericarp; sc, seed coat; scl, sclerenchymatous pitted cells; sty, style; styp, stylopodium; vb, vascular bundle; vi, vittae.) (E, From Gupta, S.C. 1964. *Phytomorphology* 14: 530–547. With permission.)

3.4.7 PEDALIACEAE (FIGURE 3.6C)

(Singh, 1960; Vaughan, 1970)

External: Seeds small-medium, flattish ovate, beaked; black, brown or white; surface smooth or weakly reticulate with faint marginal line, hilum and micropyle inconspicuous.

Internal: Seeds thinly albuminous, seed coat exotestal; epidermal cells radially elongated and those at the edges larger, forming ridges, crystalliferous (calcium oxalate crystals), crystals in outer part of cells in *Sesamum indicum* and in inner region in *S. radiatum;* mesophyll — four to six layers, thin-walled, compressed; inner epidermis — endothelium in developing seed, cells unspecialized and bound on inner side by cuticle.

Embryo: Erect, axile, broadly spatulate, cotyledons two, multi-layered, hypodermis on adaxial side of palisade cells, remaining layers of isodiametric cells; reserve food material, oil and aleurone grains.

Endosperm: Two to five layers, epidermis covered by cuticle, cells thin-walled, oil and aleurone grains as reserve food material.

3.4.8 SOLANACEAE (FIGURE 3.10A TO K)

(Saxena and Singh, 1969; Saxena, 1970; Vaughan, 1970; Sharma, 1976)

External: Seeds small to medium, flattened and subcircular or discoid (*Lycopersicon, Solanum* and *Capsicum*), minute and globose, subglobose or cubical (*Nicotiana*); white to yellowish, brown-black; smooth, reticulate or hairy; hilum in discoid seeds marginal in notch, but in globose and cubical seeds subterminal and flush; micropyle inconspicuous.

Internal: Seed coat one, exotestal, seed epidermis main mechanical layer, cells radially elongated with inner tangential and radial walls thickened, thickening heavy at base and tapering outwards, lignified, outer tangential wall remains thin (*Lycopersicon* and some *Solanum* spp.); cells flattened with rodlike thickenings on radial walls emerging from inner tangential wall (*Capsicum,* some *Solanum* spp.), epidermal cells with radial and tangential walls thickened, lumen full of pigmented contents (*Nicotiana*); mesophyll present or absent, when present cells thin-walled, compressed, if multi-layered those in outer zone distinct (*Capsicum*); inner epidermis (endothelium) — cells small, narrow, thin-walled with striate or reticulate thickenings, full of pigmented material.

Note: Freshly harvested seeds in *Lycopersicon* and *Solanum* are enclosed in a succulent sac (arillode of placental origin). It loses water on exposure to the atmosphere, reduced to a thin membrane. The dried arillode and the outer tangential wall of seed epidermis get detached, and the thick radial walls assume the appearance of hairs, silky outgrowths or rodlike fibrous thickenings. Corner (1976) called them *spurious hairs* and Rick (1978), *pseudohairs.* No true hairs occur on seeds of Solanaceae.

Embryo: Axile straight, bent or curved (*Nicotiana*), coiled, annular (*Capsicum,* some *Solanum* spp.), spirally coiled with tips incurved (*Lycopersicon,* some *Solanum* spp.), coiled with tips recurved (*Datura* spp.); cotyledons two, flat or folded, hypocotyledonary root axis well developed, cells with oil and aleurone grains.

Endosperm: Five- or six-layered; in seeds with coiled embryo with comma head and comma stem; cells thin or thick-walled, parenchymatous, reserve food material as oil and aleurone grains.

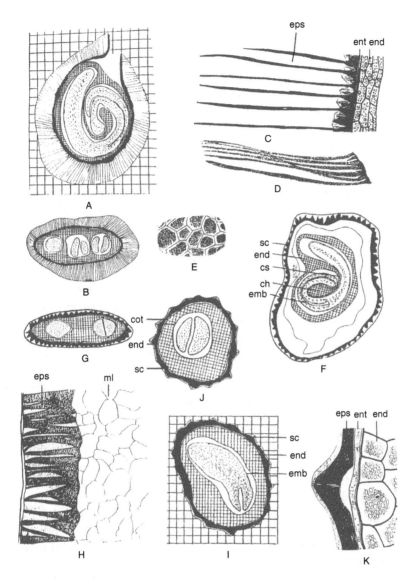

FIGURE 3.10 Structure of seed in Solanaceae. A to E, *Lycopersicon esculentum*. A, B, Ls and Ts of seed. C, Ls seed coat with endosperm. D, Single epidermal cell from maceration; note elongated thick radial walls with pointed tips. E, Endothelial cells in surface view showing pigmented contents and thick cell walls. F to H, *Capsicum annuum*. F, G, Ls and Ts seed. H, Ls part of seed coat with strong rodlike thickenings on radial walls and persistent thin-walled middle layers. I to K, *Nicotiana tabacum*. I, J, Ls and Ts seed. K, Ls part of seed coat, epidermal cells tangentially elongated and full of pigmented contents. (Abbreviations: ch, comma head and cs, comma stem of endosperm; cot, cotyledon; emb, embryo; end, endosperm; ent, endothelium; eps, seed epidermis; ml, middle layers; sc, seed coat.) (A to H, From Saxena, T. 1970. Ph.D. thesis, University of Rajasthan, Jaipur, India; I to K, From Sharma, R.C. 1976. Ph.D. thesis, University of Rajasthan, Jaipur, India.)

3.4.9 ASTERACEAE (COMPOSITAE) (FIGURE 3.11A TO C)

(Borthwick and Robbins, 1928; Vaughan, 1970; Kaul, 1972; Chopra and Singh, 1976); Kaul and Singh, 1982)

The disseminule or propagule is one-seeded indehiscent fruit known as cypsela. The description of fruit is given with reference to important crop plants.

External: Cypsils are straight, slightly or strongly curved (*Calendula*); pointed at both ends (*Lactuca* and *Guizotia*), obconical pointed at basal end and rounded at top (*Carthamus* and *Cichorium*), elongate-oblong, compressed roughly four-sided in cross-section (*Helianthus*); terete, angled, longitudinally ribbed; white, cream, light to dark brown, black; glabrous, hairy, bristled or barbed; dull or lustrous; fruit scar basal, depressed, straight or oblique, at the top annular ring often present; pappus present or absent, when present numerous fine bristles (*Lactuca*), in place of pappus stiff barbs, papery scales or stubby scales present (*Carthamus*).

Internal: Pericarp of epicarp, mesocarp and endocarp, epicarp one-layered, cells rectangular or elongated along the long axis of the fruit; rarely, cell ends forming spiny excrescences (*Lactuca* and *Cichorium*), with or without pigmented contents; mesocarp — a few or multi-layered — with or without split, schizogenous space filled with phytomelanin (*Helianthus, Carthamus,* and *Guizotia*), cells thick-walled or thin-walled; thick-walled sclerosed cells uniformly distributed (*Helianthus* and *Carthannus*) or occur in fibrovascular strands (*Lactuca* and *Guizotia*), hypodermal, three to four layers with calcium oxalate crystals (*Cichorium*). Endocarp one-layered, thin-walled, inconspicuous.

Presence of phytomelanin is of particular interest as it is shown to deter insect predation in the cultivated sunflower (Carlson and Witt, 1974; Rogers and Kreitner, 1983). It is hard and highly resistant to alkali and acids. Little is known about its chemical nature; Vries (1948) and Hegnauer (1977) consider it to be of highly unsaturated acetylenes in *Helianthus* and *Tagetes*. Hegnauer (1977) has suggested that such compounds are effective against nematodes and bacteria.

Seed Coat: Exotestal, epidermis conspicuous, cells rectangular, tangentially stretched, radially or obliquely elongated, thin- or thick-walled, thickening, uniform without or with pits (*Helianthus*) or confined to radial walls; striate or wedge-shaped (*Cichorium* and *Lactuca*), one or more hypodermal layers also present, stretched, compressed, usually thin-walled, rarely subdermal layer thick-walled (*Carthamus*).

Embryo: Axile, straight, spatulate, cotyledons two, appressed, palisade layer demarcated on adaxial side; procambial strands differentiated, oil and aleurone grains as reserve food material.

Endosperm: One- or two-layered, cells rectangular covered by thick cuticle, oil drops and aleurone grains present.

3.4.10 AMARANTHACEAE (FIGURE 3.12H TO J)

(Woodcock, 1931; Kowal, 1954; Taneja, 1981)

External: Seeds small to medium, rounded (circular), lenticular, reniform or oblong, cylindrical, usually with a distinct marginal rim, notched, brown to black,

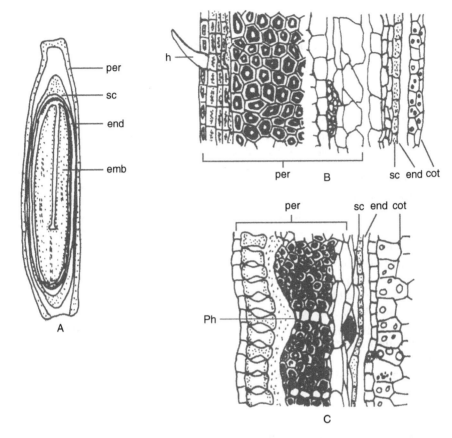

FIGURE 3.11 Cypsil structure in Asteraceae. A, Ls cypsil of *Lactuca longifolia*. B, Ts part of cypsil of *Helianthus annuus* showing hair, pericarp with phytomelanin layer, seed coat, and cotyledon. C, Ts part of cypsil of *Guizotia abyssinica*. (Abbreviations: cot, cotyledon; emb, embryo; end, endosperm; per, pericarp; ph, phytomelanin layer; sc, seed coat.) (A, From Kaul, V. and Singh, D. 1982. *Acta Biologica Cracoviensia* 24: 19–30; B, C, Vaughan, J.A. 1970. *The Structure and Utilization of Oil Seeds*. Chapman & Hall, London.)

smooth, shining; hilum inconspicuous at or near marginal notch, micropyle inconspicuous.

Internal: Seed coats two, exotestal, testa (outer coat) crustaceous, epidermis of cuboid or radially elongated cells, thick-walled, thickening as *stalactites* (long pillar-like or wedge-shaped projections with pointed ends from the outer tangential wall), stalactites rarely absent (*Digeria, Achyranthes,* and *Alternanthera*); mesophyll of one or two layers of thin-walled cells. Inner coat membranous consisting of inner epidermis of inner integument, cells narrow, elongated, full of pigmented contents and with bands of thickenings on cell walls.

Embryo: Peripheral, large, curved-lenticular to annular, rarely tips of cotyledons incurved (*Achyranthes*), cotyledons two, equal or larger than the hypocotyledonary axis.

Endosperm: Scanty, remnants around embryo and at the micropylar and chalazal ends, cell layers distinct only at ends.

Perisperm: Massive, main tissue having reserve food material, enclosed in the concavity of embryo, cells thin-walled, rich in starch grains.

3.4.11 CHENOPODIACEAE (FIGURE 3.12A TO G)

(Artschwager, 1927; Bhargava, 1936; Taneja, 1981)

External: Seed small to medium, circular-lenticular to reniform, notched, rarely ovoid and beaked (*Saeuda*) or flat and narrowly obovate (*Kochia*), brown to black, shining. Di- and polymorphism of seeds known in *Atriplex* spp. Achenes with attached perianth (seed ball) form planting unit in *Beta* and *Spinacia*.

Internal: Seed coats two, exotestal, seed epidermis radially or horizontally elongated, usually thick-walled with *stalactites*, cells with uniformly thickened outer tangential and radial walls (*Spinacia*) or full of pigmented material (*Beta*). Stalactites absent in both of them. Mesophyll one or two layers, thin-walled, compressed; innercoat membranous formed by inner epidermis, cells narrow tangentially stretched, full of pigmented contents, cell walls with bands of thickenings. In *Kochia* and *Salsola* the seed coat is membranous and uniseriate.

Embryo: Peripheral, curved-horse-shoe shaped, annular or coiled, large, cotyledons two, hypocotyledonary root axis usually longer than the cotyledons.

Endosperm: Scanty, remnants of one or two layers at micropylar and chalazal ends.

Perisperm: Massive, enclosed in the concavity of embryo; cells thin-walled, full of starch grains, main storage tissue. Perisperm absent in seeds of *Saeuda* and *Salsola*.

3.4.12 POACEAE (GRAMINAE) (FIGURE 3.13A TO I)

(Kiesselbach and Walker, 1952; Narayanswami, 1953, 1955a,b; Sanders, 1955; Bradbury et al. 1956a,b; Chandra, 1963, 1976)

One-seeded indehiscent fruit, caryopsis usually with pericarp and testa fused, form the commercial grain or seed for planting. There is considerable variability in exomorphic and internal features of the caryopsis. This account mainly concerns cultivated cereals and millets.

External: Caryopsis naked (*Triticum, Zea, Sorghum,* and *Pennisetum*) or covered with lemma and palea, which are chaffy (*Hordeum* and *Avena*) or hard horny (*Oryza* and *Paspalum*); pericarp papery, hyaline, white, easily separating and seed coat well developed, tough (*Eleusine*); elongated-ovate, elliptic, fusiform, rounded or squarish; smooth, faintly or prominently longitudinally striate, rough, pubescent or spinescents, rarely median groove on ventral surface (*Triticum*); pale white, yellow, various shades of red, brown, purplish, grey or black; basal lateral embryo area distinctive and easily identified in naked types (*Zea, Triticum, Sorghum,* and *Pennisetum*); caryopsis scar basal, opposite embryo on ventral side, inconspicuous or conspicuous, rarely floret stalk (rachilla) attached at the base (*Hordeum, Avena,* and *Oryza*); in naked grains at the apical end a blunt pointed outgrowth marks the remains of style

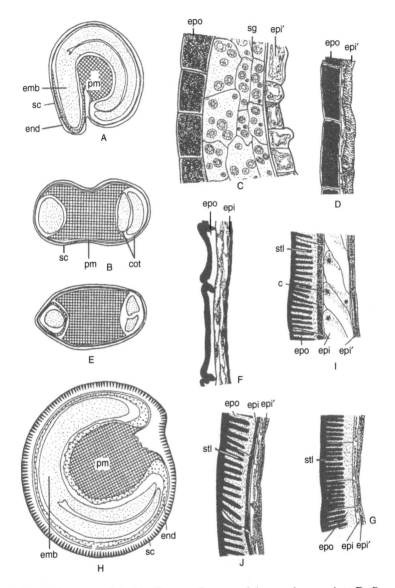

FIGURE 3.12 Structure of seed in Chenopodiaceae and Amaranthaceae. A to D, *Beta vulgaris*. A, B, Ls and Ts seed. C, Ls part of seed coat at early globular stage of embryo. Note deposition of pigmented material in outer epidermis of outer integument and inner epidermis of inner integument. D, Ls part of mature seed coat. E, F, *Spinacea oleracea* Ts seed and Ls part of seed coat, respectively. G, Ls part of seed coat of *Chenopodium album* showing stalactitis in epidermis. H to J, *Amaranthus viridis*. H, Ls seed. I, J, Ls part of immature and mature seed coat respectively. (Abbreviations: c, crystals, cot, cotyledon; end, endosperm; emb, embryo; epi, inner epidermis of outer integument; epi′, inner epidermis of inner integument; epo, outer epidermis of outer integument; pm, perisperm; sc, seed coat; sg, starch grains; stl, stalactitis.) (From Taneja, C.P. 1981. Ph.D. thesis, University of Rajasthan, Jaipur, India.)

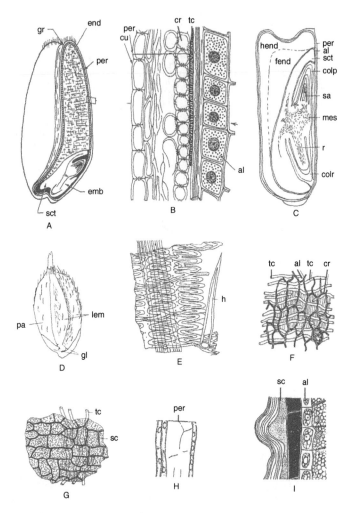

FIGURE 3.13 Structure of caryopsis in Poaceae. A, B, Ls of caryopsis and part of pericarp respectively in *Triticum*. Note differentiation of pericarp and structure of cross cells and tube cells. C, Ls caryopsis of *Zea mays*. D to G, *Oryza sativa*. D, Caryopsis covered with glumes (lemma with awn and palea). E, Macerated part of husk showing longitudinal rows of epidermal cells, mesophyll cells, and hair. F, Surface view of part of pericarp showing tube cells and cross cells. G, Tube cells and cells of pigmented testa. H, I, *Eleusine indica*. H, Ls part of pericarp. Note thinwalled cells, which form thin, loose, papery membrane around the seed. I, Ls part of seed coat, aleurone layer, and endosperm. Seed coat formed by both the layers of inner integument, outer thick-walled, and inner with pigmented contents. (Abbreviations: al, aleurone layer; colp, coleoptile; colr, coleorhiza; cr, cross cells; cu, cuticle; emb, embryo; end, endosperm; fend, floury endosperm; gl, glume; gr, groove; h, hair; hend, horny endosperm; lem, lemma; mes, mesocotyl; pa, palea; per, pericarp; r, radicle; sa, shoot apex; sc, seed coat; sct, scutellum; tc, tubular cells.) (A, B, From Esau, K. 1974. *Anatomy of Seed Plants*. Wiley Eastern, New Delhi. With permission; C to G, Redrawn from Vaughan, J.A. 1970. *The Structure and Utilization of Oil Seeds*. Chapman & Hall, London; H, I, From Chandra, N. 1963. *Proc. Indian Acad. Sci.* 58: 117–127.)

(*Zea* and *Sorghum*), a tuft of hairs in *Triticum*. Caryopsis covered with lemma and palea have base of awn on the terminal end and in *Oryza* lips of lemma and palea they become hard and pointed and are called *apiculus*.

Internal: Caryopsis coat thin to coriaceous or crustaceous, loose and membranous (*Eleusine* and *Eragrostis*); embryo lateral and endosperm abundant.

Pericarp protective and usually consists of (1) epicarp, (2) mesocarp, (3) cross cells, and (4) tube cells. Epicarp, single layer, cells elongated along long axis, walls thick and pitted, covered by a thick cuticle; mesophyll, number of layers vary, cells of outer layers usually thick-walled, thickening gradually, decreasing from outside to inside; cross cells, inner hypodermis, cells elongated at right angles to those of epicarp or transversely, walls thick and pitted; tube cells, inner epidermis of pericarp, cells elongated along the long axis of the grain or at right angles to cross cells, intercellular spaces large, walls relatively thin and pitted. Pericarp zones one to four clearly recognized in *Triticum*, *Zea*, *Sorghum*, *Pennisetum*, and *Secale*; zones recognized in *Oryza* but mesocarp greatly reduced; undifferentiated in *Hordeum* covered by multi-layered persistent adherent lemma and palea; and membranous, thin-walled in *Eleusine and Eragrostis*. In *Oryza*, hull or husk, formed by lemma and palea, epidermis of almost squarish cells, deeply sinuate walls, pointed hairs and stomata present; hypodermis of two or three layers of thick-walled elongated fibers; mesophyll spongy parenchyma; inner epidermis cells thin-walled, isodiametric, stomata present; vascular bundles present in inner zone of husk.

Seed Coat: Usually adherent to pericarp, formed by inner epidermis of inner integument, cells without or with tanniferous contents (*Sorghum, Pennisetum, Oryza,* and *Echinochloa*), crushed and only cuticle distinct (*Zea, Triticum, Euchlaena,* and *Secale*); seed coat well developed, formed by two layers of inner integument, outer thick-walled and horny, inner thin-walled with pigmented contents (*Eleusine* and *Eragrostis*); rarely pigmented contents present in outer layer also.

Placento-Chalazal Region: The placental tissue of the carpel and the adjoining tissue of the campylotropous ovule merge without the intervention of the funiculus, the region is called *placento-chalazal region*. It is multi-layered differentiated into three or more zones, cells below embryo-endosperm thin-walled, compressed; followed by a zone of compressed, nearly structureless cells with brown contents, closing zone; cells of the remaining layers thin or thick-walled, peripheral layers with suberised walls.

Nucellus: Usually as a thin cuticular membrane; when present, one- or two-layered, cells thin-walled.

Embryo: Peripheral, basal, lateral, monocotyledonous — scutellate with typical features of grass embryo — coleorhiza, coleoptile, and epiblast present or absent; radicle with primordia of lateral roots, epicotyl with primordia of foliage leaves in cultivated species.

Endosperm: Abundant, starchy, floury or floury and horny; cells of epidermis or outer two to three layers with abundant aleurone grains and poor or with no starch grains form the aleurone layer(s).

3.5 CONCLUDING REMARKS

Seed (*sensu lato*) includes true seeds as well as one-seeded dry indehiscent fruits. A viable seed has a protective covering, embryo, or germ and reserve food material, which may be stored separately or in the cotyledons of the embryo. The color, shape, size, and surface features of seeds vary considerably. The micropyle may be closed or open; the hilum varies in size, shape, and position. The organization of the hilum precedes seed maturity and partial abscission observed during seed development. The size, shape, and position of embryo in seed are variable. The axile embryo may be surrounded by endosperm or lie below the seed coat, while the peripheral embryo, despite copious endosperm, has close contacts with the protective covering. The seed coat in true seeds and pericarp in one-seeded fruits have thick-walled mechanical tissue and often in colored seeds pigmented phenolic compounds. The seed structure is variable in different taxa, but broadly follows a common pattern in a particular family.

The features of seed, particularly the nature of cuticles, position and nature of mechanical tissue, position of embryo, nature of micropyle (open or closed), and features of the seed surface (micropores, stomata, hairs, cracks, and waxy deposits) play an important role in imposing functional attributes of seeds. They act as preformed passages or barriers to water absorption and infection by microorganisms.

The seeds may show di- or polymorphism with respect to the above microfeatures. Bhattacharya and Saha (1992) have identified two categories of seeds in *Cassia tora:* seeds with (1) smooth surface and closed micropyle and (2) rough surface with pores and micropyle open. Seeds of the first category are dormant, while those of the second category are nondormant. Similarly, Russi et al. (1992) observed differences in cuticle thickness (thin and thick), which accounts for differences in seed dormancy. Thickness of the cuticle, light line, palisade cells, and presence or absence of pores and closed or open micropyle are features that have been found to affect water permeability and as infection by microorganisms in fabaceous seeds (Marouani, 1990; Kulik and Yaklich, 1991).

The presence of pigmented contents, phenols, or phenol-like substances in seed coat and pericarp may impart resistance to infection by providing biochemical barriers (Glueck and Rooney, 1978; Singh and Singh, 1979). In *Helianthus annuus*, phytomelanin in the pericarp reduces insect predation (Carlson and Witt, 1974; Rogers and Kreitner, 1983).

REFERENCES

Arora, K. 1976. Morphological and Embryological Studies in Umbelliferae. Ph.D. thesis, University of Rajastha, Jaipur, India.

Artschwager, E. 1927. Development of flower and seed in the sugar beet. *J. Agric. Res.* 34: 1–25.

Azegami, K., Tabei, H., and Fukuda, T. 1988. Entrance into rice grains of *Pseudomonas plantarii*, the causal agent of seedling blight of rice. *Ann. Phytopathol. Soc. Japan* 54: 633–636.

Baker, D.M. and Mebrahtu, T. 1990. Scanning electron microscopy examination of soybean hilum development. *Am. J. Bot.* 68: 544–550.

Barthlott, W. and Wollenweber, E. 1981. Zur Feinstruktur, Chemie und taxonomischen Signifikanz epicuticulurer Wachse und ähnlicher Sekrete. *Trop. Subtrop. Pflanzenwelt* 32: 1–67.

Bechtel, D.B. and Pomeranz, Y. 1978. Ultrastructure of the mature ungerminated rice (*Oryza sativa*) caryopsis and the germ. *Am. J. Bot.* 65: 75–85.

Bellue, M.K. 1952. *Weed Seed Handbook*. U.S. Department of Agriculture Bulletin, Sacramento, CA.

Bhargava, H.R. 1936. The life-history of *Chenopodium album* Linn. *Proc. Indian Acad. Sci.* 4: 179–200.

Bhattacharya, A. and Saha, P.K. 1992. SEM studies on morphological diversities in the seeds of *Cassia tora* L. *Seed Sci. Technol.* 20: 85–91.

Boesewinkel, F.D. 1980. Development of ovule and testa of *Linum usitatissimum* L. *Acta Bot. Neerl.* 29: 17–32.

Boesewinkel, F.D. 1984. A comparative SEM study of the seed coats of recent and 900–1100 years old, subfossil linseed. *Ber. Dtsch. Bot. Ges. Bd.* 97: 443–450.

Borthwick, H.A. and Robbins, W.W. 1928. Lettuce seed and its germination. *Hilgardia* 3: 275–305.

Bouman, F. 1974. Developmental Studies of the Ovule, Integument and Seed in Some Angiosperms. Ph.D. thesis, University of Amsterdam, Amsterdam, the Netherlands.

Bouwer, W. and Stahlin, A. 1955. *Handbuch der Samenkunde fur Landwirtschaft*. DLG Verlag, Frankurt.

Bradbury, D., Cull, J.M., and MacMasters, M.M. 1956a. Structure of the mature wheat kernel. I. Gross anatomy and relationships of parts. *Cereal Chem.* 33: 329–342.

Bradbury, D., MacMasters, M.M., and Cull, J.M. 1956b. Structure of mature wheat kernel. II. Microscopic structure of pericarp, seed coat, and other coverings of the endosperm and germ of hard red winter wheat. *Cereal Chem.* 33: 342–360.

Breggren, G. 1969. *Atlas of Seeds. Part 2: Cyperaceae*. Swedish National Science Research Council, Stockholm.

Britton, F.A. and Fuller, T.C. 1957. *Weed Seed Handbook*. U.S. Department of Agriculture Bulletin, Sacramento, CA.

Calero, E., West, S.H., and Hinson, K. 1981. Water absorption of soybean seeds and associated causal factors. *Crop Sci.* 21: 926–933.

Carlson, E.C. and Witt, R. 1974. Moth resistance in armored layer sunflower seeds. *Calif. Agric.* 28: 12–14.

Chandra, N. 1963. Morphological studies in the Gramineae. IV Embryology of *Eleusine indica* Gaertn. and *Dactyloctenium aegyptium* (Desf.) Beauv. *Proc. Indian Acad. Sci.* 58: 117–127.

Chandra, N. 1976. Embryology of some species of *Eragrostis*. *Acta. Bot. Indica* 4: 36–43.

Chopra, S. and Singh, R.P. 1976. Effect of gamma rays and 2,4-D on ovule, female gametophyte, seed and fruit development in *Guizotia abyssinica*. *Phytomorphology* 26: 240–249.

Cochrane, M.P. and Duffus, C.M. 1979. Morphology and ultrastructure of immature cereal grains in relation to transport. *Ann. Bot.* 44: 67–72.

Corner, E.J.H. 1976. *The Seeds of Dicotyledons*. Vols. 1 and 2. Cambridge University Press, Cambridge, U.K.

Dobrokhotov, V.N. 1961. *Seeds of Weed Plants*. Agricultural Literature, Moscow.

Esau, K. 1974. *Anatomy of Seed Plants*. Wiley Eastern, New Delhi.

Fukuda, T., Azegami, K., and Tabei, H. 1990. Histological studies on bacterial black node of barley and wheat caused by *Pseudomonas syringae* pv. *japonica*. *Ann. Phytopathol. Soc. Japan* 56: 252–256.

Glueck, J.A. and Rooney, L.W. 1978. Chemistry and structure of grain in relation to mold resistance. *Proceedings of the International Workshop on Sorghum Diseases*. Hyderabad, India.

Gunn, C.B. 1970a. A key and diagrams for the seeds of one hundred species of *Vicia* (Leguminosae). *Proc. Inter. Off. Seed Analysts* 35: 773–790.

Gunn, C.B. 1970b. Seeds of the tribe Vicieae (Leguminosae) in North American Agriculture. *Proc. Assoc. Off. Seed Analysts* 60: 48–70.

Gunn, C.B. 1971. *Seeds of Native and Naturalised Vetches of North America*. U.S. Department of Agriculture Handbook, p. 392.

Gupta, S.C. 1964. The embryology of *Coriandrum sativum* L. and *Foeniculum vulgare* Mill. *Phytomorphology* 14: 530–547.

Hegnauer, R. 1977. The chemistry of the Compositae. In *The Biology and Chemistry of the Compositae*. Heywood. V.H., Harborne, J.B., and Turner, B.L., Eds. Academic Press, London, pp. 283–335.

Heinisch, O. 1955. *Samenatlas*. Deutsche Akademische Landwirtschaft, Berlin.

Heywood, V.H. and Dakshini, K.H.M. 1971. Fruit structure in the Umbelliferae. Caucalideae. In *Biology and Chemistry of the Umbelliferae*. Heywood, V.H., Ed. Linnean Society of London, Academic Press, London, pp. 215–232.

Isley, D. 1947. Investigation in seed classification by family characteristics. *Iowa Agr. Exp. Sta. Res. Bull.* 351: 315–380.

Jha, S.S. and Pandey, A.K. 1989. Seed coat structure in *Melilotus* (Fabaceae). *Phytomorphology* 39: 273–285.

Johri, B.M. and Tiagi, B. 1952. Floral morphology and seed formation in *Cuscuta reflexa* Roxb. *Phytomorphology* 2: 162–180.

Joshi, P.C., Wadhwani, A.M. [nee Ramchandani, S.], and Johri, B.M. 1967. Morphological and embryological studies of *Gossypium* L. *Proc. Natl. Inst. Sci. India,* 33B: 37–93.

Kapil, R.N., Bor, J., and Bouman, F. 1980. Seed appendages in angiosperms. *Bot. Jahrb. Syst.* 101: 555–573.

Kaul, V. 1972. Cytology and Embryology of Indian Cichorieae (Compositae). Ph.D. thesis, University of Rajasthan, Jaipur, India.

Kaul, V. and Singh, D. 1982. Embryology and development of fruit in Cichorieae. *Lactuca* Linn. *Acta Biologica Cracoviensia* 24: 19–30.

Kiesselbach, T.A. and Walker, E.R. 1952. Structure of certain tissues in the kernel of corn. *Am. J. Bot.* 39: 561–569.

Korsmo, E. 1935. *Unkrautsamen* [Weed Seeds]. Glyndendal Norsk Forlag, Oslo.

Kowal, T. 1954. Morphological and anatomical features of the seeds of the genus *Amaranthus* and keys for their identification. *Monogr. Bot.* 21: 162–193.

Kulik, M.M. and Yaklich, R.W. 1991. Soybean seed coat structures: relationship to weathering resistance and infection by the fungus *Phomopsis phaseoli*. *Crop Sci.* 31: 108–113.

Kumar, P. and Singh, D. 1990. Development and structure of seed coat in *Hibiscus* L. *Phytomorphology* 40: 179–188.

Kumar, P. and Singh, D. 1991. Development and structure of seed coat in Malvaceae. V. *Althea, Pavonia, Abelmoschus*. *Acta Bot. India* 19: 62–67.

Lavialle, P. 1912. Researches sur le development de levaine en fruit chez les composees. *Ann. Soc. Nat. Ser.* 915: 39–152.

Lersten, N.R. 1981. Testa topography in Leguminosae subfamily Papilionoideae. *Proc. Iowa Acad. Sci.* 88: 180–191.

Maheshwari Devi, H., Johri, B.M., Rau, M.A., Singh, D., Dathan, A.S.R., and Bhanwra, R.K. 1995. Embryology of angiosperms. In *Botany in India: History and Progress.* Johri, B.M., Ed. Oxford and IBH Publishing Co., New Delhi, Vol. 2: 59–146.

Marouani, A. 1990. Studies in seed coat anatomy and composition, seed germination and rhizobium seed inoculation of annual *Medicago sativa. Dissertation Abs. Int. B. Sci. Eng.* 51: 484B.

Martin, A.C. 1946. The comparative internal morphology of seeds. *Am. Midl. Nat.* 36: 513–660.

Martin, A.C. 1954. Identifying *Polygonum* seeds. *J. Wildl. Mgmt.* 18: 514–520.

Martin, A.C. and Barkley, W.D. 1961. *Seed Identification Manual.* University of California Press, Berkeley, CA.

McClure, D.S. 1957. Seed characters of selected plant families. *Iowa State Coll. J. Sci.* 31: 649–682.

Murley, M.R. 1944. A seed key to fourteen species of Geraniaceae. *Proc. Iowa Acad. Sci.* 51: 241–246.

Murley, M.R. 1946. Umbelliferae in Iowa, with seed keys. *Iowa State Coll. J. Sci.* 20: 349–364.

Murley, M.R. 1951. Seeds of Cruciferae of Northeastern North America. *Am. Midl. Nat.* 46: 1–81.

Musil, A.F. 1963. *Identification of Crop and Weed Seeds.* U.S. Department of Agriculture Handbook, p. 219.

Narayanswami, S. 1953. The structure and development of the caryopsis in some Indian millets. I. *Pennisetum typhoideum* Rich. *Phytomorphology* 3: 98–112.

Narayanswami, S. 1955a. The structure and development of the caryopsis in some Indian millets. III. *Panicum miliare* Lamk. and *P. miliaceum* Linn. *Lloydia* 18: 61–73.

Narayanswami, S. 1955b. The structure and development of the caryopsis in some Indian millets. IV. *Echinocholoa frumentacea. Phytomorphology* 5: 161–171.

Netolitzky, F. 1926. *Anatomie der Angiospermen – Samen.* G. Borntraeger, Berlin.

Newell, C.A. and Hymovitz, T. 1978. Seed coat variation in *Glycine* Willd. Subgenus *Glycine* (Leguminosae) by scanning electron microscope (SEM). *Brittonia* 30: 76–88.

Pamplin, R.A. 1963. The anatomical development of the ovule and seed in the soybean. Ph.D. dissertation, University of Illinois, Urbana. *Diss. Abstr.* 63: 5128.

Prasad, K. 1974. Studies in the Cruciferae gametophytes, structure and development of seed in *Eruca sativa* Mill. *J. Indian Bot. Soc.* 53: 24–33.

Rathore, R.K.S. and Singh, R.P. 1968. Embryological studies in *Brassica campestris*. L. var. Yellow Sarson Prain. *J. Indian Bot. Soc.* 47: 341–349.

Rick, C.M. 1978. The tomato. *Sci. Am.* 239: 67–76.

Ramchandani, S., Joshi, P.C., and Pundir, S. 1966. Seed development in *Gossypium. Indian Cotton J.* Vol. 20: 97–106.

Rogers, C.E. and Kreitner, G.L. 1983. Phytomelanin of sunflower achenes: a mechanism for pericarp resistance to abrasion by larvae of the sunflower moth (Lepidoptera: Pyralidae). *Environ. Entomol.* 12: 277–285.

Russi, L., Cocks, P.S., and Roberts, E.H. 1992. Coat thickness and hard-seededness in some *Medicago* and *Trifolium* species. *Seed Sci. Res.* 2: 243–249.

Sanders, E.H. 1955. Development and morphology of the kernel in grain sorghum. *Cereal Chem.* 32: 12–25.

Saxena, T. 1970. Studies on the Development and Structure of Seed in Solanaceae. Ph.D. thesis, University of Rajasthan, Jaipur, India.

Saxena, T. and Singh, D. 1969. Embryology and seed development in tetraploid form of *Solanum nigrum. J. Indian Bot. Soc.* 48: 148–157.

Sehgal, C.B. 1965. The embryology of *Cuminum cyminum* L. and *Trachyspermum ammi* (L.) Sprague (= *Carum copticum* Clarke). *Proc. Natl. Inst. Sci. India* B 35: 175–201.

Sharma, R.C. 1976. Studies on the Structure and Development of Seed in Solanaceae with Special Reference to Medicinal Plants. Ph.D. thesis, University of Rajasthan, Jaipur, India.

Singh, B. 1953. Studies on the structure and development of seeds of Cucurbitaceae. *Phytomorphology* 3: 224–239.

Singh, B. 1964. Development and structure of angiosperm seed. I. *Bull. Natl. Bot. Gard. Lucknow* 89: 1–115.

Singh, D. 1965. Ovule and seed of *Sechium edule* S.W. A reinvestigation. *Curr. Sci.* 34: 696–697.

Singh, D. 1968. Structure and development of seed coat in Cucurbitaceae. III. Seeds of *Acanthosicyos* Hook f. and *Citrullus* Schrad. *Proc. Indian Sci. Congr.* Pt. 3: 347.

Singh, D. 1971. Structure and development of seed coat in Cucurbitaceae. II. Seeds of *Luffa* Mill. *J. Indian Bot. Soc.* 50A: 208–215.

Singh, D. and Dathan, A.S.R. 1972. Structure and development of seed coat in Cucurbitaceae. VI. Seeds of *Cucurbita* L. *Phytomorphology* 22: 29–45.

Singh, D. and Dathan, A.S.R. 1973. Structure and development of seed coat in Cucurbitaceae. IX. Seeds of Zanonioideae. *Phytomorphology* 23: 138–148.

Singh, D. and Dathan, A.S.R. 1974. Structure and development of seed coat. *Cucumis* L. *New Bot.* 1: 8–22.

Singh, D. and Dathan, A.S.R. 1978. Structure and development of seed coat in Cucurbitaceae XII. Seed of subtribes Benincasineae and Trochomerineae. In *Physiology of Sexual Reproduction in Flowering Plants*. Malik, C.P., Ed. Kalyani Publishers, New Delhi, pp. 292–299.

Singh, D. and Dathan, A.S.R. 1990. Seed coat anatomy of the Cucurbitaceae. In *Biology and Utilization of the Cucurbitaceae*. Bates, D.M., Robinson, R.W., and Jeffrey, C., Eds. Comstock Publishing Associates, Cornell University Press, Ithaca, NY, pp. 225–238.

Singh, S.P. 1960. Morphological studies in some members of the family Pedaliaceae. I. *Sesamum indicum* DC. *Phytomorphology* 10: 65–81.

Singh, T. and Singh, D. 1979. Anatomy of penetration of *Macrophomina phaseoli* in seeds of sesame. In *Recent Research in Plant Sciences*. Bir, S.S., Ed. Kalyani Publishers, New Delhi, pp. 603–606.

Singh, U., Manohar, S., and Singh, A.K. 1984. The anatomical structure of desi and kabuli chickpea seed coats. *Int. Chickpea Newslett.* 10: 26–27.

Stasz, T.E., Harman, G.E., and Marx, G.A. 1980. Time and site of infection of resistant and susceptible germinating pea seeds by *Pythium ultimum*. *Phytopathology* 70: 730–733.

Sulbha, K. 1957. Embryology of *Brassica juncea* Czern and Coss. *J. Indian Bot. Soc.* 36: 292–301.

Taneja, C.P. 1981. Structure and Development of Seed Coat in some Centrospermae. Ph.D. thesis, University of Rajasthan, Jaipur, India.

Thompson, R.C. 1933. A morphological study of flower and seed development in cabbage. *J. Agr. Res.* 27: 215–237.

U.S. Department of Agriculture (USDA). 1948. *Woody Plant Seed Manual*. Publication 654.

U.S. Department of Agriculture (USDA). 1952. *Manual for Testing Agricultural and Vegetable Seeds*. Handbook 30.

U.S. Department of Agriculture (USDA). 1961. *Seeds*. The Yearbook of Agriculture.

Vaughan, J.A. 1970. *The Structure and Utilization of Oil Seeds*. Chapman & Hall, London.

Vaughan, J.G. and Whitehouse, J.M. 1971. Seed structure and taxonomy of the *Cruciferae*. *Bot. J. Linn. Soc.* 64: 383–409.

Vries, M.A. de 1948. Over de vorming von phytomelaan by *Tagetes patula* L. an enige andere Composieten. H. Burman, Leiden.

Winton, A.L. and Winton, K.B. 1932–1939. *The Structure and Composition of Foods.* Vols. 1–4. Wiley, New York.

Woodcock, E.F. 1931. Seed development in *Amaranthus caudatum. Papers Mich. Acad. Sci. Arts Lett.* 15: 173–178.

Wolf, W.J. and Baker, F.L. 1980. Scanning electron microscopy of soybeans and soybean protein products. *Scanning Electron Microsc.* 3: 621–634.

Wolf, W.J., Baker, F.L. and Bernard, R.L. 1981. Soybean seed-coat structural features: pits, deposits and cracks. *Scanning Electron Microsc.* 3: 531–544.

Yaklich, R.W., Vigil, E.L., and Wergin, W.P. 1984. Scanning electron microscopy of soybean seed coat. *Scanning Electron Microsc.* 2: 991–1000.

Yaklich, R.W., Vigil, E.L., and Wergin, W.P. 1986. Pore development and seed coat permeability in soybean. *Crop Sci.* 26: 616–624.

Zeleznak, K. and Varriano-Marston, E. 1982. Pearl millet (*Pennisetum americanum* (L.) Leeke) and grain sorghum (*Sorghum bicolor* (L.) Moench.) ultrastructure. *Am. J. Bot.* 69: 1306–1313.

4 Penetration and Establishment of Fungi in Seed

Fungal infection of seed-borne pathogens may reach the ovule and seed at any stage from the initiation of ovule to mature seed. The previous chapters provided an account of the stages in ovule development, ovule contacts with the mother plant, probable passages and barriers to infection, and the structure of the mature seed. The developing ovule and seed are enclosed in the ovary and in contact with the mother plant while the disseminated or threshed seed is an independent unit. Different factors determine the entry of the pathogen under these two conditions.

The major pathogen groups, namely fungi, bacteria, and viruses, differ in their modes of multiplication and attack on the host. Fungal propagules germinate, and the hyphae grow. Germination of propagule and initiation of hyphal growth are important factors that determine the entry of fungal pathogens in any tissue, including the fruit and seed. Phytopathogenic bacteria multiply while the viruses replicate intracellularly, and both lack the phenomenon of growth. This chapter discusses penetration and establishment of fungal pathogens in the ovule and seed alone.

4.1 ENVIRONMENT OF OVULE AND SEED

The ovule and seed develop in the pistil, which is enclosed by other floral appendages in the hypogynous flower, but the pistil is exposed to the environment in the epigynous flowers. Bracts, bracteoles, and other accessory structures may also limit direct exposure of pistil in the hypogynous as well as epigynous flowers. These constitute the immediate environment for the ovule and seed. The physical environment, which is also important, is greatly affected by the immediately surrounding protective tissues. The developing ovules and seeds are actively growing structures, and water and humidity inside the ovary and developing fruit cannot be the limiting factors. The microclimate in enclosures formed by the fruit wall and enveloping accessory structures, if present, has been little studied. Apparently, these may provide favorable conditions for infection, establishment, and spread of fungi. However, physiological and biochemical factors inside the fruits in general and the fleshy fruit in particular may further control the establishment of infection. This is borne out by the fact that seeds of fleshy fruits such as cucumber, squash, melon, and tomato are usually remarkably free from fungi.

Seeds after harvest and threshing or after natural dissemination are devoid of the above surroundings and are directly exposed to the environment that they come to occupy such as storage areas and soil. Seeds have a well-formed cuticule, surface wax deposits (when present), seed coat, or seed coat and pericarp (one-seeded dry indehiscent fruits) with protective layers, hilum, pits or micropores, and cracks. The micropyle is narrow or open to various extents. Exposed hilum, micropyle, and raphe (when present) are features of true seeds. The latter two do not occur on the surface of one-seeded fruits, and the scar left upon separation of the fruit is at best analogous to the hilum. Since seeds are stored under dry conditions, no free moisture is available. The chief determinants for the development of fungi in and on seeds during storage are temperature and available moisture (water) of the grain. Storage fungi can grow with restricted water availability (Christensen and Kaufman, 1969; Jain et al., 1994). The soil environments of seeds are highly variable, ranging from dry to water-logged soils.

4.2 NATURE OF THE PATHOGEN

The seed-borne fungi may be parasitic or saprophytic and, according to Dickinson and Lucas (1977), may be biotrophs or necrotrophs. Biotrophs cause minimal damage to the host, including seed tissues, and are in fair harmony with the host. Biotrophs have a narrow host range and are usually obligate parasites. Necrotrophs cause apparent damage to the host cells and have a wide host range. They secrete enzymes and bring about the disintegration of cell components, resulting in cell death. The released cell contents are used by such pathogens for their growth. Basically, the mode of nutrition is like that of saprophytes. The necrotrophic fungi, depending upon time of infection and humidity, cause superficial or deep infection, whereas the biotrophs generally establish in deeper tissues including the embryo. The majority of seed-borne fungi are known to be necrotrophs. The obligate parasites that belong to Peronosporaceae, Albuginaceae, Erysiphales, Ustilaginales, and Uredinales are biotrophs. Many intermediate conditions occur between the true necrotrophs and biotrophs. Maude (1996) believes that necrotrophs, which degrade tissues as they spread, are rarely transmitted to the embryo through the mother plant.

Another quality of the pathogen that may determine its passage during infection of the ovule and seed in the field is the nature of disease in the plant and the mechanism of transmission for becoming seed-borne. Neergaard (1979) has listed eight disease cycles for seed-borne pathogens taking into consideration the location of primary inoculum in seed, course of disease development, and reinfection of ovule and seed. The infection may be systemic, local, or organospecific. The systemic infection may follow a vascular or a nonvascular course predominantly.

4.3 INFECTION IN DEVELOPING SEEDS

Since the ovules and developing seeds are present inside the ovary, the passages for their invasion need to be recognized at two levels: (1) routes leading to internal ovary infection and (2) ovary to ovule and seed infection. Infection passages of a pathogen to reach inside the ovary may be exclusive or follow a particular course predominantly together with other alternate courses. The former condition seems to occur only rarely.

4.3.1 ROUTES FOR INTERNAL OVARY INFECTION

The infection may either reach the pistil directly from the mother plant through the vascular supply or the parenchyma, primarily through the intercellular spaces of the pedicel, or take place indirectly from outside using stigma-style, ovary or fruit wall, and other floral parts, including nectaries, as sites for the receipt of inoculum. Investigations carried out using artificial inoculation and histological techniques, including SEM, have improved our understanding of the course of hyphae during penetration and growth in the tissues of the pistil (Marsh and Payne, 1984; Chikuo and Sugimoto, 1989; Neergaard, 1989; Kobayashi et al., 1990).

4.3.1.1 Direct Infection from Mother Plant

Systemic plant infection of most vascular and nonvascular pathogens enters the flower and fruit through the pedicel (Figure 4.1A, B). Local infection below the flower, if it becomes systemic, can also cause infection via the pedicel (Lawrence, Nelson, and Ayers, 1981).

4.3.1.1.1 Entry through Vascular Supply

Vascular infection of wilt pathogens, *Fusarium* and *Verticillium* species, reaches the pistil via the vascular supply. Rudolph and Harrison (1945) isolated *F. moniliforme, F. oxysporum,* and *F. scirpi* from vascular bundles from all parts of the cotton plant, including boll and seed. Snyder and Wilhelm (1962) found that *V. albo-atrum* moved through the vascular elements of the mother plant into the flower and fruit stalk in sugar beet and spinach. *Verticillium dahliae* also follow a similar mode of flower and fruit infection in these crops (Van der Spek, 1972). Parnis and Sackston (1979) found mycelium of *V. albo-atrum* in vessels of the testa of *Lupinus luteum* seeds and believed that it spread through the funicular tissue. Infection via the vascular tissue (xylem) of the mother plant is the usual route for infection of garden stock (*Mathiola incana*) seed by *F. oxysporum* f. sp. *mathiolae* (Baker, 1948).

Kingsland and Wernham ((1962) report that *F. moniliforme* invades the rudimentary ears in corn through vascular tissues of the stalk. But Lawrence, Nelson, and Ayers (1981) have found that sweet corn plants showing systemic or local infection of *F. moniliforme* and *F. oxysporum* carry hyphae of the two fungi in intercellular spaces. The xylem vessels were found occluded in stem and leaves, but no hyphae were seen. The fungus moved through the parenchyma tissue of the stalk into the cob and subsequently the pedicels of florets. Klisiewicz (1963) has observed mycelial tufts of *F. oxysporum* f. sp. *carthami* in infected receptacles of safflower heads. Hyphae traversed through the abscission zone of the cypsil and were associated with them, but not limited to the xylem. Halfon-Meiri, Kunwar, and Sinclair (1987) have found colonization of achenes of *Ranunculus asiaticus* by *Alternaria* from the mother plant through the vascular system.

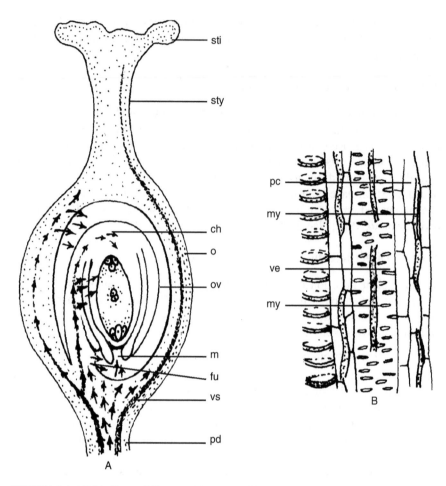

FIGURE 4.1 **(Color figure follows p. 146.)** Direct infection of ovary and ovule from mother plant. A, Diagram showing direct movement of fungal infection (arrows) via vascular elements (xylem) and nonvascular tissues (parenchyma and intercellular spaces) in pedicel. Infection may pass through the placenta and funiculus into the ovule or after escaping from the placenta, it may pass through the funiculus and ovary wall into the locule (ovary cavity) and enter the ovule via the micropyle and/or the surface of the ovule. B, Diagram of part of the pedicel region to show the spread of infection via vascular and nonvascular tissues. (Abbreviations: ch, chalaza; fu, funiculus; m, micropyle; my, mycelium; o, ovary; ov, ovule; pc, parenchyma cells; pd, pedicel; sti, stigma; sty, style; ve, vascular elements; vs, vascular supply.)

4.3.1.1.2 Entry through Nonvascular Tissues

In plants systemically infected by smuts, downy mildew, and endophytes in cereals and grasses, the mycelium moves through the intercellular spaces and enters the ear from the mother plant. In smuts and downy mildews, this infection may lead to malformation and disruption of reproductive structures (Mathre, 1978; Safeeulla, 1976). In systemically infected plants of pearl millet by *Sclerospora graminicola,* finger millet by *S. macrospora,* and sorghum by *Peronosclerospora sorghi,* the inflorescence axis is directly infected from the mother plant. The mycelium subsequently invades floral primordia (Figure 4.2A) and enters anthers and ovaries. However, secondary infection by conidia may take place through the stigma, style, or ovary (Prabhu, Safeeulla, and Shetty, 1983). Muralidhara Rao, Prakash, and Shetty (1987) also found the fungal mycelium of *P. sorghi* in systemically infected maize plants at all growth stages, including seeds. Fungal infection through silk and its establishment in seeds also occurred.

Plasmopara halstedii infection in systemically infected plants of sunflower proceeds from the receptacle to the ovary through its base. The hyphae reach the funiculus and cause seed infection (Cohen and Sackston, 1974; Doken, 1989). Doken (1989) has shown that *P. halstedii* hyphae move from the receptacle to the ovary through intercellular spaces in the pedicel at an early ovule development stage.

The mycelium of endophytic fungi also spreads intercellularly (Sampson, 1933; Philipson and Christey, 1986; Siegel, Leach, and Johnson, 1987). Using light and electron microscopy, Philipson and Christey (1986) have provided an excellent account of endophyte infection in *Lolium perenne*. The endophyte progresses intercellularly into inflorescence primordium and shows acropetal growth of its hyphae into successively formed primordia of spikelets, florets, ovary, placenta, and ovule. In the ovary wall, hyphae rarely invade the vascular system.

In barley seedlings and plants systemically infected by *Drechslera graminea*, the fungal hyphae spread through intercellular spaces, finally causing ear, floret, and pistil infection (Figure 4.2B to F) while the ear is still in the boot leaf stage (Thakkar, 1988). The infection reaches the base of the ovary and ovule prior to the differentiation of vascular elements in these structures (Figure 4.2E, F).

4.3.1.2 Indirect Infection from Outside

When transferred from other infected plants or from a local infection on the same plant to the ovary or fruit, fungal spores and conidia result in indirect infection. The inoculum may be transferred through various dispersal agencies such as wind, water (rain, irrigation), and insects. The routes for internal infection vary, depending on the site of receipt of inoculum.

4.3.1.2.1 Stigma and Style

During the first quarter of the 20th century, several authors reported fungal infection through the stigma (Figure 4.3A) and its establishment in the seed, viz., loose smuts of barley and wheat (Lang, 1910, 1917) and anther mold of red clover, caused by *Botrytis anthophila* (Silow, 1933). This has been denied by later workers (Batts, 1955; Jung, 1956; Campbell, 1956; Malik and Batts, 1960; Bennum, 1972). Jung (1956) inoculated stigmas of 61 plant species with numerous fungi but never found hyphal entry into the ovary. Bennum (1972) reinvestigated *B. anthophila* infection in *Trifolium* and found that hyphae from the stem enter the ovary and reach the funiculus. She also observed hyphae growing acropetally from the receptacle into the style but never saw this in the stigma.

Ergot disease of grasses caused by *Claviceps* spp. is a classic example of blossom infection. *Claviceps purpurea* infects floral tissues prior to fertilization or within the

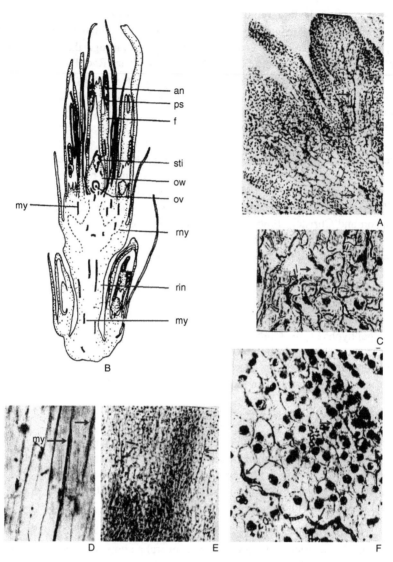

FIGURE 4.2 Direct infection from mother plant. A, Ls of part of systemically infected spike showing spread of *Sclerospora graminicola* mycelium (arrows) in floral buds of pearl millet. B to F, Pathway of infection in systemically infected barley plants by *Drechslera graminea*. B, Diagrammatic Ls of part of developing spikelets from 80-day-old plant with mycelium in node, internode, and different parts of the floret, including base of the ovary. C, D, Parts of node and internode, respectively, showing intercellular mycelium (arrows). E, F, Parts of the base of the ovary and the ovule showing intercellular mycelium (arrows). (Abbreviations: an, anther; fu, funiculus; my, mycelium; o, ovary; ov, ovule; ps, pollen sac; rin, rachis internode; rn, rachis node; sti, stigma.) (A, From Safeeulla, K.M. 1976. *Biology and Control of the Downey Mildews of Pearl Millet.* Downy Mildew Research Laboratory, Manasagangotri, Mysore University, India; B to F, Thakkar, R. 1988. Ph.D. thesis, University of Rajasthan, Jaipur, India.)

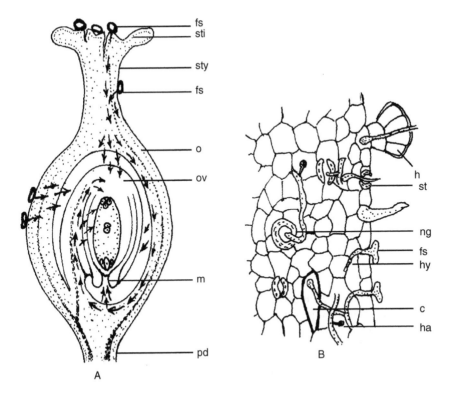

FIGURE 4.3 (Color figure follows p. 146.) Indirect infection via stigma, style, and ovary wall. A, Diagram to show infection movement from stigma, style, and ovary wall (arrows) to the ovule. Hyphae may enter ovule via funiculus, micropyle, or ovule surface. B, Diagram of part of ovary wall showing various avenues (cuticle, epidermal cells, stomata, cracks or wounds, nectary, and hairs) for hyphae to penetrate the ovary wall. (Abbreviations: c, crack or wound; fu, funiculus; fs, fungal spore; h, hair; ha, haustorium; hy, hypha; m, micropyle; ng, nectary gland; o, ovary; ov, ovule; pd, pedicel; st, stomata; sti, stigma; sty, style; vs, vascular supply.)

first few days after fertilization. On reaching the stigma, ascospores germinate, and the germ tubes penetrate the stigma-style and grow down to the ovary (Agrios, 1988; Shaw and Mantle, 1980). Campbell (1958) and Webster (1986) have doubted this infection route. However, Luttrell (1977) has clearly shown that the primary infection by ascospores and secondary infection by conidia of *Claviceps paspali* on *Paspalum dilatatum* occur through the stigma and the style (Figure 4.4A to C). The hyphae grow downward inside the style to the ovary and permeate in its inner layers. A similar path of ovary infection is observed in sorghum for *C. sorghi* (Bandyopadhyay et al., 1990) and for pearl millet by *C. fusiformis* (Thakur, Rao, and Williams, 1984; Willingale and Mantle, 1987). The infection in pearl millet by *C. fusiformis* occurs on fresh receptive stigmas, and hyphae follow the path normally taken by pollen tubes. Colonization of the ovary by the fungus predominantly takes place along the abaxial wall toward the vascular trace supplying the ovary (Willingale and Mantle, 1987). Prakash, Shetty, and Safeeulla (1980), using artificially inoculated pistils of

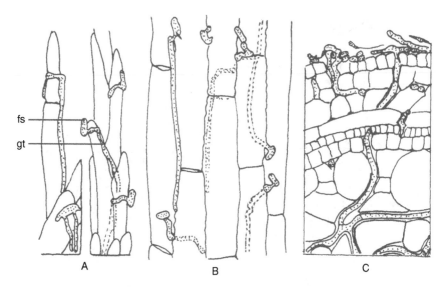

FIGURE 4.4 Artificially inoculated stigma and style of *Paspalum dilatatum* by ascospores and conidia of *Claviceps paspali*. A, Germ tubes from conidia penetrating between cells of stigma. B, Conidia germinating on surface of style and germ tubes penetrating cells and growing downward toward ovary. C, Ts of infected ovary 2 days after inoculation, intercellular mycelium in nucellus (lower) and ovary wall (upper) with hyphal tips emerging between cells of ovary epidermis. (Abbreviations: fs, fungal spore; gt, germ tube.) (From Luttrell, E.S. 1977. *Phytopathology* 67: 1461–1468. With permission.)

pearl millet by conidial suspension of *C. fusiformis*, noted that the germ tubes enter through the stigma, style, and ovary wall.

Neergaard (1989) traced the infection path of *Didymella bryoniae,* a cause of internal fruit rot in cucumber. The spores adhere to stigmatic papillae and germinate, and the germ tubes penetrate the stigma and invade the ovary through the style (Figure 4.5A, B). The preferred route taken by the hyphae is the transmitting tissue. Initially the hyphae grow intercellularly, but subsequently they also become intracellular.

Marsh and Payne (1984) artificially inoculated silk of corn at different stages (green, yellow-brown, and brown). Conidia on yellow-brown silks germinated in 4 to 8 hours, and hyphae entered directly or indirectly through cracks and intercellular gaps. The hyphae reached inside the parenchyma cells and grew parallel to the silk axis.

Halfon-Meiri and Rylski (1983) reported fruit infection in pepper by *Alternaria alternata,* following stigmatic infection. In a comparative study to determine features associated with the degree of resistance to tomato fruit rot by *F. oxysporum* f. sp. *lycopersici,* Kabayashi et al. (1990) observed germination of fungal conidia on the stigma of both susceptible and resistant cultivars, and the hyphae grew in their styles for the first four days. The growth was retarded in the styles of the resistant cultivar, but continued in those of the susceptible cultivar.

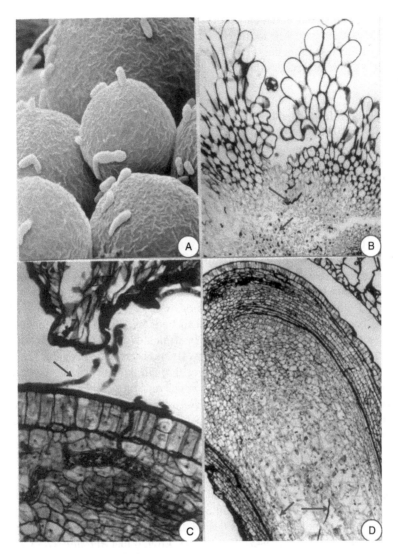

FIGURE 4.5 Infection of *Cucumis sativus* stigma and ovule by *Didymella bryoniae* after artificial inoculation. A, SEM photomicrograph of stigma showing papillae and germinating conidia, 4 hours after inoculation. B, Ls of stigma showing papillae and hyphae (arrows) after 48 hours of inoculation. C, Part of pericarp and chalazal region of ovule showing hyphae transversing the gap between the pericarp and ovule, and pentrating the epidermis of the outer integument. D, Ls of part of ovule with hyphae (arrow) in nucellus. (From Neergaard, E. 1989. *Can. J. Plant. Pathol.* 11: 28–38. With permission.)

4.3.1.2.2 Ovary and Fruit Wall

Fungal inoculum may reach the ovary and developing fruit wall from outside after the flower has bloomed in hypogynous flowers. It may reach them easily when the nonessential whorls (sepals and petals) have dropped. The inferior ovary, which remains exposed from initiation, can get infected at any stage during the development. The infection may be localized, forming lesions or progressive discoloration, or it may remain symptomless. The pathogens may penetrate the ovary wall through the cuticle, epidermal cells, stomata, cracks, hairs, and nectaries, if any (Figure 4.3A, B).

The course of infection of loose smut of wheat and barley seems to exemplify this type of infection, although there is no general agreement (Brefeld, 1903; McAlpin, 1910; Lang, 1910, 1917; Ruttle, 1934; Simmonds, 1946; Batts, 1955; Pederson, 1956; Shinohara, 1972). Using artificially inoculated and naturally infected ears of wheat by *Ustilago tritici,* Batts (1955) concluded that despite plentiful spores on the stigma, the spores caught on the ovary germinated, and hyphae penetrated the ovary wall directly. Stomata occur in the ovary wall, but entry of the hyphae through them was not seen. Malik and Batts (1960), who studied infection of barley ears by *U. nuda,* also found numerous spores on the stigma and style, but observed that spores caught on the ovary surface alone caused infection.

Localized infection of the fruit wall causing internal infection of the locule is common (Neergaard, 1979; Agarwal and Sinclair, 1997). This has been reported for siliqua and seed infection of cabbage by *Phoma lingam* (Bonman and Gahrielson, 1981), pod infection of soybean by *Phomopsis longicola* (Roy and Abney, 1985), and infection of sugar beet flower and seed bolls by *Colletotrichum dematium* f. *spinaciae* (Chikuo and Sugimoto, 1989). In artificially inoculated cotton bolls with *Colletotrichum capsici,* Roberts and Snow (1984) observed that conidia germinated on the fruit wall and penetrated directly or through the stomata and hairs (Figure 4.6A to D). Hyphae invaded the pericarp parenchyma, the vascular system, endocarp, and lint fibers. Conidia of *Colletotrichum gloeosporioides,* after artificial inoculation of *Carica papaya* green fruits, germinated and penetrated the cuticle and epidermal cells (Chau and Alvarez, 1983).

4.3.1.2.3 Other Floral Parts Including Nectary

Although the necrotrophs often attack senescing nonessential floral parts (Jarvis, 1977), the exact role of this type of infection, including that of floral nectaries in causing fruit and seed infection, is little understood. *Colletotrichum lini* and *Aureobasidium lini* infect senescing petals in linseed flowers and developing fruits (Lafferty, 1921; Johansen, 1943). Davis (1952) found that *Acronidiella eschscholtziae* on *Eschscholtzia californica* penetrates the capsule from infected petals through the fruit stalk. Infection of barley caryopsis by *Rhynchosporium secalis* occurs commonly on the inside of the lemma just below the base of the awn and rarely on the inside of the palea. This infection subsequently spreads to the portion of the pericarp in contact with the infected regions of the lemma and palaea (Skoropad, 1959). Neergaard (1989) has observed the invasion of cucumber ovaries through the natural openings in the nectary by *D. bryoniae* and presumed that this infection may spread to the transmitting tissue inside the ovary.

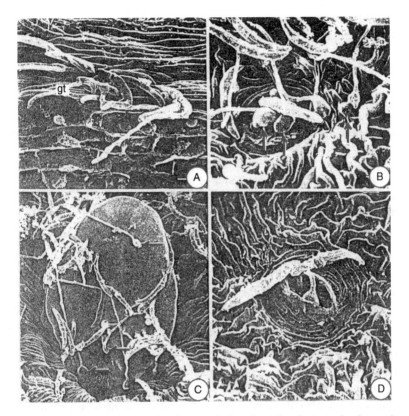

FIGURE 4.6 SEM photomicrographs of cotton boll surface showing germination and penetration of *Colletotrichum capsici* conidia. A, Three-septate conidia and germinating conidia — one of the germ tubes appears to penetrate the cuticle without producing an appressorium. B, Appressoria (arrows) produced on guard cells. C, Appressoria (arrows) on a multicellular trichome. D, Direct penetration by germ tube and formation of appressorium in the cavity of the stomata. (Abbreviation: gt, germ tube.) (From Roberts, R.G. and Snow, J.P. 1984. *Phytopathology* 74: 390–397. With permission.)

4.3.2 ROUTES FOR INFECTION FROM OVARY TO OVULE AND SEED

Various options open to the hyphae to invade the ovule and seed after reaching inside the ovary are (1) funicular vascular supply for systemic or localized systemic, vascular pathogens; (2) funicular parenchyma for systemic nonvascular pathogens and also for pathogens entering through stigma-style and reaching the placentae in the ovary; (3) ovule and seed surface consisting of integuments and seed coat, chalaza, and raphe (when present) for pathogens causing localized ovary and fruit wall infection that permeates on to ovules and seeds beneath infection courts, and also for fungal hyphae that grow out of the placentae, funiculus, and ovary wall in the locule and reach the ovule and seed surface; and (4) the micropyle for hyphae of fungal pathogens following strictly the path of the pollen tube and also for free-growing mycelium in the locule (Figures 4.1A and 4.3A). The ovule and seed surface may be invaded directly through the cuticle and epidermal cells, or entry may take place through the stomata, micropores, and pits and natural cracks in the seed surface. Micropores and cracks develop during ripening and desiccation of the seed (Yaklich, Vigil, and Wergin, 1986).

Rudolph and Harrison (1945) isolated *F. moniliforme, F. oxysporum* and *F. scirpi* from xylem isolated from seeds and concluded that the infection reached through the xylem. The hyphae of *Alternaria* sp. in achenes of *Ranunculus asiaticus* invade the ovule and seed through the ovular vascular supply via the funiculus, raphe, and chalaza, but the hyphae that escape from the pericarp and/or funiculus in the locular cavity penetrate the seed coat directly (Halfon-Meiri, Kunwar, and Sinclair, 1987).

Vaughan et al. (1988) determined the routes of entry of *A. alternata* into ovules and seeds in artificially inoculated pods of soybean cultivars Union, PI 181550 and Williams. The fungal hyphae cover the seed surface and enter the micropyle. Entry directly through the cuticle and seed epidermis and in the case of cultivar Williams, through micropores on the seed coat, also occur. Variation occurs in seeds of soybean cultivars for the presence and nature of pitting, and also the features of the micropyle, closed or open type (Yaklich, Vigil, and Wergin, 1986; Wolf, Baker, and Bernard, 1981). Kulik and Yaklich (1991) have found that hyphal infection of *Phomopsis phaseoli* via the micropyle is far more prevalent in seeds of cultivars with high incidence of infection. Seeds of cultivars with high infection have open micropyle and seed coats with multiple pits, whereas those with low infection have closed micropyle, and the seed coats lack pits. The hyphae of *P. longicolla* from localized pod infection in soybean enter the seed through the funiculus or directly through the seed coat (Roy and Abney, 1988).

D. bryoniae infection in the pistil of *Cucumis sativus* follows the path of the pollen tube and reaches the ovary mainly in the transmitting tissue. Invasion of the ovules takes place through the funiculus, and occasionally the hyphae transverse directly from the inner ovary wall to the ovule and seed surface in the chalazal region and enter through the epidermis (Figure 4.5C). Thereafter, the mycelium spreads throughout the ovule (Figure 4.5D).

The mycelium in infected ovaries of wheat and barley by *U. tritici* and *U. nuda,* respectively, grows centripetally to enter the testa and other seed tissues (Batts, 1955; Malik and Batts, 1960). The young ovule in barley receives infection of *Drechslera graminea* through the funiculus, hyphae moving in intercellular spaces. However, during grain development, the hyphae may also trespass from the inside of the ovary on to the seed surface, which they penetrate directly (Thakkar, 1988).

Styer and Cantliffe (1984) observed penetration of the pericarp through small random cracks in the surface of cultivar Sh_2 of sweet corn by *F. moniliforme,* and Smart, Wicklow, and Caldwell (1990) observed the same in Dekalb hybrid $\times$ L12 of corn by *A. flavus.*

4.4 AVENUES OF INFECTION IN THRESHED SEEDS

After harvest, seeds are generally stored under dry conditions. Fungal inoculum of pathogens (field fungi) infesting the seed surface rarely develop further and remain quiescent during storage. However, storage fungi, particularly species of *Aspergillus* and *Penicillium,* can tolerate low available moisture and are more thermotolerant.

In a favorable environment, such fungi are able to grow and cause internal infection. In true seeds, avenues for infection include (1) seed surface through the cuticle, natural openings, cracks, or injuries caused during threshing; (2) hilum that is covered by the cuticle but with fissures; (3) micropyle, particularly the open type; and (4) accessory structures, e.g., hairs, wings, aril, and caruncle. Little information is available on the role of accessory structures in seed infection.

Seenappa, Stobbs, and Kempton (1980) found infection of *Aspergillus halophilicus* in dried red peppers, stored at 70% relative humidity. The fungal conidia germinated, germ tubes entered the stalks through the stomata and pericarp through crevices caused by mechanical injury. It colonized the inner fruit wall, produced conidiophores and conidia, which reached the seed surface. These conidia germinated forming hyphae with appressoria and infection pegs entering the seed coat. The germination of spores of *Alternaria brassicicola* on the seed coat, hilum, and micropyle took place on artificially inoculated cabbage seeds (Knox-Davies, 1979). Extensive mycelial growth occurred on a seed with damaged testa.

One-seeded dry fruits such as caryopsis, cypsils, achenes, and cremocarp may also get infected during the postharvest period. In these seeds (*sensu lato*), the pericarp, adhering bracts, separation scar (analogous to hilum), remnants or scars of the style and stigma, and other persistent structures such as the seed cap in sugar beet and spinach provide points of entry. Such seeds lack the micropyle, hilum, raphe and chalaza as exposed areas. Mycock, Lloyd, and Berjak (1988) noted that *A. flavus* var. *columnaris* penetrates maize caryopsis through lesions in the pericarp and peduncle (caryopsis base) under suitable storage conditions.

4.5 MECHANISM OF PENETRATION OF OVARY, FRUIT, AND SEED SURFACES

The penetration of fungal hyphae into the ovary and fruit and the ovule and seed surfaces is similar to the processes observed during their entry into vegetative parts (see Dodman, Barker, and Walker, 1968; Heath, 1980; Kulik, 1987). It may be mechanical or enzymatic or both. The germ tube may enter directly without any special manifestation or penetrate after forming appressoria, cushions, and pegs. The limited information based on artificial inoculations or natural infection of the fruit wall reveals that the germ tubes from fungal propagules may invade the surface using more than one mode of entry. Similarly, there is evidence that there may be different modes of entry on surfaces of different cultivars (Bassi, Moore, and Batson, 1979).

The formation of appressorium at the site of infection on the fruit surface seems to be quite common. Batts (1955) and Malik and Batts (1960) report that as the wheat and barley ovaries were inoculated with spores of *U. tritici* and *U. nuda*, respectively, the tips of the promycelia, upon coming into contact with the epidermal cells of the ovary wall, developed appressorium-like swellings. Beneath the appressorium a bulbous swelling developed, through which the penetrating hypha passed and entered the cell immediately below it. Binyamini and Schiffmann-Nadel (1972) found that in avocado fruits infected by *C. gloeosporioides*, the spores germinate and germ tubes enter the thick wax layers above the cuticle and form dark appressoria

(Figure 4.7A, B), which persist in fruits until the fruits harden. During softening of the fruits, thin hyphae develop from the appressoria and penetrate the cuticle and cell wall. Appressoria formation and subsequent infection by narrow germ tubes have been reported in ethylene-degreened Robinson tangerines by *Colletotrichum gloeosporioides* (Brown, 1977) and sugar beet flowers and seed bolls by *C. dematium* f. *spinaciae* (Chikuo and Sugimoto, 1989).

Penetration of the fruit surface of *Carica papaya* by *Colletotrichum gloeosporioides* takes place after appressoria formation (Figure 4.7C, D) or directly by the germ tube. The penetration by appressoria and infection peg formation is the most common mode of entry (Chau and Alvarez, 1983). Similar mode of entry of hyphae of *Capsicum capsici* occurs on cotton bolls in which appressoria are formed on the cuticle of the epidermal cells, multicellular hairs and stomata (Figure 4.6 A to D) (Roberts and Snow, 1984). The hyphae of *Fusarium moniliforme* enter directly through small cracks and by appressoria formation on intact areas in sweet corn (Styer and Cantliffe, 1984).

The stigmatic surface is usually penetrated directly by fungal hyphae (Luttrell, 1977; Bandyopadhyay et al. 1980; Prakash, Shetty, and Safeeulla, 1980; Neergaard, 1989). Snow and Sachdev (1977) have observed direct penetration of wax, epidermal cells, and epidermal hairs on cotton bolls by *Diplodia gossypina*.

Bassi, Moore, and Batson (1979) observed differential behavior of *Rhizoctonia solani* on tomato fruits of a susceptible cultivar, C-28, and those of a resistant cultivar, PI 193407. The mycelial growth was extensive with cushion formation on C-28 fruits (Figure 4.7E), but it was sparse on those of PI 193407. The penetration was by multiple infection pegs under infection cushions in the former (Figure 4.7F) and rare with individual hyphae in the latter.

Appressoria formation takes place on seeds of *Capsicum* before the seed coat is penetrated by *Aspergillus halophilicus* (Seenappa, Stobbs, and Kempton, 1980).

The direct or indirect (through the development of appressoria or cushions) penetration of the fruit or seed surface by fungal pathogens may be mechanical or enzymatic. Histological studies suggest that if the cuticle and/or cell wall at the point of penetration is depressed inwards, the penetration is accomplished by physical force. Lack of such depressions and the symptoms of digestion of cuticle are taken as evidence of an enzymatic penetration (Kolattukudy, 1985). Enzymatic hydrolysis of the cuticle and cell wall will certainly weaken these barriers and permit the entry of infection hyphae. Secretion of cutinase by fungi during surface penetration of plant parts has been demonstrated (Shaykh. Soliday, and Kolattukudy, 1977; Dickman, Patil, and Kolattukudy, 1982). *Fusarium solani* f. sp. *pisi, C. capsici* and *C. gloeosporioides,* which are common pathogens infecting fruits and seeds, secrete cutinase. Dickman, Patil, and Kolattukudy (1982) isolated cutinase from papaya fruits infected by *C. gloeosporioides* and further demonstrated that exogenous application of cutinase helped fungal pathogens, which are incapable of infection because of their inability to penetrate the cuticle, to cause infection. *Mycosphaerella* sp., which infects papaya fruits only when the cuticular barrier is mechanically breached, infected fruits with an intact cuticle when the surface was pretreated with cutinase from *C. gloeosporioides* (Dickman, Patil, and Kolattukudy, 1982). Similar information on the penetration of the ovule and seed surface is lacking.

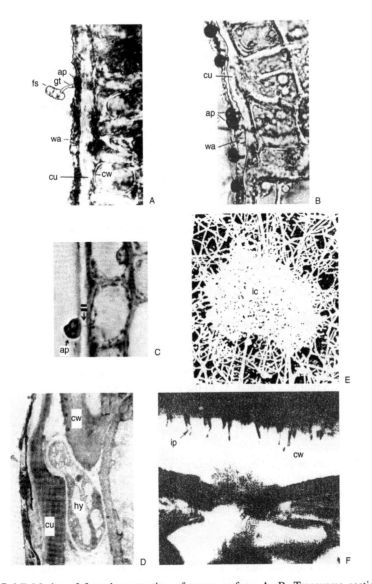

FIGURE 4.7 Modes of fungal penetration of ovary surface. A, B, Transverse section of avocado fruit from periphery showing *Colletotrichum gloeosporioides* spore germination and formation of appressorium. Many appressoria are seen in the wax layer in B. C, D, Section of papaya fruit wall showing infection of *C. gloeosporioides* forming appressorium and infection peg. E, F, Infection of tomato fruit wall of susceptible cultivar C-28 by *Rhizoctonia solani* showing infection cushion in E and numerous infection pegs in F. (Abbreviations: ap, appressorium; cu, cuticle; cw, cell wall; fs, fungal spore; gt, germ tube; hy, hyphae; ic, infection cushion; ip, infection peg; wa, wax.) (A, B, From Binyamini, N. and Schiffman-Nadel, M. 1972. *Phytopathology* 62: 569–594; C, D, From Chau, K.F. and Alvarez, A.M. 1983. *Phyto-pathology* 73: 1113–1116; E, F, From Bassi, A., More, E.L., and Batson, W.E. 1979. *Phyto-pathology* 69: 556–559. With permission.)

4.6 CONCLUDING REMARKS

The infection course pathogens take and the passages they use to reach the ovule and seed differ in developing seeds and seeds in storage. They also vary with the nature of pathogens, necrotroph or biotroph, and the course of disease development and reinfection of the plant. In flower and fruit, the pathogen may follow predominantly one path of infection or it may also use alternate routes. After reaching the interior of the ovary, the nonvascular pathogens may become more opportunistic and enter the ovule and seed through more than one site.

The avenues of infection in threshed seeds including one-seeded fruits differ from those of developing seeds. Field fungi usually remain quiescent. Storage fungi, which may show activity under low available moisture, may cause internal infection if they are thermotolerant.

The penetration of the ovary and fruit and the ovule and seed surfaces may be mechanical or enzymatic, or both. The invasion may be direct or may take place after the formation of appressoria, cushions, and pegs. A fungus may invade the surface using more than one mode of entry. Entry may also differ on surfaces of different cultivars. If the cuticle or cell wall at the point of entry is depressed inward, the penetration is mechanical and evidence of weakening or digestion of cuticle is considered to support an enzymatic penetration. The resistant reactions between pathogen and host tissues are seldom described, but the information on specific and quick resistant reactions in tissues forming passages for entry and spread in the ovary and fruit and the ovule and seed could be of practical importance in breeding cultivars with inhibitory reactions to the invasion of fungi from the outside.

REFERENCES

Agarwal, V.K. and Sinclair, J.B. 1997. *Principles of Seed Pathology*, 2nd ed. CRC Press, Boca Raton, FL.

Agrios, G.N. 1988. *Plant Pathology*, 3rd ed. Academic Press, San Diego.

Baker, K.F. 1948. Fusarium wilt of garden stock (*Mathiola incana*). *Phytopathology* 38: 399.

Bandyopadhyay, R., Mughogho, L.K., Manohar, S.K., and Satyanarayana, M.V. 1990. Stroma development, honeydew formation and conidial production in *Claviceps sorghi*. *Phytopathology* 80: 812–818.

Bassi, A., More, E.L., and Batson, W.E. 1979. Histopathology of resistant and susceptible tomato fruit infected with *Rhizoctonia solani*. *Phytopathology* 69: 556–559.

Batts, C.C.V. 1955. Observation on infection of wheat by loose smut (*Ustilago tritici* [Per.]) Rosta). *Trans. Br. Mycol. Soc.* 38: 465–475.

Bennum, A. 1972. *Botrytis anthophila* Bondarzew, med sacrligt henblik på patologisk anatomi. Ph.D. Thesis, Royal Veterinary and Agricultural University, Copenhagen, Denmark.

Binyamini, N. and Schiffman-Nadel, M. 1972. Latent infection in avocado fruit due to *Colletotrichum gloeosporioides*. *Phytopathology* 62: 592–921.

Bonman, J.M. and Gabrielson, R.L. 1981. Localized infections of siliques and seed of cabbage by *Phoma lingam*. *Plant Dis.* 65: 868–869.

Brefeld, O. 1903. Neue Untersuchungen und Ergebnisse über natürliche Infektion und Verbreitung der Brandkrankheiten des Getreides. *Klub Landw. Berlin. Nachr.* 466: 4224–4234.

Brown, G.E. 1977. Ultrastructure of penetration of ethylene-degreened Robinson tangerines by *Colletotrichum gloeosporioides*. *Phytopathology* 67: 315–320.

Campbell, W.P. 1958. Infection of barley by *Claviceps purpurea*. *Can. J. Bot.* 36: 615–619.

Chau, K.F. and Alvarez, A.M. 1983. A histological study of anthracnose on *Carica papaya*. *Phytopathology* 73: 1113–1116.

Chikuo, Y. and Sugimoto, T. 1989. Histopathology of sugar beet flowers and seed bolls infected with *Colletotrichum dematium* f. *spinaciae*. *Ann. Phytopathol. Soc. Japan* 55: 404–409.

Christensen, C.M. and Kaufmann, H.H. 1969. *Grain Storage. The Role of Fungi in Quality Loss*. University of Minnesota Press, Minneapolis, MN.

Cohen, Y. and Sackston, W.E. 1974. Seed infection and latent infection of sunflowers by *Plasmopara halstedii*. *Can. J. Bot.* 22: 231–238.

Davis, L.H. 1952. The Heterosporium disease of California poppy. *Mycologia* 44: 366–376.

Dickinson, C.H. and Lucas, J.A. 1977. Basic microbiology. In *Plant Pathology and Plant Pathogens*. Wilkinson, J.E., Ed. Vol. 6, Blackwell Scientific, London.

Dickman, M.B., Patil, S.S., and Kolattukudy, P.E. 1982. Purification and characterization of an extracellular cutinolytic enzyme from *Colletotrichum gloeosporioides* on *Carica papaya*. *Physiol. Plant Pathol.* 20: 333–347.

Dodman, R.L., Barker, K.R., and Walker, J.C. 1968. A detailed study of the different modes of penetration by *Rhizoctonia solani*. *Phytopathology* 58: 1271–1276.

Doken, M.T. 1989. *Plasmopara halstedii* (Farl.) Berl et de Toni in sunflower seeds and the role of infected seeds in producing plants with systemic symptoms. *J. Phytopathol.* 124: 23–26.

Halfon-Meiri, A., Kunwar, I.K., and Sinclair, J.B. 1987. Histopathology of achenes and seeds of *Ranunculus asiaticus* infected with an *Alternaria* sp. *Seed Sci. Technol.* 15:197–204.

Halfon-Meiri, A. and Rylski, I. 1983. Internal mold caused in sweet pepper by *Alternaria alternata*. *Phytopathology* 73: 67–70.

Heath, M.C. 1980. Fundamental questions related to plant fungal interactions: Can recombinant DNA technology provide the answers? *In Biology and Molecular Biology of Plant Pathogen Interactions*. Bailey, J., Ed. NATO NSI Series. Springer-Verlag, Berlin. Vol. 3, pp. 15–27.

Jain, P.C., Shukla, A.K., Agarwal, S.C., and Lacey, J. 1994. Spoilage of cereal grains by thermophilous fungi. In *Vistas in Seed Biology*. Vols. 1 and 2. Singh, T. and Trivedi, P.C., Eds. Printwell, Jaipur. Vol. 1, pp. 353–365.

Jarvis, W.R. 1977. *Botryotinia and Botrytis species: Taxonomy, Physiology and Pathogenicity.* Monograph No.15, Information Division, Canada Department of Agriculture, Ottawa.

Johansen, G. 1943. Horsygdomme. *Tidsskr. Plant Avl.* 48: 187–298.

Jung, J. 1956. Sind Narbe und Griffel Eintrittspforten für Pilzinfektionen? *Phytopathol. Z.* 27: 405–426.

Kingsland, G.C. and Wernham, C.C. 1962. Etiology of stalk rot of corn in Pennsylvania. *Phytopathology* 52: 519–523.

Klisiewicz, J.M. 1963. Wilt-incitant *Fusarium oxysporum* f. *carthami* present in seed from infected safflower. *Phytopathology* 53: 1046–1049.

Knox-Davies, P.S. 1979. Relationships between *Atlernaria brassicicola* and *Brassica* seeds. *Trans. Br. Mycol. Soc.* 73: 235–248.

Kobayashi, I., Sakurai, M., Tomikawa, A., Yamamoto, T., Yamaoka, N., and Kunoh, H. 1990. Cytological studies of tomato fruit rot caused by *Fusarium oxysporum* f. sp. *lycopersici* race 13(II) ultrastructural differences in infected styles of susceptible and resistant cultivars. *Ann. Phytopathol. Soc. Japan* 56: 235–242.

Kolattukudy, P.E. 1985. Enzymatic penetration of the plant cuticle by fungal pathogens. *Ann. Rev. Phytopathol.* 23: 225–250.

Kulik, M.M. 1987. Observations by scanning electron and bright-field microscopy on the mode of penetration of soybean seedlings by *Phomopsis phaseoli*. *Plant Dis.* 72: 115–118.

Kulik, M.M. and Yaklich, R.W. 1991. Soybean seed coat structures: Relationship to weathering resistance and infection by the fungus *Phomopsis phaseoli*. *Crop Sci.* 31: 108–113.

Lafferty, H.A. 1921. The "browning" and "stem-break" disease of cultivated flax (*Linum usitatissimum*) caused by *Polyspora lini* n. gen. et sp. *Sci. Proc. R. Dubl. Soc.* 16 (N.S.): 248–278.

Lang, W. 1910. Die Bluteninfektion von Weizenflugbranol. *Zentralbl. Bakteriol. Parasitenkd. Abt.* 2, 25: 86–101.

Lang, W. 1917. Zur Ansteckung der Gerste durch *Ustilago nuda*. *Ber. Dtsch. Bot. Ges.* 35: 4–20.

Lawrence, E.B., Nelson, P.E., and Ayers, J.E. 1981. Histopathology of sweet corn seed and plants infected with *Fusarium moniliforme* and *F. oxysporum*. *Phytopathology* 67: 1461–1468.

Luttrell, E.S. 1977. The disease cycle and fungus — host relationships in dalligrass ergot. *Phytopathology* 67: 1461–1468.

Malik, M.M.S. and Batts, C.C.V. 1960. The infection of barley by loose smut (*Ustilago nuda* (Jens.) Rostr.). *Trans. Br. Mycol. Soc.* 43: 117–125.

Marsh, S.F. and Payne, G.A. 1984. Scanning EM studies on the colonization of dent-corn by *Aspergillus flavus*. *Phytopathology* 74: 557–561.

Maude, R.B. 1996. *Seed-Borne Diseases and Their Control*. CAB International, Wallingford, U.K.

McAlpin, D. 1910. *The Rusts of Australia, Their Structure, Life History, Treatment and Classification*. Melbourne.

Mathre, D.E. 1978. Disrupted reproduction. In *Plant Disease: An Advanced Treatise*. Horsfall, J.G. and Cowling, E.B. Eds. Academic Press, New York. Vol. 3, pp. 257–275.

Muralidhara Rao, B., Prakash, H.S., and Shetty, H.S. 1987. Colonization of maize seed by *Peronosclerospora sorghi* and its significance. *Geobios* 14: 237–240.

Mycock, D.J., Lloyd, B.L., and Berjak, P. 1988. Micropylar infection of post-harvest caryopses of *Zea mays* by *Aspergillus flavus* var. *columnaris* var. nov. *Seed Sci. Technol.* 16: 647–653.

Neergaard, E. 1989. Histological investigation of flower parts of cucumber infected by *Didymella bryoniae*. *Can. J. Plant Pathol.* 11: 28–38.

Neergaard, P. 1979. *Seed Pathology*. Vols. 1 and 2. Macmillan Press, London.

Parnis, E.M. and Sackston, W.E. 1979. Invasion of lupin seed by *Verticillium albo-atrum*. *Can. J. Bot.* 57: 597–601.

Pedersen, P.N. 1956. Infection of barley by loose smut, *Ustilago nuda* (Jens.) Rostr. *Friesia* 5: 341–348.

Philipson, M.N. and Christey, M.C. 1986. The relationships of host and endophyte during flowering, seed formation and germination of *Lolium perenne*, *N.Z. J. Bot.* 24: 125–134.

Prabhu, S.C.M., Safeeulla, K.M., and Shetty, H.S. 1983. Penetration and establishment of downy mildew mycelium in sorghum seeds and its transmission. *Proc. Indian Natl. Sci. Acad.* 49B: 459–465.

Prakash, H.S., Shetty, H.S., and Safeeulla, K.M. 1980. Histology of carpel infection of *Claviceps fusiformis* in pearl millet. *Proc. Indian Natl. Sci. Acad.* 46B: 708–712.

Roberts, R.G. and Snow, J.P. 1984. Histopathology of cotton boll rot caused by *Colletotrichum capsici*. *Phytopathology* 74: 390–397.

Roy, K.W. and Abney, T.S. 1988. Colonization of pods and infection of seeds by *Phomopsis longicolla* in susceptible and resistant soybean lines inoculated in the greenhouse. *Can. J. Plant Pathol.* 10: 317–320.

Rudolph, B.A. and Harrison, G.J. 1945. The invasion of the internal structure of cotton seed by certain Fusaria. *Phytopathology* 35: 542–546.

Ruttle, M.L. 1934. Studies on barley smuts and on loose smut of wheat. *N.Y. Agr. Exp. Sta. Tech. Bull.* 221: 1–31.

Safeeulla, K.M. 1976. *Biology and Control of the Downey Mildews of Pearl Millet.* Downy Mildew Research Laboratory Manasagangotri, Mysore University, Mysore, India.

Sampson, K. 1933. The systemic infection of grasses by *Epichloe typhina* (Pers.) Tul. *Trans. Br. Mycol. Soc.* 18: 30–47.

Seenappa, M., Stobbs, L.W., and Kempton, A.G. 1980. *Aspergillus* colonization of Indian red pepper during storage. *Phytopathology* 70: 318–322.

Shaw, B.I. and Mantle, P.G. 1980. Host infection by *Claviceps purpurea*. *Trans. Br. Mycol. Soc.* 75: 77–90.

Shaykh, M., Soliday, C.L., and Kolattukudy, P.E. 1977. Proof for the production of cutinase by *Fusarium solani* f. pisi during penetration into its host, *Pisum sativum*. *Plant Physiol.* 60: 170–172.

Shinohara, M. 1972. Anatomical studies on barley loose smut (*Ustilago nuda* [Jens] Rostrup) I. Path of embryo-infection in the developing caryopsis. II. Host-parasite morphological features at the point of entry and in the tissue of the developing caropsis. *Bull. Coll. Agric. Vet. Med. Nihan Univ.* 6: 437–440.

Siegel, M.R., Latch, G.C.M., and Johnson, M.C. 1987. Fungal endophytes of grasses. *Ann. Rev. Phytopathol.* 25: 298–315.

Silow, R.A. 1933. A systemic disease of red clover caused by *Botrytis anthophila* Bond. *Trans. Br. Mycol. Soc.* 18: 239–248.

Simmonds, P.M. 1940. Detection of the loose smut fungi in embryos of barley and wheat. *Sci. Agric.* 26: 51–56.

Skoropad, W.P. 1959. Seed and seedling infection of barley by *Rhynchosporium secalis*. *Phytopathology* 49: 623–626.

Smart, M.C., Wieklow, D.T., and Caldwell, R.W. 1990. Pathogenesis in *Aspergillus* ear rot of maize: light microscopy of fungal spread from wounds. *Phytopathology* 80: 1287–1294.

Snow, J.P. and Sachdev, M.G. 1977. Scanning electron microscopy of cotton ball invasion by *Diplodia gossipina*. *Phytopathology* 67: 589–591.

Snyder, W.C. and Wilhelm, S. 1962. Seed transmission of Verticillium wilt of spinach. *Phytopathology* 52: 365.

Styer, R.C. and Cantliffe, D.J. 1984. Infection of two endosperm mutants of sweet corn by *Fusarium moniliforme* and its effect on seedling vigour. *Phytopathology* 74: 189–194.

Thakkar, R. 1989. Studies on Some Seed-Borne Infection of Barley (*Hordeum vulgare* L.) Grown in Rajasthan. Ph.D. thesis, University of Rajasthan, Jaipur, India.

Thakur, R.P., Rao, V.P., and Williams, R.J. 1984. The morphology and disease cycle of ergot caused by *Claviceps fusiformis* in pearl millet. *Phytopathology* 74: 201–205.

Van der Spek, J. 1972. Internal carriage of *Verticillium dahliae* by seeds and its consequences. *Meded. Fac. LandWet. Gent* 37: 567–573.

Vaughan, D.A., Kunwar, I.K., Sinclair, J.B., and Bernard, R.L. 1988. Routes of entry of *Alternaria* sp. into soybean seed coats. *Seed Sci. Technol.* 16: 725–731.

Webster, J. 1986. *Introduction to Fungi*, 2nd ed. Cambridge University Press, Cambridge, U.K.

Willingale, J. and Mantle, P.G. 1987. Stigmatic constriction in pearl millet followng infection by *Claviceps fusiformis*. *Physiolog. Mol. Plant Pathol.* 30: 247–257.

Wolf, W.J., Baker, F.L., and Bernard, R.L. 1981. Soybean seed-coat structural features: pits, deposits and cracks. *Scanning Electron Microsc.* 3: 621–624.

Yaklich, R.W., Vigil, E.L., and Wergin, W.P. 1986. Pore development and seed coat permeability in soybean. *Crop Sci.* 26: 616–626.

5 Location of Fungal Hyphae in Seeds

Contamination of seeds by smut propagules (Remmant, 1637; Tull, 1733), and accompanying infection of the ergot pathogen, *Claviceps purpurea* (Hellwig, 1699) were reported long before internal seed infection was observed. Frank (1883) was probably the first to report that the mycelium of *Colletotrichum lindemuthianum*, the cause of anthracnose in beans, often penetrated the bean (*Phaseolus vulgaris*) seed cotyledons. The embryo infection of wheat seed by *Ustilago tritici*, loose smut fungus, was recorded by Maddox in 1896 and confirmed by Brefeld (1903). Heald, Wilcox, and Pool (1909) reported the internal seed-borne mycelium of *Diplodia zeae* in endosperm and embryo of maize (*Zea mays*). Neergaard (1979) provided such information under seed and fruit component subheadings, namely (1) embryo infection; (2) endosperm infection; (3) seed coat infection; (4) pericarp infection; (5) bract infection; and (6) contamination of seed coat and pericarp. He pointed out that the location of pathogen within the embryo is dependent on the species, perhaps even the race or strain of the pathogen and the variety or cultivar of the host species. Agarwal and Sinclair (1997) followed Neergaard's categories in their book. Maude (1996), while describing the internal inoculum of seed, stressed the nutritional status of the fungus, necrotroph or biotroph. Accordingly, the necrotrophic fungi are generally located in the seed and fruit coat, and deeper penetration, i.e., endosperm and embryo infection, is infrequent. Biotrophs, on the other hand, are located in the embryo.

5.1 SEVERITY OF INFECTION AND LOCATION

Histopathology of infected categorized seeds of different crops by pathogens ranging from well-known necrotrophs to biotrophs has shown that the extent and amount of fungal inoculum in seed tissues are directly correlated with the severity of seed infection. In symptomatic infected seeds (Figure 5.1), severity can be determined by the coverage of seed surface by fungal inoculum, such as number and distribution of microsclerotia, pycnidia, aceruli, dormant mycelium (with or without propagules), and degree of discoloration (Figure 5.1) and shriveling. Asymptomatic infection cannot be evaluated by visual observation. In symptomatic infected seeds, weak infections are usually confined to seed coat and pericarp, whereas heavy infection may invade all parts including the embryo.

The severity of infection of any pathogen in an affected cultivar may depend on the stage at which the ovule and seed become infected and also on its environment after the infection has taken place. Late infections, when the seed is reaching

FIGURE 5.1 (Color figure follows p. 146.) Seeds of soybean (*Glycine max*) showing slight to very severe symptoms, purple seed stain, caused by *Cercospora kikuchii*.

maturity, may remain superficial unless the environment (humidity and temperature) is favorable to pathogens for a long period. Early infections of necrotrophs and nonsystemic pathogens may cause failure of ovule and seed development or deep infections. Pirson (1960) found a relationship between the time of inoculation of *Stagnospora nodorum* in winter wheat and the amount of infection, expressed by the number of conidia produced in the spikes and the reduction of grain weight. Early inoculation, June 10, led to a reduction in grain weight of about 40% while late inoculation, July 10, had little or almost no effect. Djerbi (1971), using infection of *Fusarium culmorum* to the caryopsis of wheat, also observed that the extent of invasion reflects the actual time of inoculation or infection. When infection takes place at an earlier stage, colonization may be deep, reaching the integument, endosperm, and the surroundings of the embryo, but if infection takes place near maturity of the caryopsis, only the outer layers of the pericarp will be invaded.

Ponchet (1966) reported the occurrence of the mycelium of *S. nodorum* beneath the testa of diseased wheat kernels. However, Agarwal et al. (1985), while examining the seeds of cultivars Svenno and Starke-II infected by *S. nodorum* that were categorized as bold, loose and cracked pericarp, shriveled, and discolored, found mycelium restricted to the outer layers of the pericarp in bold infected seeds, but in seeds with loose and cracked pericarp, shriveled and discolored types, profuse mycelium occurred in pericarp, and it extended to other parts in order of severity (Figure 5.2A). The discolored seeds carried the mycelium in all parts, including the embryo.

A similar trend in spread of infection has been observed in wheat kernels infected by *Bipolaris sorokiniana* (Figure 5.2B, C) and *Drechslera tetramera* (Figure 5.2D, E) (Yadav, 1984), maize by *Botryodiplodia theohromae* (Singh et al., 1986b), chili by *Colletotrichum dematium* (Chitkara, Singh, and Singh, 1990), barley by *Drechslera graminea* (Thakkar et al., 1991), sorghum by *Curvularia lunata* and *Phoma sorghina* (Rastogi, Singh, and Singh, 1990, 1991), sunflower by *Rhizoctonia bataticola* (Godika, Agarwal, and Singh, 1999), and mustard by *Albugo candida* (Sharma, Agarwal, and Singh, 1997).

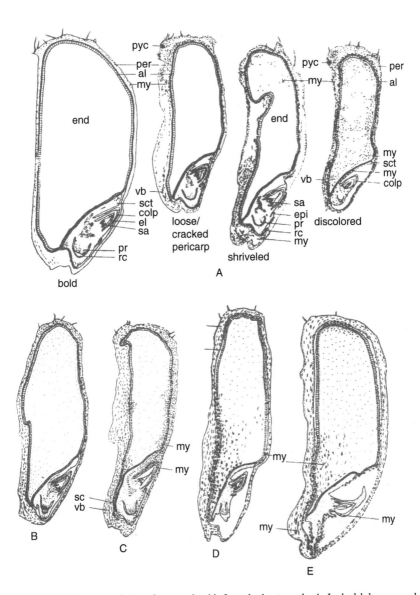

FIGURE 5.2 Diagrammatic Ls of categorized infected wheat seeds. A, Ls bold, loose cracked pericarp, shriveled and discolored types of *Stagnospora nodorum*-infected wheat seeds showing the expanse of mycelium. B, C, Ls of moderate and heavily infected wheat seeds by *Bipolaris sorokiniana*. D, E, Ls wheat seeds moderate and heavily infected by *Drechslera tetramera*. The embryo damage is more by *D. tetramera* than *B. sorokiniana* or *S. nodorum* (Abbreviations: al, aleurone layer; colp, coleoptile; el, embryonic leaf; end, endosperm; epi, epiblast; my, mycelium; per, pericarp; pr, primary root; pyc, pycnidium; rc, root cap; sa, shoot apex; sc, seed coat; sct, scutellum; vb, vascular bundle.) (A, From Agarwal, K. et al. 1985. *Phytomorphology* 35: 87–94; B to E, From Yadav, V. 1984. Ph.D. thesis, University of Rajasthan, Jaipur, India.)

5.2 PRIMARY SITES OF COLONIZATION

Developmental histopathological studies suggest that there is a primary site of colonization by the fungal mycelium in seed, depending on the course of penetration and the availability of space or soft tissue. The infection through the micropyle spreads readily in spaces between the components of the ovule and developing seed (Singh and Singh, 1979; Singh, Mathur, and Neergaard, 1980). This inoculum subsequently infects cells in different components. The infection coming through the funiculus spreads first in the integument or developing seed coat or raphe, if present. The hyphae further spreads from outside to the inside in different components of seeds, including the spaces between the components. The fungal mycelium entering directly through seed epidermis via natural openings, such as stomata and micropores, or through cracks and wounds during late stages of seed development, predominantly occupies the cells with prominent air spaces and parenchymatous cells. In leguminous seeds, this infection is readily observed in hourglass cells with prominent air spaces (Ilyas et al., 1975; Singh and Sinclair, 1985, 1986; Mathur, 1992; Sharma, 1992). If the fungus penetrates the ovule and seed through the funiculus or micropyle, the infected seeds, particularly weakly infected ones, often lack hyphae or propagules of the fungus on the seed surface, whereas when infection takes place through the seed surface, the aggregation of mycelium on the seed surface is common. The fungal hyphae may follow more than one course of entry, as seen in soybean seeds infected with *Alternaria* sp. (Vaughan et al., 1988) and *Cercospora sojina* (Singh and Sinclair, 1985), *Phomopsis longicolla* (Roy and Abney, 1988), and *Ranunculus asiaticus* by *Alternaria* sp. (Halfon-Meiri, Kunwar, and Sinclair, 1987).

5.3 HOST–PATHOGEN INTERACTIONS

The reaction of seed tissues to different pathogens is variable. Usually heavy infection of fungi, particularly necrotrophs, causes distortion and weakening of tissues, including thickenings of cuticles and cell wall (Singh, Mathur, and Neergaard, 1977, 1980; Singh, 1983) and poor storage of reserve food material in the endosperm and/or embryo, resulting in deformed and shriveled seeds. Several fungi, *Alternaria brassicicola, A. tenuis, Trichothecium roseum* in rape and mustard (Sharma, 1989), *T. roseum* in maize (Singh, Singh, and Singh, 1985), *D. tetramera* in wheat (Yadav, 1984), *C. lunata* in sorghum (Rastogi, Singh, and Singh, 1990), *R. bataticola* in mothbean (*Vigna aconitifolia*) (Varma, Singh, and Singh, 1992b), and *Fusarium oxysporum* in mothbean and cowpea (*Vigna unguiculata*) (Varma, 1990), cause lysis of the cells of the embryo, forming cavities.

    Macrophomina phaseolina (*R. bataticola*) infection occurring in spaces between seed components in sesame seeds induces cell division in the endosperm and embryo (Singh and Singh, 1979). The divisions are conspicuous on the adaxial side of cotyledons (Figure 5.3A, B). Cell divisions are also incited by *C. dematium* in the embryo of chili (Chitkara, Singh, and Singh, 1990), the epithelium of scutellum in wheat by *B. sorokiniana* (Yadav, 1984), and in the seed coat of rape and mustard by *A. brassicicola* (Sharma, 1989). Premature xylogenesis in the mesocotyl of sorghum embryo (Figure 5.3C, D) is incited by *C. lunata* (Rastogi, Singh, and Singh, 1990) and in that of wheat by *B. sorokiniana* (Yadav, 1984). In rubber (*Hevea brasiliansis*) seeds infected with *B. theobromae*, some cells in mesophyll of cotyledons became hypertrophied and thick-walled (Varma, Singh, and Singh, 1990). The embryo in heavily infected seeds shows conspicuous axial elongation, comprising narrow elongated cells in sorghum and wheat infected by *C. lunata* (Rastogi, Singh, and Singh, 1990) and *D. tetramera* (Yadav, 1984), respectively.

Common histological manifestations at the subcellular level in heavily affected cells of cotyledons in soybean by *F. oxysporum* and *R. bataticola* (Mathur, 1992; Sharma, 1992) are the loosening of cell wall fibrils, the broadening of plasmodesmata, the depletion of plasma membrane, the degradation of cytoplasm and cell organelles, and the deformation of protein bodies, lipid bodies, and nuclear membrane. The soybean cotyledons infected with *R. bataticola* also revealed the appearance of striated excretory bodies not found in uninfected cotyledons (Figure 5.3E, F). Thus, the changes at subcellular level show symptoms of disturbed metabolic activity, including triggering of some new pathways.

5.4. MIXED INFECTIONS

Seeds usually do not have unifungal infection, but information on histopathology of seeds infected with more than one fungus is limited (Wilson, Noble, and Gray, 1945; Kunwar, Singh, and Sinclair, 1985; Sharma, 1992). Wilson, Noble, and Gray (1945) distinguished the hyphae of *Gloetinia granigena* (blind seed fungus) from that of the endophyte in rye grains on the basis of hyphal morphology. *Gloetinia granigena* occurred in different parts of the seed and its expanse depended on the stage at which infection took place. But the mycelium of the endophyte was usually confined to a layer immediately outside the aleurone layer (probably the seed coat). In germinating caryopsis, the endophyte spread to the plumule, while *G. granigena* was usually restricted to the endosperm and scutellum.

Kunwar, Singh, and Sinclair (1985) found *Colletotrichum truncatum* and *Phomopsis* sp. or *Cercospora sojina* in soybean seeds. Based on hyphal color (unstained), width, presence or absence of oil globules, and reaction to stains (safranin-light green and trypan blue), the mycelium of these fungi could be distinguished from one another (Table 5.1). These pathogens compete with one another for colonization of seed tissues. The hyphae of *Phomopsis* sp., an aggressive pathogen, are restricted to seed coat layers when *C. truncatum* is present. Similarly, *C. sojina* hyphae are more numerous in the seed coat than *C. truncatum* hyphae in seeds having mixed infection of these pathogens. An additive effect is seen on deterioration of tissues of seed coat when infected with *C. truncatum* and *Phomopsis* sp.

Seeds of soybean can be infected with *F. oxysporum* and *Papulaspora coprophila*, a saprophyte. Hyphae of *F. oxysporum* are hyline, branched, septate, 2 to 3.5 μm broad and stained green or light red, whereas those of *P. coprophila* are brown to dark brown, much branched, shortly septate, 4 to 10.5 μm broad and dark red brown in safranin-fast green stained preparations. With trypan blue the hyphae of *F. oxysporum* are stained blue, but those of *P. coprophila* retained the natural brown color. Both the fungi invade all the components, but *F. oxysporum* is more

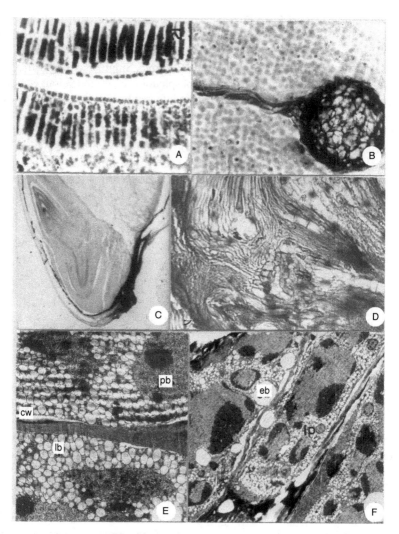

FIGURE 5.3 Photomicrographs of normal and infected parts of embryo showing reaction of host tissue to infection. A, B, Ts part of cotyledons from adaxial surface of healthy and infected sesame seeds by *Macrophomina phaseolina*. Note mycelium and microclerotia between cotyledons in B and pronounced cell divisions in cells. C, D, Ls embryo from normal and infected sorghum seeds by *Curvularia lunata* showing induced premature xylogenesis in the mesocotyl region in D. E, F, TEM of cells from normal (E) and infected cotyledons of soybean infected by *Rhizoctonia bataticola,* the latter showing weakening of cell wall, dilated protein bodies, and ergastic bodies. The ergastic bodies are absent in cells of normal cotyledons. (Abbreviations: cw, cell wall; eb, ergastic bodies; lb, lipid bodies; pb, protein bodies.) (A, B, Singh, T. and Singh, D. 1979. In *Recent Research in Plant Science*. Bir, S.S., Ed. Kalyani Publishers, New Delhi; C, D, Rastogi, 1984; E, F, Mathur, R. 1992. Ph.D. thesis, University of Rajasthan, Jaipur, India.)

TABLE 5.1
Characteristics of Hyphae of *Cercospora sojina, Colletotrichum truncatum,* and *Phomopsis* sp. in Seed Tissues Carrying Mixed Infection

	Hyphal Morphology			Stain Reaction	
Pathogen	Color	Width (µm)	Oil Globule	Safranin–Light Green	Trypan Blue
Cercospora sojina	Dark brown	0.8–1.6	+	Light green	Blue
Colletotrichum truncatum	Brown	3.0–11.0	+	Green	Blue
Phomopsis sp.	Hyaline	3.8–8.7	–	Red or green	Blue

Note: + = presence of oil globules; – = absence of oil globules.

Based on Kunwar, I.K., Singh, T., and Sinclair, J.B. 1985. *Phytopathology* 75: 489–592.

extensively ramified in deeper tissues. Hyphae of both the fungi occurred on the surface of the plumule and radicle, but only *F. oxysporum* was seen in inner layers (Sharma, 1992).

5.5 COLONIZATION OF SEED TISSUES

The histopathology of seeds infected by fungi belonging to Oomycetes, Ascomycetes, Basidiomycetes, and Deuteromycetes is described for each genus separately. Seed infection of endophytes is given separately.

5.5.1 OOMYCETES

Terrestrial Oomycetes, mostly members of Peronosporales, are either facultative or obligate plant parasites. The hyphae of facultative parasites grow indiscriminately into and through the affected cells of the host, whereas those of obligate parasites are usually intercellular forming haustoria into the host cells. Species of *Phytophthora, Plasmopara, Peronospora, Peronosclerospora, Sclerospora, Sclerophthora,* and *Albugo* are known to be seed-borne (Richardson, 1990). The information on location of Oomycetes in seeds is presented in Table 5.2.

5.5.1.1 *Phytophthora*

Aseptate and branched mycelium without haustoria of *Phytophthora parasitica* var. *sesami,* causing sesamum blight, occur in the seed coat, endosperm, and embryo of infected sesame seeds (Sehgal and Prasad, 1966). Dubey and Singh (1999) reported *P. parasitica* var. *sesami* mycelium, chlamydospores, and oospores in seed components. The distribution of mycelium in seed varies and depends on the severity of infection. It is superficial, confined to the seed coat in seeds with weak infection, but seeds with deep infection carry mycelium in the seed coat, endosperm, and embryo (Figure 5.4 A, B). Severely infected seeds fail to germinate, but weak infection is transmitted to seedling and plant.

TABLE 5.2
Location of Oomycetes in Seeds of Crop Plants

Pathogen	Host	Part(s) of Seed and Fruit	Important References
Albugo bliti	*Amaranthus* sp.	Seed coat	Melhus, 1931
A. candida	*Brassica juncea*	Seed coat, endosperm, embryo	Sharma, 1989; Sharma et al., 1997
Phytophthora parasitica var. *sesami*	*Sesamum indicum*	Seed coat, endosperm, embryo	Sehgal and Prasad, 1966; Dubey and Singh, 1999
Phytophthora sp.	*Theobroma cacao*	Seed coat, endosperm, embryo	Kumi et al., 1996
P. cinnamomi	*Persea americana*	Embryo	Neergaard, 1979
Peronospora ducomati	*Fagopyrum esculentum*	Persistent calyx, pericarp, seed coat	Zimmer et al., 1992
P. farinosa (P. schachtii)	*Beta vulgaris*	Seed coat	Leach, 1931
P. manshurica	*Glycine max*	Seed coat	Roongruangsree et al., 1988
P. viciae	*Pisum sativum*	Seed coat	Melhus, 1931
Plasmopara halstedii (*P. helianthii*)	*Helianthus annuus*	Pericarp, seed coat, endosperm, embryo	Novotelnova, 1963; Cohen and Sackston, 1974; Doken, 1989
Peronosclerospora heteropogoni	*Zea mays*	Aleurone layer, under scutellum	Rathore et al., 1987
P. maydis (*Sclerospora maydis*)	*Zea mays*	Embryo	Purakusumah, 1965; Semangoen, 1970
P. phillippinensis (*Sclerospora phillippinensis*)	*Zea mays*	Endosperm, embryo	Weston, 1920; Miller, 1952
P. sacchari (= *Sclerospora sacchari*)	*Zea mays*	Embryo	Singh et al., 1968; Semangoen, 1970
P. sorghi (*Sclerospora sorghi*)	*Sorghum vulgare*	Pericarp, endosperm	Jones et al., 1972; Kaveriappa and Safeeulla, 1975, 1978; Safeeulla, 1976; Prabhu et al., 1984
	Zea mays	Pericarp, endosperm, embryo	Murlidhara Rao et al., 1984, 1985
Sclerospora graminicola	*Pennisetum typhoides*	Pericarp, endosperm, embryo-scutellum	Safeeulla, 1976; Shetty et al., 1978, 1980
Sclerophthora macrospora (*Sclerospora macrospora*)	*Zea mays*	Embryo	Ullstrup, 1952
	Eleusine coracana	Pericarp, endosperm, embryo	Safeeulla, 1976; Raghvendra and Safeeulla, 1979
Sclerophthora rayssaiae var. *zeae* (*Sclerospora rayssiae* var. *zeae*)	*Zea mays*	Embryo	Singh et al., 1968

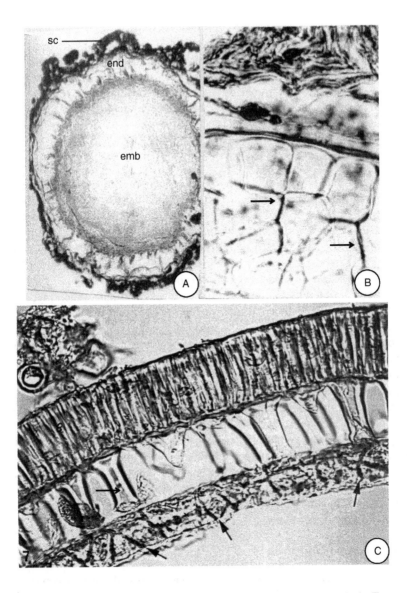

FIGURE 5.4 A, B, *Phytophthora parasitica* f. sp. *sesami* in sesame seed. A, Transverse section of seed having oospores and hyphae in seed coat, and hyphae in endosperm and embryo. B, Part from A magnified to show intercellular mycelium in endosperm. C, Ts part of soybean seed coat with hyphae of *Peronospora manshurica* in various layers (arrows) and oospores and mycelium on seed surface. (Abbreviations: emb, embryo; end, endosperm; sc, seed coat.) (A, B, From Dubey, A.K. 2000. Ph.D. thesis, University of Rajastan, Jaipur, India; C, From Roongruangsree, U-Tai et al. 1988. *J. Phytopathol.* 123, 1326–1328. With permission).

Neergaard (1979) reported the occurrence of *Phytophthora cinnamomi* myce-lium in the embryo of avocado pear seed. Kumi et al. (1996) also observed that cacao seeds from symptomatic (black) and asymptomatic pods from infected trees of *Theobroma cacao* yielded *Phytophthora* from more than 90% of embryos, but the recovery was only 4% from seed coat. Such seeds either failed to germinate or produced seedlings that usually died soon after germination or, if they survived, were stunted and developed leaf spots.

5.5.1.2 *Peronospora*

Leach (1931) found mycelium and oospores of *Peronospora farinosa* (*Peronospora schachtii*) in the seed coat of sugar beet and Melhus (1931) found mycelium and oospores of *P. viciae* in the seed coat of pea. The presence of oospore crust on soybean seeds, first described by Johnson and Lefebvre (1942), is well known, Such seeds are produced in infected pods (Hildebrand and Koch, 1951; Mckenzie and Wyllie, 1971) and cause systemic infection in seedlings (Lehman, 1953; Zad, 1989). Roongruangsree, Olson, and Lange (1988) found that seeds encrusted by *P. man-shurica*, contain, in addition to the oospores, thick- and thin-walled mycelium on the seed surface. Only thin-walled mycelium occurred in the seed coat between palisade cells, hourglass cells, and thin-walled parenchyma layers (Figure 5.4 C). The oospores, thick-walled mycelium on seed surface and mycelium in seed coat, retain viability. Hildebrand and Koch (1951) also observed fragments of mycelium of *P. manshurica* within the palisade and hourglass cells, but considered them to be senescent.

Peronospora ducomati, the causal agent of downy mildew of buckwheat (*Fagopyrum esculentum*), is also internally seed-borne (Zimmer, McKeen, and Campbell, 1992). The commercial seed is in fact an indehiscent, one-seeded fruit, the achene. According to Zimmer, McKeen, and Campbell (1992), oospores of *P. ducomati* occur in the remnant of persistent calyx on the inside of the seed coat (probably the pericarp) and in the spermoderm layer (the seed coat). The seed-borne oospores constitute the primary inoculum that causes systemic invasion of the seed-lings at the time of germination.

5.5.1.3 *Plasmopara*

Plasmopara halstedii (*P. helianthii*) is an obligate parasite that causes downy mildew of sunflower. Seeds (cypsils) with severe infection contain mycelium in all parts including the embryo (Novotelnova, 1963; Cohen and Sackston, 1974). Doken (1989) observed hyphae of *P. halstedii* in the pericarp and testa (seed coat) of seeds from systemically infected sunflower plants. No infection was seen in the embryo in these seeds. The infection entered the ovary from the receptacle, and the fungal colonization was especially dense in the hilum region (the point of fruit separation).

5.5.1.4 *Sclerospora*, *Peronosclerospora*, and *Sclerophthora*

In 1920, Weston was the first to show that the mycelium of *Sclerospora* is internally seed-borne, followed by Arya and Sharma in 1962. Weston (1920) reported the mycelium of *Sclerospora phillippinensis* in the pericarp and endosperm of maize kernels. Arya and Sharma (1962) found dormant mycelium of *S. graminicola*, the causal organism of the green ear disease of pearl millet, in seeds collected from the infected portion of the ears. However, Suryanarayana (1962), who also observed mycelium in seed, did not attribute it to *S. graminicola*. It was not until recently, during the last two decades, that convincing evidence of the occurrence of hyphae of *S. graminicola* in pearl millet seed and its role in disease transmission has been revealed. The dormant mycelium of *S. graminicola* is reported in the pericarp, endosperm, and embryo (Figure 5.5A to C) of infected grains (Safeeulla, 1976; Shetty et al., 1978; Shetty, Mathur, and Neergaard, 1980; and Subramanya, Safeeulla, and Shetty, 1981). The mycelium is coenocytic, thick, and netlike with constrictions and forked haustoria (Figure 5.5D). In embryos, scutellum alone is infected, and the radicle and plumule are not invaded. This embryal mycelium alone causes systemic infection of seedlings and plants (Shetty, Mathur, and Neergaard, 1980).

Several *Sclerospora* species have been placed under *Peronosclerospora* and *Sclerophthora* in recent years. The mycelium and oospores of *P. sorghi* occur in the glumes and pericarp of seeds from systemically infected sorghum plants (Kaveriappa and Safeeulla, 1975; Safeeulla, 1976). The fungal mycelium occasionally invades the endosperm, but neither mycelium nor oospores are found in embryos. Kernels collected from maize plants systemically infected by *Peronosclerospora sorghi* oospores and mycelium occurred in all parts of infected kernels, i.e., the pericarp, endosperm, and embryo (Muralidhara Rao et al., 1984, 1985). The infected seeds produced diseased plants in large numbers.

In seeds from partially malformed finger millet (*Eleusine coracana*) earheads, *P. sorghi* mycelium is reported in the pericarp, endosperm, and embryo. Similarly, Rathore, Siridhana, and Mathur (1987) found mycelium of *Peronosclerospora heteropogoni* in most parts including the embryo in kernels of *Zea mays*.

Sclerophthora macrospora (*Sclerospora macrospora*), the cause of crazy top in maize, is internally seed-borne. Ullstrup (1952) reported coenocytic mycelium in cob tissues including seeds. In a few kernels from severely infected plants, the mycelium was found in the scutellum and coleorhiza. Singh, Joshi, and Chaube (1968) found the presence of *Sclerophthora rayssiae* var. *zeae* mycelium in the plumule and the adjoining part of coleoptile in corn kernels from infected plants. The aseptate mycelium in the embryo was swollen and irregular with prominent vacuoles and granular protoplasm. Thin hyphae extended from the broad swollen mycelium to the adjoining tissues. Ullstrup (1952) reported the transmission of *Sclerophthora* mycelium to the seedlings and plants, including aerial vegetative and floral primordia.

5.5.1.4 *Albugo*

Albugo species cause white blisters (white rust) and hypertrophy of vegetative and floral parts in crucifers. Alcock (1931) reported minute white spots of *A. candida* (*Cystopus candidus*) on turnip seeds. Oospore contamination is common in *Brassica* seeds (Petrie, 1975; Verma and Bhowmik, 1988). According to Sharma, Agarwal, and Singh (1997), the white rust-affected seeds may be symptomless or symptomatic

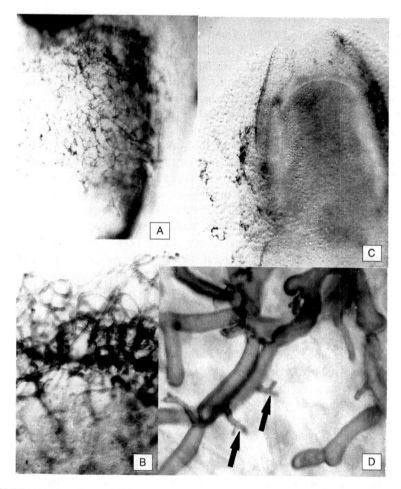

FIGURE 5.5 Mycelium of *Sclerospora graminicola* in pearl millet seed. A to C, Cleared whole-mount prepartions of pericarp, endosperm and scutellum, respectively, showing the mycelium. D, Mycelium from seed tissues showing constrictions and forked haustoria (arrow). (From Shetty, H.S. et al. 1978. *Seed Sci. Technol.* 6: 935–941. With permission.)

(Figure 5.6A). The latter can be subclassified into (1) seeds partly covered with white crust, (2) bold-discolored with white mycelium and oospores, and (3) shriveled discolored seeds. The symptomatic seeds are usually smaller and lighter than the symptomless seeds. Coenocytic, branched, and intercellular mycelium of *A. candida* occurs in seed epidermis, hypodermis, and rarely, superficial layers of cotyledons in symptomatic seeds (Figures 5.6B to D). Fungal mycelium was observed in about 4% of symptomless seeds from seed lots of infected crop plants. Symptomatic seeds carried the mycelium in seed coat together with sex organs (oogonia and antheridia) and immature to mature oospores (Figure 5.6E to G). Occasionally the mycelium also invaded the endosperm and embryo. Embryonal infection took place only rarely in heavily infected seeds belonging to the last two categories (Sharma, Agarwal, and

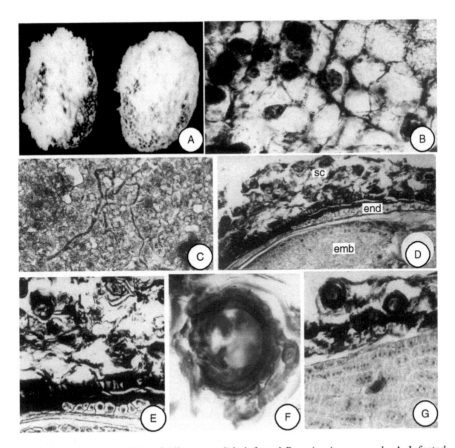

FIGURE 5.6 Histopathology of *Albugo candida*-infected *Brassica juncea* seeds. A, Infected symptomatic seeds with white crust on seed surface. B, C, Cleared whole-mount preparations of epidermis and cotyledon, respectively, showing mycelium and oospores. D, E, Ls part of bold-discolored seed with mycelium and oospore in seed coat with slight increase in number of its layers. F, A part of mature seed coat magnified to show oogonium and antheridium. G, Ls part of shriveled discolored seed showing abundant mycelium and oospores in seed coat. (Abbreviations: emb, embryo; end, endosperm; sc, seed coat.) (Sharma, J., Agarwal, K., and Singh, D. 1997. *J. Phytol. Res.* 10: 25–29.)

Singh, 1997). The seed-borne mycelium and oospores remain viable and can cause seedling infection (Sharma, Agarwal, and Singh, 1994).

5.5.2 ASCOMYCETES

Table 5.3 gives an account of Ascomycetes in seeds of crop plants.

5.5.2.1 *Protomyces*

Protomyces macrosporus is the most widespread species causing discoloration and gall formation on members of the Apiaceae (Umbelliferae). *Coriandrum sativum* is

TABLE 5.3
Infection of Ascomycetes and Basidiomycetes in Seeds of Crop Plants

Pathogen	Host	Part(s) of Seed and Fruit	Important References
Ascomycetes			
Protomyces macrosporus	Coriandrum sativum	Persistent calyx, stylopodium, pericarp, carpophore	Gupta, 1962; Rao, 1972; Pavgi and Mukhopadhyay, 1972; Singh et al., 2001
Claviceps fusiformis	Pennisetum typhoides	Seed sclerotia	Prakash et al., 1980; Roy, 1984
C. paspali	Paspalum dilatatum	Seed sclerotia	Luttrell, 1977
C. purpurea	Secale secale, Triticum aestivum, Hordeum vulgare	Seed sclerotia	Campbell, 1958
Didymella bryoniae	Cucurbita pepo	Seed coat, embryo	Lee, Mathur, and Neergaard, 1984
D. lycopersici	Lycopersicum esculentum	Seed coat	Fischer, 1954
	Helianthus annuus	Pericarp, testa	Tollenaar and Bleiholder, 1971
Sclerotinia sclerotiorum	Eruca sativa	Seed coat	Sharma, 1992
Basidiomycetes			
Melanopsichium eleusinis	Eleusine coracana	Seed cavity	Thirumalachar and Mundkur, 1947
Tolyposporium panicillariae	Pennisetum typhoides	Affects all parts except pericarp	Bhat, 1946; Mitter and Siddiqui, 1995
Ustilago tritici	Triticum aestivum	Pericarp, endosperm, embryo	Batts, 1955; Popp, 1959; Ohms and Bever, 1956
Ustilago nuda	Hordeum vulgare	Pericarp, endosperm, embryo	Malik and Batts, 1960a
Tilletia controversa	Triticum aestivum	Colonizes developing ovary, affects all parts within pericarp	Mathur and Cunfer, 1993
T. indica (Neovossia indica)	Triticum aestivum	Partial or complete infection, all parts except aleurone layer and pericarp	Goates, 1988; Cashion and Luttrell, 1988
T. laevis (T. foetida)	Triticum aestivum	Colonizes ovary, affects all parts except pericarp	Singh, 1991
T. tritici (T. caries)	Triticum aestivum	Colonizes ovary, affects all parts except pericarp	Mathur and Cunfer, 1993
Puccinia calcitrapae var. centaureae (P. carthami)	Carthamus tinctorius	Uredospore and teleutospore infestation of seed surface	Halfon-Meiri, 1983
Uromyces betae	Beta vulgaris	Uredospores in seed balls	Alcock, 1931

the common host, and the disease appears in the form of tumor-like swellings on aerial parts including flowers. The infection is partial and scattered on different parts. The infected fruits are hypertrophied, and they are called galls. The fungal mycelium and chlamyspores are found only in infected tissues (Gupta, 1962; Pavgi and Mukhopadhyay, 1972; Rao, 1972; Singh, 1991).

The healthy cremocarp is symmetrical and comprises two mericarps. The hypertrophied and galled fruits are symmetrical or asymmetrical depending on whether the infection is complete or partial. The ridges on the surface of these fruits are inconspicuous, and the affected pericarp lacks differentiation. Its cells are uniformly enlarged with prominent air spaces and abundant chlamydospores. The vittae are not recognized and the vascular bundles are dispersed with poor thickening of xylem elements. The cells in the inner epidermis are tangentially stretched and thin-walled.

The cells in carpophore are also hypertrophied, and its vittae and vascular bundles are inconspicuous. The pedicel and stylopodium, which remain with fruit, show hypertrophy of cells, prominent air spaces, distinct but reduced amount of vascular zone, indistinct vittae, and abundant mycelium and chlamydosphores. In partially infected fruits, the uninfected part shows normal anatomical features (Figure 5.7).

The size of locules and the development of ovules are directly correlated with the intensity of infection. In severely infected and fully hypertrophied fruits, the ovules abort. In moderately hypertrophied fruits, the ovules develop to varying extents and have endosperm and embryo, but the latter is weakly developed. Fruits with chlamydospores confined only to the pericarp usually have well-formed seeds with a developed embryo. Such seeds germinate normally (Gupta, 1962).

5.5.2.2 *Claviceps*

Species of *Claviceps* cause ergot disease in diverse graminaceous crops and grasses. The disease cycle (Figure 5.8) follows an almost uniform pattern. The path of infection and development of stroma and sclerotia under natural and experimental conditions are known for *Claviceps purpurea* in rye (Tulsane, 1853; Campbell, 1958), *C. paspali* in *Paspalum dilatatum* (Brown, 1916; Luttrell, 1977), *C. fusiformis* in pearl millet (Thakur, Rao, and Williams, 1984; Roy, 1984), and *C. sorghi* in sorghum (Bandyopadhyay et al., 1990). The development of sclerotia follows two different courses.

Initial infection by ascospores (primary) and conidia (secondary) takes place through the stigma. The invading hyphae grow down the style (Figure 5.9A) and colonize the ovary. In ovaries of rye infected by *C. purpurea* and pearl millet by *C. fusiformis*, the development begins at the base of ovary (Figure 5.9B, C) with the appearance of the sphacelial stage (Weihing, 1956; Roy, 1984). The affected tissues of the ovary produce sweet viscid fluid in which conidia remain embedded. It is exuded as honeydew. The apical part of the ovary resists disintegration, and it is carried at the top of the sphacelial stroma. The downward colonization of the host does not proceed beyond the plate of cells at the upper limit of rachilla (Figure 5.9C), where a tissue (plate) of four to six layers separates the infected ovary from the

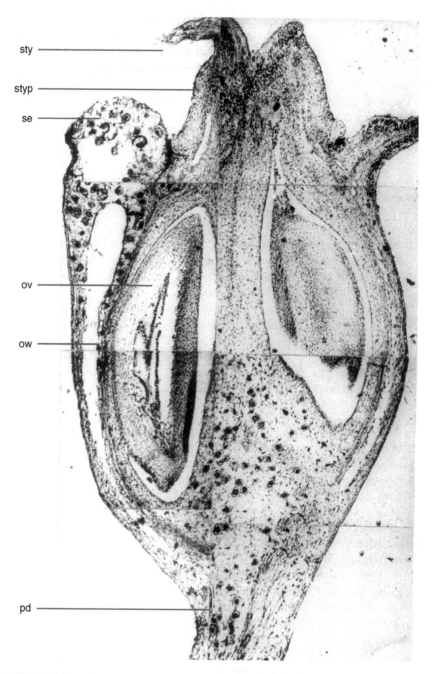

sty

styp

se

ov

ow

pd

FIGURE 5.7 Reconstructed Ls of partially hypertrophied fruit of coriander showing infection of *Protomyces macrosporus* in different parts. (Abbreviations: ov, ovule; ow, ovary wall; pd, pedicel; se, sepal; sty, style; styp, stylopodium.) (From Singh, B.K. 1991. Ph.D. thesis, University of Rajasthan, Jaipur, India.)

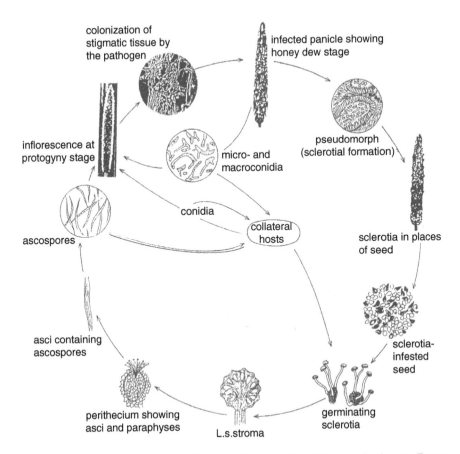

FIGURE 5.8 Disease cycle of ergot of pearl millet caused by *Claviceps fusiformis*. (Reproduced with permission from Chahal, S.S., Thakur, R.P. and Mathur, S.B. 1994. *Seed-Borne Diseases and Seed Health Testing of Pearl Millet*. Danish Government Institute of Seed Pathology for Developing Countries, Copenhagen, Denmark.)

uninfected part of the floret. The development of this *cell plate* is completed within 8 days of infection in pearl millet (Roy, 1984).

The ovary wall is also infected, and intercellular hyphae emerge in between the epidermal cells to form the extramatrical sphacelial stage (Campbell, 1958; Roy, 1984). The development of ergot sclerotium begins with the development of pseudo-parenchymatous or plectenchymatous sclerotic tissue in the lower half of the ovary (Roy, 1984; Bandyopadhyay et al., 1990). This tissue gradually increases, carrying the conidial pouches at the top. Finally the sclerotium formation is completed and the entire ovary is filled with pseudoparenchymatous hyphal mass.

After invading stigma and style, *C. paspali* permeates the ovary wall in *P. dilatatum* and spreads outward, emerging between the cells of the outer epidermis and forming a plectenchymatous extrametrical stroma around the ovary (Figure 5.9D). The fungus grows to a lesser extent toward the inside. The hyphae on the surface transform into conidiophores forming conidia (Figure 5.9D) and the

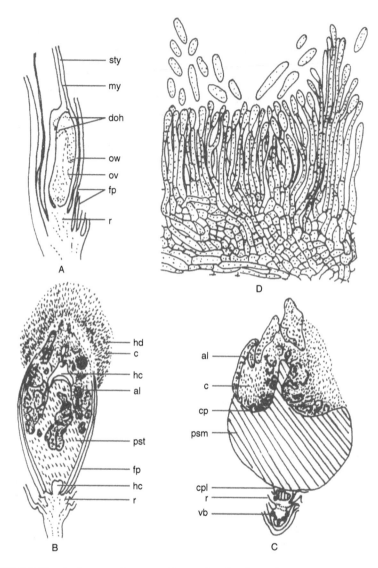

FIGURE 5.9 Developing gramineaceous florets showing formation of sclerotia of *Claviceps* species. A to C, *Claviceps fusiformis* in *Pennisetum typhoides*. A, Ls ovary after 5 days of inoculation with mycelium in stylar region. B, C, Ls infected ovaries showing development of sclerotium, conidial pouches, and position of cell plate separating infected ovary from rachilla. D, Cross section of infected ovary of *Paspalum dilatatum* by *Claviceps paspali* producing conidia from extramatrical stroma on surface. (Abbreviations: al, remnants of anther lobes; c, conidia; cp, conidial pouch; cpl, cell plate separating ovary from rachilla; doh, areas showing dissolution of host cells; fp, floral parts; hc, host cells; hd, honey dew; my, mycelium; ov, ovule; ow, ovary wall; psm, pseudoparenchymatous mass; pst, presclerotic tissue; r, rachilla; sty, style; vb, vascular bundle.) (A to C, Roy, S. 1984. Ph.D. thesis, University of Rajasthan, Jaipur, India; D, Luttrell, E.S., 1977, *Phytopathology*, 67: 1461–1468. With permission.)

affected tissues secrete honeydew (Luttrell, 1977). The extramaterical stroma expands, fills the space between the lemma and palea, and protrudes above their tips. In the base of the sclerotium in *P. dilatatum*, the outline of the ovary is distinct. A loose plectenchyma fills the space occupied by the ovary.

The mature sclerotia in all cases are dark colored, and the remnants of the apical part of the ovary, particularly the fragments of stigmatic cells together with conidial pouches, persist (Luttrell, 1977; Roy, 1984). Anatomically, the mature sclerotium is made up of dark colored (brown) rind or cortex of relatively thick-walled irregular plectenchyma on the surface and uniform compact plectenchyma inside in the case of *C. paspali*. The sclerotia are also composed of dark cortex in *C. purpurea*, but internally it is differentiated into an outer region of fine hyphal cells and an inner region of isodiametric cells (Campbell, 1958).

5.5.2.3 *Sclerotinia*

Sclerotinia spp., particularly *Sclerotinia sclerotiorum*, cause destructive diseases of vegetable and flower crops. The early symptoms on infected plants are a white fluffy mycelial growth in which sclerotia, white at first but ultimately black and hard, are produced. *Sclerotinia sclerotiorum* sclerotia may be attached to the seeds or may occur mixed with them (Neergaard, 1979). Tollenaar and Blieholder (1971) observed mycelial strands of *S. sclerotiorum* in the thick-walled and thin-walled layers of the pericarp and in the seed coat of sunflower kernels.

Sharma (1992) detected *S. sclerotiorum* in seeds of *Eruca sativa* from Rajasthan, India. Infected seeds are symptomatic with black mycelium and sclerotia on the seed surface. The infection is confined to the seed coat, usually the seed epidermis, and only rarely on the subepidermal cells. The infection occurs as thick mycelium and microsclerotia, which are either submerged in the epidermis or emergent.

5.5.2.4 *Didymella*

Didymella bryoniae, the cause of internal fruit rot in *Cucumis* and *Cucurbita*, invades the pistil through the stigma and style in *Cucumis sativus*. The fungus infects the ovules and developing seeds superficially or internally (Neergaard, 1989). In pumpkin seeds, the fungus is located in the seed coat and rarely in cotyledons. It occurs in all the layers of the seed coat and is usually prominent in the chlorenchyma and the inner epidermis (Lee, Mathur, and Neergaard, 1984).

5.5.3 BASIDIOMYCETES

5.5.3.1 Ustilaginales (Smuts and Bunts)

The seed-borne smuts and bunts attack the ovary and sometimes other parts of florets or the inflorescence (Table 5.3). Three main disease cycles are found in these fungi. The primary infection is intraembryal and causes systemic, latent infection in the plant, expressing itself in the ovary or developing grains in which abundant teliospores (chlamydospores) are formed. The exposed teliospores are disseminated by the wind and germinate on the stigma and ovary wall in florets on healthy plants

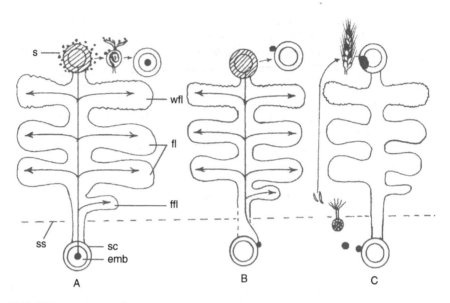

FIGURE 5.10 Schematic diagrams to show different pathways of development of seed-borne infections of smuts and bunts in cereals. A, *Ustilago tritici* (loose smut) of wheat — intra-embryal seed inoculum followed by systemic infection. B, *Ustilago hordei* (covered smut) of barley — externally seed-borne inoculum causing systemic infection. C, *Tilletia indica* (karnal bunt) — externally seed as well as soil-borne inoculum, spores cause blossom infection of individual florets. (Abbreviations: mb, embryo; ffl, first foliar leaf; fl, foliar leaves; s, seed; sc, seed coat; ss, soil surface; wfl, weakened foliar leaves.) (Adapted and redrawn from Neergaard, P. 1979. *Seed Pathology.* Vols. 1 and 2. Macmillan, London.)

to form promycelium. They cause secondary infection to developing caryopsis, result-ing in embryal infection, which remains dormant in seeds, e.g., *Ustilago tritici* and *Ustilago nuda* (loose smuts of wheat and barley), until germination (Figure 5.10A). The primary inoculum is externally seed-borne or, as in some *Tilletia* spp., externally seed- as well as soil-borne, and causes seedling (coleoptile) infection at the time of seed germination. Infection is systemic and latent in the vegetative phase and expresses in ears. All ears of tillers of a plant and every seed in the infected ear develop teliospores held by the persistent membrane and pericarp (Figure 5.10B), e.g., *Ustilago hordei* (covered smut of barley). Sometimes, as in the common bunt, individual seeds or ears may remain healthy (*Tilletia tritici and Tilletia laevis*). The primary inoculum is externally seed-borne and soil-borne. Sporidia formed in soil-become airborne, cause infection of individual florets, and invade ovary and ovule forming teliospores. Grain infection is either total or partial, covered by pericarp (Figure 5.10C), e.g., *Tolyposporium penicillariae* (smut of pearl millet), *Tilletia indica* (*Neovossia indica,* Karnal bunt).

5.5.3.1.1 Type 1 Disease Cycle

The embryal infection in wheat grains by *Ustilago tritici*, causal organism of loose smut of wheat, was first reported by Maddox (1896) in Australia and later confirmed by Brefeld (1903) in Germany. Subsequently, many studies have described the course

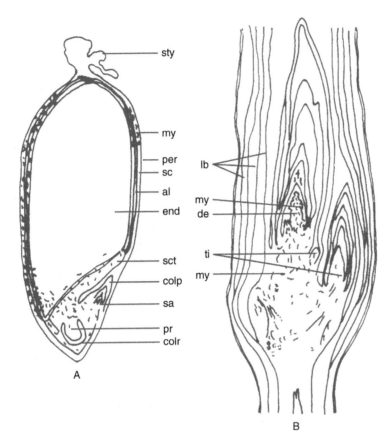

A

B

FIGURE 5.11 Location of loose smut mycelium in seed and young seedling of barley. A, Diagrammatic representation (Ls) of the distribution of mycelium of *Ustilago nuda* in an infected barley grain. B, Diagrammatic representation of *U. nuda* mycelium in crown node of 5-week-old seedling. (Abbreviations: al, aleurone layer; colp, coleoptile; colr, coleorhiza; de, developing ear; end, endosperm; lb, leaf bases; my, mycelium; per, pericarp; pr, primary root; sa, shoot apex; sc, seed coat; sct, scutellum; sty, stylar tissue; ti, tillers.) (Redrawn from Malik, M.M.S. and Batts, C.C.V. 1960a. *Trans. Br. Mycol. Soc.* 43: 112–125 and Malik, M.M.S. and Batts, C.C.V. 1960b. *Trans. Br. Mycol. Soc.* 43: 126–131.)

of infection and location of mycelium of *U. tritici* and *U. nuda* in wheat and barley grains (Ruttle, 1934; Vanderwalle, 1942; Simmonds, 1946; Batts, 1955; Pedersen, 1956; Malik and Batts, 1960a; Loiselle and Shands, 1960; Kozera, 1968; Shinohara, 1972). Batts (1955), working with artificially as well as naturally infected wheat grains, and Malik and Batts (1960a), working with barley grains, have shown that entry and location of *U. tritici* and *U. nuda* mycelium in wheat and barley grains are similar. In grains obtained from plants inoculated at the time of anthesis, the mycelium occurs in the pericarp, testa, aleurone layer, other layers of endosperm, and embryo (Figure 5.11A). It is mainly intracellular in the pericarp and testa, and usually intercellular in the aleurone layer, the other layers of endosperm, and the embryo. The mycelium has not been found in the radicle. In cases where the infection occurs just before the ripening of the grains, hyphae are confined to the endosperm, especially at the base (Pedersen, 1956). Ohms and Bever (1956) also observed the highest percentage of infected embryos with a large amount of hyphae when inoculations were made at anthesis in winter wheat. Inoculation made after anthesis showed a decreased percentage of infected embryos with a small amount of hyphae.

The embryo count method is commonly used in routine seed health testing for the detection of loose smut mycelium in wheat seed lots. In extracted embryos, the mycelium takes trypan blue stain, but the older mycelium is brown and does not stain. The smut mycelium is broad and branched (Khanzada et al., 1980; Mathur and Cunfer, 1993).

Loiselle and Shands (1960) compared the location of *U. nuda* in grains of resistant and susceptible barley cultivars. They found that in grains of resistant plants mycelium was confined to the chalaza and the parenchyma associated with the vascular bundle while in grains of susceptible cultivars, it occurred in the integument (seed coat and pericarp), aleurone, endosperm, and embryo. Batts and Jeater (1958) and Popp (1959) found three different reactions when resistant and susceptible cultivars were artificially inoculated by different races of *U. tritici*. Mycelium spread in all tissues of the embryo of susceptible cultivars. The scutellum and plumular bud were regularly infected, but the coleoptile, epiblast, radicle, coloerhiza, and suspensor were less frequently infected. In resistant varieties, only the scutellum was consistently infected and the plumule bud had no infection. Other structures were less frequently infected. All parts of the embryo were free of infection in totally resistant or immune varieties. Popp (1959) further demonstrated that only infection of plumule bud is directly correlated with smut infection in plants grown from infected seed (Table 5.4). Based on the above criteria, a reasonably accurate prediction of percentage of smut infection in field plants can be made. Based on recordings of scutellum infection, however, Khanzada et al. (1980) found that the laboratory counts of the embryo infection in wheat seed closely correlated to the number of smutted plants in the field.

Batts and Jeater (1958) and Malik and Batts (1960b) determined the development of the fungus in infected germinating seed, seedling, and plant in susceptible cultivars of wheat and barley, respectively. The mycelium becomes active at the time of seed germination and is carried upward by the elongation of the plumule (epicotyl). The mycelium spreads in the crown node in which the development of all parts of the adult plant (nodes, internodes, and ears of each tiller with its leaf sheath) takes place. The smut mycelium permeates the whole structure including the ears (Figure 5.11B). During further development, stems elongate and the mycelium already present in the ears is carried up. By the time the ears emerge out of the leaf sheath, all of its parts except the rachis are completely replaced by spores.

Wallen (1964) has reported that the cultivar Keystone barley in Canada is extremely embryo susceptible but mature plant resistant. The embryal infection is never transferred to the plant.

5.5.3.1.2 Type 2 Disease Cycle

Common bunt or stinking smut is caused by two closely related pathogens, *T. tritici* (*T. caries*) and *T. laevis* (*T. foetida*). Bunt balls are seed- as well as soil-borne. The

TABLE 5.4
Estimates of Loose Smut Infection in Wheat Varieties of Different Adult Plant Reaction as Determined by Scutellum, Plumular Bud, and Adult Plant Tests

| | Percentage Infection | | |
| | Embryos | | |
Varietal Reaction	Scutellum	Plumular Bud	Adult Plants
Artificially inoculated			
Highly susceptible	85	85	86
Moderately susceptible	68	48	49
Moderately resistant	83	8	7
Highly resistant	68	0	0
Immune	0	0	0
Naturally infected			
Highly susceptible	8	5	5
Highly resistant	5	5	5

From Popp, W. 1959. *Phytopathology* 49: 75–77. With permission.

teliospores germinate at the time of seed germination and directly cause seedling coleoptile infection. The plant infection is systemic and latent in the vegetative phase. Microscopic examination prior to ear emergence from the boot leaf shows that the pistil in affected ears is larger and deep green and the ovary is double the length of the normal ovary. The stamens are reduced in length and breadth. The anthers are pale yellow instead of green, and they possess defective pollen grains. The bunted grains are full of black spore mass (teliospores) covered with pericarp. The number of teliospores in affected grains varies even in a susceptible cultivar (Griffith, Zscheile, and Oswald, 1955; Zscheile and Anken, 1956). Infected grains with a limited number of teliospores contain normal-appearing tissue surrounding the sorus. The spore-bearing region decreases toward the stigmatic side and increases toward the base. The spores mature first at the center of the sorus and proceed centrifugally to the periphery. The development of teliospores at more than one place may take place in a kernel, but these sori usually fuse in the mature kernel.

Several other smuts, namely covered smut of barley caused by *Ustilago hordei* (externally seed-borne), loose smut of oats caused by *U. avenae* (grain infection confined to the pericarp and external seed contamination), covered smut of oats caused by *U. kolleri* (externally seed-borne), loose smut of sorghum caused by *Sphacelotheca cruenta* (externally seed-borne), and grain smut of sorghum caused by *S. sorghi* (externally seed-borne), have disease cycles broadly similar to those of *T. tritici* (*T. caries*) and *T. laevis* (*T. foetida*). The teliospores fill the kernels and are enclosed in the pericarp membrane. No anatomical details are available.

5.5.3.1.3 Type 3 Disease Cycle

T. penicillariae is the causal organism of smut of pearl millet and the smutted grains are scattered in earheads. The primary inoculum consists of spore balls that adhere to the seed surface during harvesting and threshing or are soil-borne. In nature, smut infection occurs through air-borne sporidia produced from germinating teliospores in soil at the time of flowering of the crop. Smut infection occurs through the stigmas before anthesis and is confined to individual spikelets (Bhat, 1946; Mitter and Siddiqui, 1995). The mycelium, after reaching the ovary, branches profusely between the pericarp and aleurone layer. The hyphae are inter- and intracellular. A cavity, full of fungal mass, is formed inside the ovary. Teliospore formation follows, and finally all the structures within the ovary except the ovary wall are replaced by teliospores (Mitter and Siddiqui, 1995). The smutted grains are oval or pear-shaped, bigger than healthy grains, and project beyond the glumes. The top is bluntly rounded to conical. The enclosing membrane is tough and consists of host tissue.

Smut of finger millet (*Eleusine coracana*) due to *Melanopsichium eleusinis* is caused by air-borne sporidia. Soon after infection the cells in the ovary wall divide, making it multilayered. This is followed by the disintegration of cells at different sites, resulting in the formation of small lysigenous cavities. The cavities enlarge, become ovate or spherical, and a thick felt of mycelium borders each cavity, filled with a mucilaginous fluid. Teliospores are formed from the hyphae around the cavity in a centripetal manner. The sorus is usually multilocular due to the development of two, three, or even four cavities full of spores. Each cavity or locule remains distinct, separated from the others by the host tissue, even at maturity (Thirumalachar and Mundkur, 1947).

A similar disease cycle occurs in rice bunt, caused by *Tilletia horrida* (Chowdhury, 1946), and karnal wheat bunt, caused by *Tilletia indica* (*Neovossia indica*). In *T. indica* the primary infection, which takes place in individual florets, under favorable conditions, causes secondary and tertiary spread of the pathogen within and between spikelets through mycelium or secondary sporidia produced on initially infected florets (Dhaliwal et al., 1983; Bedi and Dhiman, 1984). In the case of *T. indica* entry of hyphae is reported through the ovary wall.

5.5.3.2 Uredinales (Rusts)

Some rust fungi that cause serious diseases in crop plants are known to be seed-borne (Alcock, 1931; Savile, 1973; Neergaard, 1979; Halfon-Meiri, 1983). Alcock (1931) reported *Uromyces betae* on seed balls of sugar beet as yellow spots, sori containing uredospores (Table 5.3). Hungerford (1920) found abundant sori of ure-dospores and teleutospores of *Puccinia graminis* var. *tritici* on wheat seeds.

Puccinia calcitrapae var. *centaureae* (=*Puccinia carthami*), a macrocyclic auto-ecious rust, has been convincingly shown to be transmitted directly from seeds (Figure 5.12). Uredo- and teleutospores of *P. calcitrapae* var. *centaureae* occur inside the crevices, scarred ends, and surface of safflower cypsils (Figure 5.13A to D). Roughness of spore surface also contributes to their adherence to cypsil surface. The teleutospores, larger in size and two-celled, are easily distinguished from ure-dospores (Figure 5.13B). The teleutospores germinate, forming sporidia, which cause seedling infection (Halfon-Meiri, 1983). Schuster and Christiansen (1952), Schuster (1956), and Zimmer (1963) have also reported that seed or soil infested by teleutospores supply the initial inoculum for seedling infection.

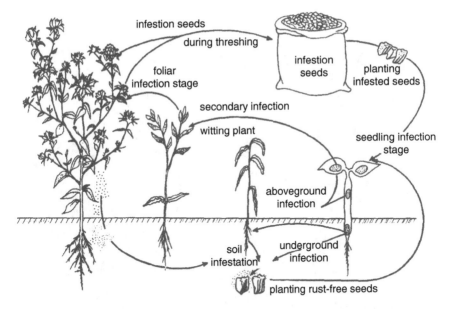

FIGURE 5.12 Disease cycle of seed-borne safflower rust, *Puccinia calcitrapae* var. *centaureae* (From Halfon-Meiri, A. 1983. *Seed Sci. Technol.* 11, 835–851. With permission.)

5.5.4 Deuteromycetes

Deuteromycetes or fungi imperfecti are very economically important because many of them are the causes of serious plant diseases. It is generally assumed that these fungi reproduce exclusively by asexual means. However, in recent years, sexual or perfect stages of several taxa have been discovered. It has been revealed that species having similar conidial form may have a sexual form belonging to different genera. Species of *Rhizoctonia* are placed in the basidiomycetous genera *Thanetophorus* and *Cerotobasidium,* while those of *Phoma* are placed in the ascomycetous genera *Didymella (Phoma lycopersici), Leptosphaeria (Phoma lingam),* and *Pleospora (Phoma betae).* Furthermore, sexual or perfect stages of many species belonging to such form genera are still unknown. The literature also reveals that the imperfect or asexual form causes disease and is associated with seed. The Deuteromycetes are subclassified into Hyphomycetes and Coelomycetes, and they are dealt with separately. Their location in seed is presented in Table 5.5.

5.5.4.1 Hyphomycetes

5.5.4.1.1 Alternaria

Plant pathogenic species of *Alternaria* usually cause leaf spots and blights in various crop plants throughout the world. They are weak to serious pathogens, and a large number of them are seed-borne and seed-transmitted (Richardson, 1990). Many of the seed-borne *Alternaria* species cause deep infection in seed (Table 5.5). The mycelium is dark colored in *Alternaria* and moderate to heavy seed infections result in symptomatic seeds.

FIGURE 5.13 *Puccinia calcitrapae* var. *centaureae* on safflower seed surface. A, Infested seed surface. B, Teleutospores (two-celled) and uredospores (one-celled). C, D, SEM photomicrographs of seed surface showing uredo- and teleutospores, respectively. (From Halfon-Meiri, A. 1983. *Seed Sci. Technol.* 11: 835–851. With permission.)

TABLE 5.5
Location of Fungi Belonging to Deuteromycetes in Seeds of Crop Plants

Fungus	Host	Seed and Fruit Part(s)	Important References
Hyphomycetes			
Alternaria alternata	*Helianthus annuus*	Pericarp, seed coat, endosperm, embryo	Singh et al, 1977
	Capsicum annuum	Seed coat, endosperm	Chitkara et al., 1986a
	Glycine max	Seed coat, endosperm	Kunwar et al., 1986a
	Triticum aestivum	Asymptomatic seed-pericarp only; symptomatic all parts	Agarwal et al., 1987
	Hordeum vulgare	Husk, lodicules, pericarp, endosperm	Thakkar et al., 1988
	Brassica sp.	Seed coat, endosperm, embryo	Sharma, 1989
	Coriandrum sativum	Pericarp, seed coat, endosperm, embryo	Singh, 1991
	Linum usitatissimum	Seed coat, endosperm, embryo	Sharma, 1992
	Eruca sativa	Seed coat, endosperm, embryo	Sharma et al., 1993
	Cyamopsis tetragonoloba	Seed coat, endosperm, embryo	Bhatia, 1995
	Cuminum cyminum	Pericarp, seed coat, carpophore, rarely endosperm	Rastogi et al., 1998a
	Trigonella foenum-graecum	Seed coat, endosperm, embryo	Rastogi et al., 1998b
A. brassicae	*Brassica rapa*	Seed coat	Chupp, 1935
A. brassicicola	*Brassica* sp.	Seed coat	Boek, 1952; Domsch, 1957; Knox-Davies, 1979
	Brassica sp.	Seed coat, endosperm embryo	Sharma, 1989; Maude and Humpherson-Jones, 1980
	Eruca sativa	Seed coat, endosperm, embryo	Sharma, 1992
A. brunsii	*Cuminum cyminum*	Pericarp, carpophore, endosperm	Rastogi, 1995
A. dauci	*Daucus carota*	Pericarp	Soteros, 1979
A. longipes	*Nicotiana tabaccum*	Seed coat	Heursel, 1961
A. padwickii (*Trichoconis padwickii*)	*Oryza sativa*	Glume, pericarp, endosperm, embryo	Cheeran and Raj, 1972
A. radicina	*Daucus carota*	Pericarp	Netzer and Kenneth, 1969; Strandberg, 1983
A. raphani	*Raphanus sativus*	Seed coat, embryo	Atkinson, 1950

(continued)

TABLE 5.5 (CONTINUED)
Location of Fungi Belonging to Deuteromycetes in Seeds of Crop Plants

Fungus	Host	Seed and Fruit Part(s)	Important References
		Hyphomycetes (continued)	
A. sesamicola	Sesamum indicum	Seed coat, endosperm, embryo	Singh et al., 1980
A. zinniae	Zinnia elegans	Usually pericarp, seed coat, endosperm (rarely)	Tarp, 1979; Parmar, 1981
Acroconidiella tropaeoli (Heterosporium tropaeoli)	Tropaeolum majus	Pericarp, seed coat, embryo-cotyledons (rarely)	Baker and Davis, 1950
Bipolaris maydis	Zea mays	Pericarp, seed coat, endosperm, embryo	Singh et al, 1986a
B. nodulosa	Eleusine coracana	Pericarp, endosperm	Ranganathaiah and Mathur, 1978
B. oryzae	Oryza sativa	Pericarp, endosperm, embryo	Paul, 1987
B. setariae	Pennisetum typhoides	Pericarp, endosperm, embryo	Shetty et al., 1982
	Setaria italica	Pericarp, endosperm	Keshavamurthy, 1990; Ranganathaiah, 1994
B. sorokiniana	Triticum aestivum	Pericarp, endosperm, embryo	Yadav, 1984
Botrytis allii	Allium cepa	Seed coat	Maude and Presly, 1977
B. anthophila	Trifolium pratense	Seed coat	Silow, 1933; Bennum, 1972
B. cinerea	Linum usitatissimum	Seed coat	Van der Spek, 1965
B. fabae	Vicia faba	Seed coat, embryo	Harrison, 1978
Cercospora kikuchii	Glycine max	Seed coat, hilar region, embryo rare	Ellis et al., 1975; Singh and Sinclair, 1986
C. sojina	Glycine max	Seed coat, hilar region, embryo rare	Singh and Sinclair, 1985
C. traversiana	Trigonella foemum-graecum	Seed coat, endosperm, embryo	Rastogi et al., 1998c
Curvularia lunata	Sorghum vulgare	Pericarp, seed coat, endosperm, embryo	Rastogi et al., 1990
Cylindrocladium crotalariae	Arachis hypogaea	Seed coat, embryo-cotyledons	Porter et al., 1991
Drechslera graminea	Hordeum vulgare	Husk, pericarp, endosperm, embryo	Thakkar et al., 1991
D. tetramera	Triticum aestivum	Pericarp, endosperm, embryo	Yadav, 1984

(continued)

TABLE 5.5 (CONTINUED)
Location of Fungi Belonging to Deuteromycetes in Seeds of Crop Plants

Fungus	Host	Seed and Fruit Part(s)	Important References
		Hyphomycetes (continued)	
Fusarium culmorum	*Triticum aestivum*	Pericarp, seed coat, endosperm, embryo	Djerbi, 1971
	Secale cereale	Pericarp, seed coat, endosperm, embryo	Djerbi, 1971
F. moniliforme	*Zea mays*	Pericarp, endosperm, embryo (rarely)	Singh and Singh, 1977; Singh et al., 1985
F. oxysporum	*Vigna* spp.	Seed coat, hilar region, embryo	Varma, 1990; Sharma, 1999
	Glycine max	Seed coat, hilar tissue, embryo	Sharma, 1992
	Cajanus cajan	Seed coat, hilar tissue, embryo	Sharma, 1996
	Cyamopsis tetragonoloba	Seed coat, hilar tissue, endosperm, embryo	Bhatia et al., 1996a
F. oxysporum var. *carthami*	*Carthamus tinctorius*	Pericarp, seed coat	Klisiewicz, 1963
F. oxysporum var. *laginariae*	*Lagenaria siceraria*	Seed coat	Kuniyasu, 1980
F. oxysporum f. sp. *mathiolae*	*Mathiola incana*	Seed coat	Baker, 1952
Helminthosporium papaveris	*Papaver somniferum*	Seed coat, endosperm	Meffert, 1950; Schmiedeknecht, 1958
Itersonilia pastinacae	*Pastinaca sativa*	Pericarp (?)	Channon, 1969
Nigrospora oryzae	*Zea mays*	Pericarp, endosperm, embryo	Singh and Singh, 1989
Pyricularia oryzae	*Oryza sativa*	Husk, pedicel, endosperm, embryo	Chung and Lee, 1983
P. grisea	*Eleusine coracana*	Pericarp, endosperm	Ranganathaiah and Mathur, 1978
Ramularia foeniculi	*Foeniculum vulgare*	Pericarp	Singh, 1991
Rhynchosporium secalis	*Hordeum vulgare*	Husk, pericarp	Mathre, 1982; Skoropad, 1959
Trichothecium roseum	*Zea mays*	Pericarp, endosperm, embryo	Singh et al., 1985
	Brassica sp.	Seed coat, endosperm, embryo	Sharma, 1989
	Eruca sativa	Seed coat, endosperm, embryo	Sharma, 1992
	Vigna sp.	Seed coat, embryo	Varma and Singh, 1991

(continued)

TABLE 5.5 (CONTINUED)
Location of Fungi Belonging to Deuteromycetes in Seeds of Crop Plants

Fungus	Host	Seed and Fruit Part(s)	Important References
Hyphomycetes (continued)			
Verticillium alboatrum	*Helianthus annuus*	Pericarp, seed coat	Sackston and Martens, 1959
	Carthamus tinctorius	Pericarp, seed coat	Klisiewicz, 1975
	Lupinus lutens	Seed coat only or seed coat, tracheid bar, funicular trace, cotyledons	Parnis and Sackston, 1979
	Medicago sativa	Seed coat	Christen, 1982; Huang et al., 1985
V. dahliae	*Carthamus tinctorius*	Pericarp, seed coat	Klisiewicz, 1975
Coelomycetes			
Colletotrichum dematium	*Glycine max*	Seed coat, hourglass cells	Schneider et al., 1974
	Capsicum annuum	Seed coat, endosperm, embryo	Chitkara et al., 1990
	Vigna sp.	Seed coat, embryo	Varma et al., 1992c
	Cyamopsis tetragonoloba	Seed coat, endosperm, embryo	Bhatia et al., 1996b
C. dematium f. sp. *spinaciae*	*Beta vulgaris*	Pericarp, seed coat	Chikuo and Sugimoto, 1989
C. graminicola	*Sorghum vulgare*	Pericarp, endosperm, embryo	Basu Chaudhary and Mathur, 1979; Prasad et al., 1985
C. lini	*Linum usitatissimum*	Seed coat	Lafferty, 1921
C. lindemuthianum	*Phaseolus vulgare*	Seed coat, embryo	Frank, 1883; Grummer and Mach, 1955; Zaumeyer and Thomas, 1957
C. truncatum	*Glycine max*	Seed coat, embryo	Kunwar et al., 1985
Ascochyta fabae f. sp. *lentis*	*Lens esculentum*	Seed coat, embryo	Singh et al., 1993
A. pinodes	*Pisum sativum*	Seed coat, embryo	Maude, 1996
A. pisi	*Pisum sativum*	Seed coat, embryo	Dekker, 1957
A. rabiei	*Cicer arietinum*	Seed coat, embryo	Maden et al., 1975
Botryodiplodia theobromae	*Zea mays*	Pericarp, closing tissue, endosperm, embryo	Singh et al., 1986b; Kumar and Shetty, 1983
	Hevea brasiliensis	Seed coat (tegmen), perisperm, endosperm, embryo	Varma et al., 1990

(continued)

TABLE 5.5 (CONTINUED)
Location of Fungi Belonging to Deuteromycetes in Seeds of Crop Plants

Fungus	Host	Seed and Fruit Part(s)	Important References
		Coelomycetes (continued)	
Diplodia zeae	*Zea mays*	Pericarp, endosperm, embryo	Heald et al., 1909; Miller, 1952
Phoma lingam	*Brassica oleracea*	Seed coat	Jacobsen and Williams, 1971
P. sorghina	*Sorghum vulgare*	Pericarp, endosperm, embryo	Rastogi et al., 1991
Phomopsis heveae	*Hevea brasiliensis*	Seed coat (tegmen) perisperm, endosperm, embryo	Varma et al., 1991
Phomopsis sp.	*Glycine max*	Seed coat, hilar tissue, embryo	Ilyas et al., 1975; Singh and Sinclair, 1986
Septoria glycines	*Glycine max*	Seed coat	Sinclair and Backman, 1989
S. linicola	*Linum usitatissimum*	Seed coat	Lafferty, 1921
Stagnospora nodorum	*Triticum aestivum*	Pericarp, seed coat, endosperm, embryo	Agarwal et al., 1985
Stagnospora sp.	*Hevea brasiliensis*	Seed coat (tegmen); perisperm, endosperm, embryo	Varma et al., 1991
Rhizoctonia bataticola	*Vigna* spp.	Seed coat, embryo	Sinha and Khare, 1977; Varma et al., 1992a,b
	Phaseolus vulgaris	Seed coat, embryo	Agarwal and Jain, 1978
	Sesamum indicum	Seed coat, endosperm, embryo	Singh and Singh, 1979
	Crotolaria juncea	Seed coat, embryo	Basu Choudhary and Pal, 1982
	Helianthus annuus	Pericarp, endosperm, embryo-cotyledons (rare)	Raut, 1983; Godika, 1996; Godika et al., 1999
	Glycine max	Seed coat, endosperm, embryo	Kunwar et al., 1986b; Mathur, 1992
	Cajanus cajan	Seed coat, embryo	Sharma, 1996
	Cyamopsis tetragonoloba	Seed coat, endosperm, embryo	Bhatia et al., 1998
	Abelmoschus esculentus	Seed coat, embryo	Agarwal and Singh, 2000
R. solani	*Capsicum annuum*	Seed coat, endosperm	Baker, 1947; Chitkara et al., 1986b

Alternaria infection, established as dormant mycelium, is reported in seed coat for *Alternaria brassicae* in *Brassica* (Chupp, 1935), *Alternaria brassicicola* in *Brassica* (Boek, 1952; Domsch, 1957), and *Alternaria longipes* in tobacco (Heursel, 1961), and in the pericarp of carrot for *Alternaria dauci* and *Alternaria radicina* (Soteros, 1979; Netzer and Kenneth, 1969). Atkinson (1950) alone recorded *Alternaria raphani* in the seed coat and embryo of radish.

Dormant mycelium of *Alternaria alternata* (*Alternaria tenuis*) occurs in the seed coat and pericarp in asymptomatic and weakly symptomatic seeds. Moderately to heavily infected seeds carry fungal mycelium in all parts — the seed coat or pericarp and seed coat (one-seeded dry fruits), endosperm, and embryo in the case of *Capsicum, Glycine, Brassica, Eruca, Helianthus,* and *Cumin* (Table 5.5). In weakly infected seeds of soybean inter- and intracellular hyphae occur in all the layers of the seed coat. Rarely do hyphae traverse the cells of endosperm and peripheral layers of the cotyledon. Moderately and heavily infected seeds carry a thick mat of hyphae in the region of seed coat parenchyma and endosperm; the two zones are indistinct. Hyphae also occur in all the parts of the embryo. In the hilar region, hyphae are seen outside as well as inside the hilium. When the hyphae are internal, they occur in the parenchyma and the tracheid bar (Kunwar, Manandhar, and Sinclair, 1986a). In *Hordeum vulgare* the husk, persistent lodicules, pericarp, and endosperm are infected (Thakkar, 1988). The mycelium usually occurs in layers of pericarp and testa outside the thick cuticle of the endosperm in sunflower seeds. Rarely in heavily infected seeds, the mycelium penetrates the cuticle and invades the endosperm and embryo (Singh, Mathur, and Neergaard, 1977; Godika, Agarwal, and Singh, 1999).

In *Alternaria sesamicola,* the cause of sesame blight, the mycelium readily spreads in soft parenchymatous layers of the seed coat. The epidermal cells, which have calcium oxalate crystals, remain free of infection in weak and moderately infected seeds (Figure 5.14A, B). In severely infected seeds, abundant mycelium occurs in all parts of the seed showing host cell disintegration (Figure 5.14C, D). Conidia formation occurs in spaces in the seed and also on the seed surface. Such seeds were probably infected early during development or had prolonged humid environments after being infected.

In *A. brassicicola,* the cause of leaf spot and blight of crucifers, infected seeds are asymptomatic and symptomatic in *Brassica juncea, B. campestris,* and *Eruca sativa.* In asymptomatic infected and weakly infected seeds, the hyphae are confined to the seed coat. However, in moderate and heavily infected seeds (bold discolored and shriveled discolored), the hyphae occur in the seed coat, endosperm, and embryo. The infection is maximum (100%) in the seed coat and gradually declines in the endosperm and embryo (Sharma, 1989; Sharma, 1992). Among the *Brassica* spp., the severity of seed infection was greater in mustard (*B. juncea*) than in rapeseed (*B. campestris*). Knox-Davies (1979) has found that in cabbage seeds naturally infected by *A. brassicicola,* the testa is usually colonized in the hilum area.

5.5.4.1.2 Curvularia

Curvularia lunata is a common seed-borne fungus, recorded in seeds of field crops all over the world (Richardson, 1990). It is one of the fungi that caused grain mold disease in sorghum (Rastogi, Singh, and Singh, 1990). The affected seeds are

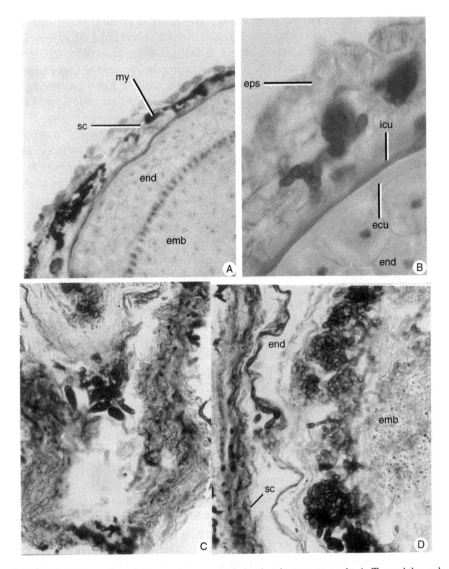

FIGURE 5.14 Location of *Alternaria sesamicola* hyphae in sesame seeds. A, Ts seed through a weakly infected seed showing dark-colored mycelium in subepidermal cells of seed coat. B, A portion from A magnified. C, Ls micropylar part of severely infected seed showing fungal conidia in the space. D, Ls part of severely infected seed. The wall layers in the seed coat are poorly differentiated, endosperm and embryo with mycelium all over. (Abbreviations: ecu, cuticle of endosperm; emb, embryo; end, endosperm; eps, seed epidermis; icu, inner enticle of seed coat; my, mycelium; sc, seed coat.) (From Singh, D., Mathur, S.B., and Neergaard, P. 1980. *Seed Sci. Technol.* 8: 85–93. With permission.)

discolored, becoming black. Thick, dark brown, septate, knotty mycelium occur exo- and endophytically in glumes of infected seeds. *Curvularia lunata* primarily colonizes the pericarp and rarely invades the aleurone layer in moderately infected seeds.

It is common in the persistent stylar tissue. Hyphae are distributed in all parts, i.e., the pericarp, aleurone layer, and other layers of the endosperm and embryo in heavily infected seeds. Embryal infection occurs in the scutellum, coleoptile, mesocotyl, primary root, and coleorhiza. Cells in infected embryos are narrow and elongated with poor contents, and those of mesocotyl show premature xylogenesis. Rarely, the embryo remains undifferentiated, made up of small parenchyma cells with clumps of mycelium. The heavily infected seeds fail to germinate.

Symptomatic seeds of onion (*Allium cepa*) infected with *C. lunata* are dull black. Asymptomatic seeds rarely (about 20%) carry mycelium on the seed surface. Thick, brown, septate, and branched mycelium occurs in the seed coat and rarely in the endosperm of symptomatic seeds. Hyphal aggregation is more toward the micropylar end (Dwivedi, 1994).

5.5.4.1.3 Bipolaris, Drechslera, and Helminthosporium

Most plant pathogenic species, initially described under *Helminthosporium*, were placed at one time under *Drechslera* (Subramanian and Jain, 1966; Ellis, 1971; Subramanian, 1971). Subsequently, these have been reclassified under *Bipolaris, Drechslera,* and *Helminthosporium*. These fungi cause leaf spots, blight, and crown or root rot in plants of Poaceae (Graminae). They are weak to potent pathogens and many of them are seed-borne and seed-transmitted. Infection of such species was primarily recorded in the pericarp and seed coat in early studies, e.g., *Bipolaris sorokiniana* in wheat (Weniger, 1925) and barley (Mead, 1942), *Bipolaris oryzae* in rice (Nisikado and Nakayama, 1943; Fazli and Schroeder, 1966), and *Drechslera graminiea* in barley (Vogt, 1923; Genau, 1928; Platenkamp, 1975). Meffert (1950) reported *Helminthosporium papaveris* in the seed coat and endosperm of poppy seeds. Fazli and Schroeder (1966) also found *B. oryzae* hyphae in the endosperm. Recent investigations on seeds infected by *Bipolaris* and *Drechslera* species of several crop plants have clearly shown that the expanse of mycelium in seed is directly correlated with the severity of infection (Yadav, 1984; Singh, Singh, and Singh, 1986a; Thakkar et al., 1991).

Yadav (1984) divided the wheat seeds affected with *B. sorokiniana* into four categories: bold seeds, seeds with loose pericarp, shriveled seeds, and discolored seeds. The mycelium is usually confined to outer layers of pericarp in bold seeds and rarely penetrates cross and tube cells (Figure 5.15H). Kernels with loose pericarp and shriveled type carry abundant mycelium in all the layers of the pericarp. Infection occurs in all components, i.e., pericarp, aleurone layer, endosperm (Figure 5.15I), and embryo in discolored seeds. The embryonal infection varies from weak to heavy. Weak infection is confined to the scutellum, but moderate to heavy infection spreads to all parts of the embryo. Rarely, the cells of scutellar epithelium undergo two to three transverse divisions, and premature xylogenesis is induced in the mesocotyle-donary node. Heavy infection of *B. sorokiniana* caused 62% pre- and postemergence losses, and 28% of the 38% of the surviving seedlings developed symptoms (Yadav, 1984).

Wheat seeds from earheads artificially inoculated by *D. tetramera* carry myce-lium in the pericarp, aleurone layer, endosperm, and rarely the embryo (Yadav, 1984).

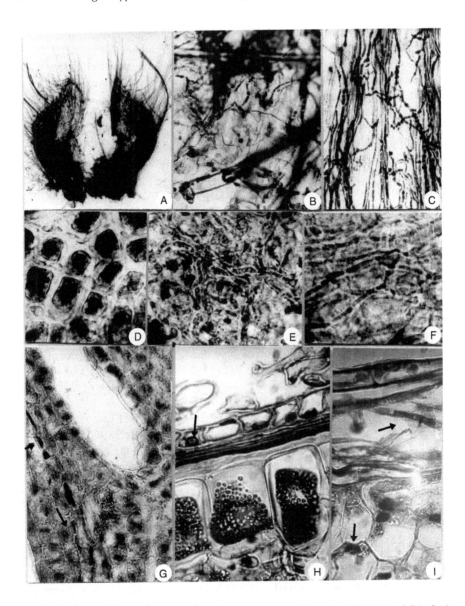

FIGURE 5.15 Photomicrographs of parts of seeds infected by *Drechslera and Bipolaris* species. A to G, *D. graminea* in barley kernels. A, B, Cleared whole-mounts of persistent lodicules showing hyphae. C to F, Whole-mounts of husk, aleurone layer, endosperm, and scutellum, respectively, having abundant hyphae. G, Ls part of embryo showing base of the coleoptile having hyphal bits (arrow). H, I, Ls part of pericarp and endosperm in wheat kernels showing inter- and intracellular hyphae (arrows) of *B. sorokiniana*. (A to G, From Thakkar, R. 1988. Ph.D. thesis, University of Rajasthan, Jaipur, India; H, I, From Yadav, V. 1984. Ph.D. thesis, University of Rajasthan, Jaipur, India.)

Infected embryos remain thin, and those with heavy infection possess mycelium in all parts. Occasionally, embryos develop induced premature xylogenesis in the mesocotyl and lysogenous cavities (Yadav, 1984). More deleterious effects on wheat embryo are caused by *Drechslera tetramera* than by *B. sorokiniana*.

The inter- and intracellular mycelium of *Bipolaris maydis* occurs in the basal cap, closing tissue, and pericarp in weakly infected maize kernels, but it is abundant in the pericarp, basal cap, closing tissue, aleurone layer, endosperm, and embryo in heavily infected seeds (Singh, Singh, and Singh, 1986a).

Drechslera graminea, which causes leaf stripe disease in barley, is seed-borne. According to Vogt (1923), Genau (1928), and Platenkamp (1975), the hyphae are confined to the pericarp. Thakkar et al. (1991) found that moderate to heavily infected seeds are symptomatic, and the expanse of mycelium and its effect on embryo are variable. In moderately infected seeds, the hyphae may be confined to the pericarp (husk) and vascular strand of seed, or they may invade persistent lodicules (Figure 5.15A, B), all layers up to the aleurone layer (Figure 5.15C, D), or penetrate the endosperm and rarely the embryo (Figure 5.15E to G). The hyphae spread inter- and intracellularly in all parts of heavily infected kernels. In very heavily infected seeds, hyphae freely traverse from one part to another, binding the tissues and making it difficult to separate husk, pericarp, aleurone layer, and even endosperm. Heavily infected seeds may have a small well-formed embryo, a shriveled poorly differentiated embryo, or undifferentiated embryonal mass. The mycelium invades all parts except seminal leaves and the plumule bud in small embryos, all parts of shriveled embryos, and formless embryos. Barley kernels with deep infection of *D. graminea* fail to germinate (Thakkar, 1988).

Bipolaris oryzae, a serious pathogen that causes brown leaf spot in rice, occurs in the pericarp, seed coat, and endosperm (Fazli and Schroeder, 1966). Paul (1987) noted *B. oryzae* in the husk (palea and lemma), pericarp on the embryal side, and hilar region in weakly infected kernels. It spread all over the pericarp and testa, but not in the endosperm and embryo in moderately infected seeds. In shriveled heavily infected seeds, *B. oryzae* occurred in all parts. The fungus formed pseudosclerotia between pericarp and testa and between testa and aleurone layer. Hyphae occurred in the peripheral layers of the endosperm and in the scutellum and other tissues of the embryo.

5.5.4.1.4 Cylindrocladium

Cylindrocladium black rot caused by *Cylindrocladium crotalariae* is widespread in peanut-growing areas in the United States. Porter et al. (1991) found a high degree of correlation between the incidence of disease in the field and seed infection. Hyphae of *C. crotalariae* ramified both inter- and intracellularly in cells of the seed coat in discolored seeds. Seeds with dark brown seed coat carried hyphae in cotyledons also. In these seeds, abundant hyphae occurred in the seed coat, between the seed coat and the cotyledons, and in cells of cotyledons (Porter et al., 1991). Porter et al. (1991) have proposed cleaning seed lots by removing shriveled and discolored seeds to check the disease transmission.

5.5.4.1.5 Cercospora

Cercospora species cause serious diseases of *Arachis hypogaea, Glycine max,* and *Beta vulgaris. Cercospora kekuchii* and *C. sojina* cause purple seed stain and seed discoloration of soybean, respectively. Ilyas et al. (1975) found hyphae of *C. kekuchii* confined to the seed coat, but Singh and Sinclair (1985, 1986) observed their occurrence in the seed coat, endosperm, and embryo. Singh and Sinclair (1985), using symptomatic seeds with gray to brown discoloration of the seed coat as well as cotyledons (collected from uninoculated and artificially inoculated plants by *C. sojina*), found that the pathogen colonized the seed coat tissues (Figure 5.16A), the space between the seed coat and embryo, and, rarely, the hypocotyl-radicle region of the embryo. Abundant hyphae occurred in the hilar region including tracheid bar (Figure 5.16B).

Rastogi, Singh, and Singh (1998c) found hyphae of *C. traversiana* in the seed coat and tissues of the hilar region in moderately infected seeds of *Trigonella foenum-graecum*. Abundant hyphae occurred in the seed coat, endosperm, and embryo in severely infected seeds. Singh (1991) has reported the presence of conidia of *C. personata* in the shells of groundnut, but there is no histopathological evidence for this.

5.5.4.1.6 Botrytis

Botrytis anthophila, which causes anther mold of *Trifolium pratense* (red clover), reduces pollen fertility and also seed formation. Bondarzew (1914) and Silow (1933) reported systemic infection of plants. The infection originated from intraseminal mycelium, particularly present in the hourglass cells of the seed coat. Using seeds incubated on agar for 3 to 5 days, Bennum (1972) observed predominantly intercellular and rarely intracellular mycelium in the seed coat and hilar region — remnants of funiculus, tracheid bar, and aerenchyma (stellate parenchyma) — but not in the embryo.

Botrytis fabae occurs in the seed coat, cotyledons, and embryonal axis in infected seeds of *Vicia faba*. Hyphae are present in the seed coat of all the infected seeds, but these occur in 40% of cotyledons and 20% of embryonic axes (Harrison, 1978). Van der Spek (1965) has reported *Botrytis cinerea* in the cells of seed coat in flax.

5.5.4.1.7 Pyricularia

Pyricularia oryzae causes the most destructive and widely distributed blast disease in rice. Chung and Lee (1983) have reported infection of *P. oryzae* within the tissues of the glumes, rachilla, pedicel, palea, lemma, and pericarp. In heavily infected seeds, the mycelium penetrates the aleurone layer and peripheral layers of the endosperm. The affected layers of endosperm turn dark brown. The embryos are free of infection. Earlier, Suzuki (1930, 1934) had reported that seeds from ears artificially inoculated before, during, and after the flowering period, carried the fungal hyphae in the endosperm and embryo.

The seeds of *Eleusine coracona* infected by *Pyricularia grisea* are asymptomatic and symptomatic (discolored). The discolored seeds exhibited two to three times higher infection than the asymptomatic seeds. Pericarp of all the infected seeds had

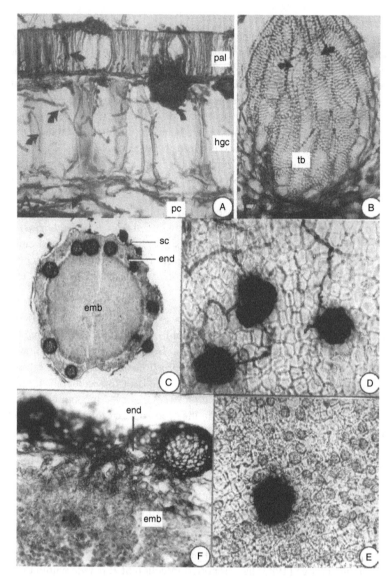

FIGURE 5.16 Distribution of fungal mycelium of *Cercospora sojina* and *Macrophomina phaseolina* in seeds of *Glycine max* (A, B) and *Sesamum indicum* (C to F), respectively. A, Ts seed showing hyphae and hyphal aggregations (arrows) in palisade layer, hourglass cells, and in parenchyma. B, Enlarged region of hilar tracheids showing hyphae. C, Ts seed showing microsclerotia in seed coat and endosperm. D, E, Whole-mount preparations of seed coat and endosperm respectively, having thick, knotty, and branched mycelium and microsclerotia. F, Ts part of seed with mycelium in endosperm and its spread to the embryo. (Abbreviations: emb, embryo; end, endosperm; hgc, hourglass cells; pal, palisade cells; pc, parenchymatous cells; sc, seed coat; tb, tracheid bar.) (A, B, From Singh, T. and Sinclair, J.B. 1985. *Phytopathology,* 75, 185–189. With permission; C to F, Singh, T. and Singh, D. 1979. In *Recent Research in Plant Science.* Bir, S.S., Ed. Kalyani Publishers, New Delhi.)

fungal mycelium, but it occurred only in 56.8% of the endosperms. The embryos were free of infection (Rangnathaiah and Mathur, 1978).

5.5.4.1.8 Fusarium

Fusarium spp. are a common associate of seeds of a large number of crop plants (Richardson, 1990). They cause vascular wilts, primarily of annual vegetable and flowering plants. Djerbi (1971) has given an excellent account of *Fusarium culmorum* infection in wheat kernels. Infection may occur only in the outer layers of the pericarp, in the layers of pericarp and seed coat, and in all parts of the caryopsis including the thick cuticle of the endosperm and the aleurone layer, where the mycelial cushions may be formed. The degree of invasion of kernel tissues reflects the time of infection. If infection takes place during the early stages of kernel development, colonization may reach deeper tissues. If infection takes place near maturity, only the pericarp is invaded (Djerbi, 1971).

Fusarium moniliforme was initially reported to be confined to the pericarp between the brown cap and the pedicel in maize kernels (Sumner, 1966). Singh and Singh (1977) observed abundant mycelium of *F. moniliforme* in the pericarp (Figure 5.17A, B) and parenchyma layers outside the brown sclerotic cap in maize seeds (Figure 5.17A, D). Only rarely does the mycelium penetrate the endosperm on the sides (Figure 5.17C) as well as around the brown cap. Singh, Singh, and Singh (1985) in a study of maize kernels from tribal areas in Rajasthan, India, found hyphae of *F. moniliforme* in the pericarp, butt, endosperm, and peripheral layers of scutellum. The hyphae also penetrated the placentochalazal region. Mathur, Mathur, and Neergaard (1975) have also reported the presence of *F. moniliforme* in the pericarp, endosperm, and embryo in sorghum.

Fusarium oxysporum is common in seeds of Fabaceae (Table 5.5). Velicheti and Sinclair (1991) have reported the hyphae of *F. oxysporum* over the seed surface, in the hilar region, and seed coat of soybean. Sharma (1992) has reported that the soybean seeds infected by *F. oxysporum* are asymptomatic or symptomatic, depending on the degree of infection. The symptomatic seeds are reddish brown. In asymptomatic seeds, the hyphae are confined to the seed coat and the hilar stellate parenchyma. In symptomatic seeds, however, the infection occurs in all the layers of the seed coat, stellate parenchyma, and rarely, in the tracheid bar, aleurone layer, and peripheral two to three layers of cotyledons on the abaxial surface in weakly infected seeds. In moderately infected seeds, inter- and intracellular mycelium occurs in all components. Heavy colonization is found in the seed coat with mycelial mat in the parenchymatous region. In cotyledons the infection is more on the abaxial than the adaxial surface. Infection in the embryonal axis is rare. In heavily infected seeds, aggregation of mycelium occurs in different components including the embryonal axis. Chlamydospores, microsclerotia, and microconidia may be present on the seed surface, in spaces between the components, and in the lysigenous cavities in cotyledons and the hypocotyl root-shoot axis. Mycelium is inter- and intracellular and also seen in the vascular elements of the seed coat and cotyledons (Sharma, 1992).

A similar trend in the spread of mycelium of *F. oxysporum* has been observed in the seeds of *Cyamopsis, Cajanus,* and *Vigna* (Varma, 1990; Bhatia, Singh, and Singh, 1996a; Sharma, 1996).

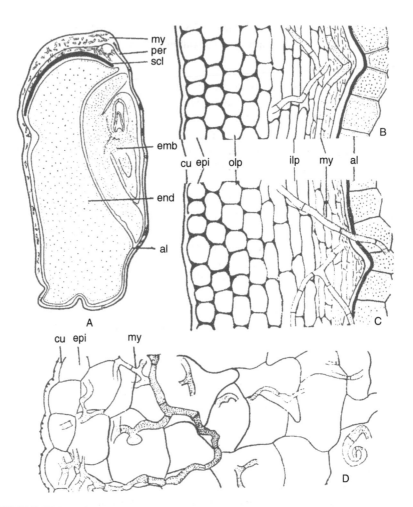

FIGURE 5.17 Location of *Fusarium moniliforme* in maize kernels. A, Ls maize kernel showing mycelium in pericarp and pedicel. B, C, Ls part of pericarp showing mycelium in the inner layers and penetration of hyphae into the aleurone layer in C. D, Ls part from pedicel region showing inter- and intracellular hyphae. (Abbreviations: al, aleurone layer; cu, cuticle; emb, embryo; end, endosperm; epi, epidermis; ilp, inner layers of pericarp; olp, outer layers of pericarp; my, mycelium; per, pericarp; scl, sclerotic layer.) (From Singh, D. and Singh, T. 1977. *J. Mycol. Plant Pathol.* 7: 32–38.)

5.5.4.1.9 Verticillium

Verticillum causes vascular wilt in plants and is found almost worldwide, but is significant in temperate zones. *Verticillium albo-atrum* and *V. dahliae* attack hundreds of hosts and are reported to be seed-borne (Richardson, 1990). However, even in cases where *Verticillium* is positively reported to be seed-borne, the infection is often found to be inconsistent (Rudolph, 1944; Sackston and Martens, 1959; Parnis and Sackston, 1979; Huang, Hanna, and Kokke, 1985). According to Van der Spek (1972), *V. dahliae* invades through the vascular system of the mother plant and establishes in seeds of beet and spinach. Huang, Hanna, and Kokke (1985) observed *V. albo-atrum* throughout the stems of alfalfa, but it occurred only sporadically in the peduncle, pedicels, pods, and seeds. Artificial inoculation of stigmas of healthy plants of cultivar Vernal of *Medicago sativa* (alfalfa) caused infection of the stigma and upper part of style. The infection remained confined to these parts throughout the fruit and seed development. Under humid conditions, the fungus in the remnants of style attached to the mature pod, rarely invaded the pod and seed coat (Huang, Hanna, and Kokke, 1985). Christen (1982) found *V. albo-atrum* hyphae in alfalfa seeds, particularly small seeds obtained from artificially inoculated greenhouse plants of susceptible cultivars. SEM photomicrographs showed the mycelium within and between hourglass cells of the seed coat.

Artificial inoculation by injecting spore-mycelium suspension of *V. albo-atrum* in stems of *Lupinus albus* and *L. luteus* showed variability in the colonization of fruit and seed. The pathogen was discontinuous in tissues of fruit, including funiculus in *L. albus*. It was confined to the seed coat, but not seen in vascular elements. In *L. luteus*, mycelium occurred in vascular bundles of the seed coat, within the tracheid bar and the funicular trace in infected seeds. Profuse mycelium occurred in the space between the seed coat and embryo and from there it spread to the cotyledons (Parnis and Sackston, 1979).

5.5.4.1.10   Trichothecium

Trichothecium roseum, a well-known saprophyte, is common in seeds of crop plants. It causes seed rot and seedling blight in incubation tests in maize kernels (Singh, Singh, and Singh, 1985), seeds of cowpea (Varma and Singh, 1991), and rape and mustard (Sharma, 1989). The seeds are asymptomatic or symptomatic. In cowpea, rape, and mustard, symptomatic seeds show discoloration and shiny pinkish mycelium and/or conidia on the seed surface. In maize kernels, thin, shining hyaline hyphae colonized the basal cap, pericarp, endosperm, and embryo. The infected embryo was well formed, but thin with lysigenous cavities containing abundant mycelium and bicelled spores of *T. roseum*.

In symptomatic seeds of cowpea, *T. roseum* occurred in the seed coat, endosperm, and embryo. The mycelium was inter- and intracellular in hourglass cells and formed a thick mat in parenchyma cells and spongy parenchyma in the hilar region. The embryonal infection occurred in the peripheral layers of cotyledons, radicle, and all over in the plumule and seminal leaves (Varma and Singh, 1991). In rape and mustard the mycelium of *T. roseum* was confined to the epidermis and subepidermis of the seed coat in symptomless and bold weakly discolored seeds. In heavily discolored seeds, however, it occurred in the seed coat, endosperm, and embryo. The mycelium was inter- as well as intracellular and caused depletion of cell contents (Sharma, 1989). The symptomatic seeds in cowpea and heavily discolored and shriveled discolored seeds of rape and mustard failed to germinate.

5.5.4.2   Coelomycetes

5.5.4.2.1   Colletotrichum

Colletotrichum species, which cause anthracnose and die-back diseases, have mostly *Glomerella* and occasionally *Physalospora* or some other genus as the sexual stage.

Colletotrichum lindemuthianum was the first seed-borne fungus whose mycelium was found to penetrate deep into the cotyledons of bean (*Phaseolus vulgaris*) by Frank (1883). *Colletotrichum* spp. are commonly encountered in seeds of legumes, solanads, and cereals (Richardson, 1990).

Chili seeds infected with *Colletotrichum dematium* have a few to many brown to black spots (acervuli) on the seed surface (Chitkara, Singh, and Singh, 1990). In moderately infected seeds, the mycelium is usually confined to the seed coat, space between seed coat and endosperm, and superficial layers of endosperm. Hyphae rarely invaded the embryo. Inter- and intracellular hyphae and acervuli occurred in the seed coat, endosperm, and embryo in heavily infected seeds (Table 5.5). Thick septate mycelium covered the seed surface and spread among spaces between different components of seed. A thick mat of mycelium was found in parenchymatous layers of the seed coat, the comma stem region of endosperm, and the radicle end and tips of cotyledons, which lie close to the comma stem region (Chitkara, Singh, and Singh, 1990). Symptomatic seed showed only 53% germination and 32% seedling survival.

Chikuo and Sugimoto (1989) observed variations in the establishment of *C. dematium* f. sp. *spinaceae* mycelium in seeds obtained from plants of *Beta vulgaris* artificially inoculated at early and late flowering stages. In the former condition, infection spread in the fruit cavity, and hyphae invaded developing seed, causing its collapse. But in the latter situation, the mycelium was usually confined to the surface of the seed ball, and hyphae were rarely seen in the apical pore and under the seed coat.

Tiffani (1951) artificially inoculated seeds of soybean with *C. dematium* and observed pre-emergence killing, seedling blight, and plants with latent infection. In fruits of plants with latent infection, fungal hyphae were traced in the carpel wall, ovary cavity, and in the cotyledons of developing seeds. Initially these pods were symptomless, but at maturity, some of them were covered with acervuli. Schneider et al. (1974) found that *C. dematium* mycelium in symptomatic soybean seeds from India is confined to the hourglass layer of the seed coat and to naturally occurring wounds.

Varma, Singh, and Singh (1992c) and Bhatia et al. (1996b) found that heavy infection of *Colletotrichum dematium* occurred in all components of categorized naturally infected seeds of *Vigna aconitifolia* and *Cyamopsis tetragonoloba*, respectively (Table 5.5). Varma, Singh, and Singh (1992c) found hyphae occurring on or in the palisade cells, hourglass cells, and, rarely, in the peripheral layers of the hilar region in moderately infected seeds. The hyphae in the heavily infected seeds were distributed all over in the seed, i.e., in the seed coat-palisade cells, the hourglass layer, and the parenchyma layers, the embryo-cotyledons and hypocotyledonary root-shoot axis, and in the hilar region in all tissues, including the tracheid bar and spongy parenchyma. Heavily infected seeds failed to germinate.

Basu Chaudhary and Mathur (1979) recorded *C. graminicola* in the pericarp (97 and 83%), horny (24.6 and 39.2%) and floury endosperms (8.6 and 14.2%), and embryos (1.4 and 1.3%) of two seed samples of sorghum. About 55% of seeds did not germinate, showing severe seed rot. Of the emerging seedlings, the majority showed drying of leaves from tip to base and died in 2 to 3 weeks. Only 2 to 3% of the plants survived.

5.5.4.2.2 Ascochyta

Ascochyta rabiei and *A. fabae* var. *lentis* cause chickpea and lentil blight, respectively. Infected seeds show discolored areas. Lesions vary from small to large with mycelium and pycnidia often seen on the seed surface of chickpea (Maden et al., 1975) and occasionally in lentil (Singh, Khare, and Mathur, 1993). In chickpea the lesions are localized, and abundant mycelium occurs in the region of the lesions, but the adjoining clean areas are free of infection. Seeds with small superficial lesions carry mycelium in the seed coat and on the surface of cotyledons, whereas those with deep lesions contain profuse mycelium in the seed coat (Figures 5.18A, B) and cotyledons (Figure 5.18C). Pycnidia are common in the seed coat (Figure 5.18B) and hyphae often become intracellular in vascular elements (Figure 5.18D). Both superficial and deep infections have equal potential to cause seedling infection (Maden et al., 1975; Vishnuawat, Agarwal, and Singh, 1985).

The hyphae of *A. fabae* f. sp. *lentis* travel vertically as well as horizontally in the palisade cells, hourglass cells, and parenchymatous cells of the seed coat in moderately infected seeds. Hyphae are inter- and intracellular in parenchyma. In heavily infected seeds, the fungal mycelium occurs in the seed coat, the space between the seed coat and the cotyledons, and, rarely, the space between the cotyledons and the embryonal axis. Heavy accumulation of mycelium also takes place in the hilar tissues (Singh, Khare, and Mathur, 1993).

Ascochyta pisi, which causes blight, leaf, and pod spot in pea, is carried as mycelium in the seed coat in infected seeds; 75% of cotyledons and 40% of plumule are also infected in such seeds. Infection may be caused by mycelium spreading in the pod cavity or directly by lesions on the pod surface (Dekker, 1957).

5.5.4.2.3 Botryodiplodia

Botryodiplodia theobromae is seed-borne in a large number of crops (Kumar and Shetty, 1983; Richardson, 1990). It causes stem rot and seed rot in maize. Singh, Singh, and Singh (1986b) found that the seed surface in weak to moderately infected maize seeds develops black streaks or pinhead-like microsclerotia, but heavily infected seeds are almost black. *B. theobromae* hyphae are confined to the pericarp and basal cap in weakly infected seeds, but occur in the pericarp, basal cap, closing tissue, aleurone layer, and, rarely, in other layers of the endosperm and placentochalazal region in moderately infected seeds. It occurs in all tissues, including the embryo in heavily infected kernels (Kumar and Shetty, 1983; Singh, Singh, and Singh, 1986b). Heavy infection causes disintegration of tissues and depletion of reserve food material.

The decoated (woody testa removed) rubber seeds infected with *B. theobromae* have brown to black discoloration progressing from micropylar to chalazal end (Varma, Singh, and Singh, 1990). The expanse of mycelium differed quantitatively in different components in seeds showing different degrees of discoloration. The fungus colonized the tegmen, perisperm, endosperm, and embryo in all the symptomatic seeds, but it was recorded only rarely in asymptomatic seeds. In moderately

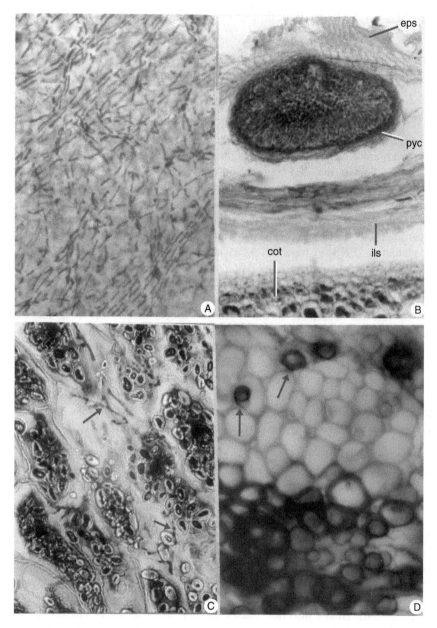

FIGURE 5.18 *Ascochyta rabiei* in seeds of *Cicer arietinum*. A, Cleared whole-mount of inner layers of seed coat showing profuse branched and septate mycelium. B, Ls part of seed with mycelium aggregated in the layers of seed coat and subepidermal pycnidium. C, Ls part of cotyledon showing inter- and intracellular mycelium (arrow). D, Ts through vascular region of seed in the position of lesion showing intracellular hyphae (arrows). (Abbreviations: cot, cotyledon; eps, seed epidermis; ils, inner layers of seed coat; pyc, pycnidium.) (From Maden, S. et al. 1975. *Seed Sci. Technol.* 3: 667–681. With permission.)

and heavily infected seeds, abundant mycelium occurred in various tissues, including vascular elements in the tegmen. It formed a thick mycelial mat in the spaces around the endosperm and embryo. In heavily infected seeds, *B. theobromae* caused disintegration of cells. Some of the mesophyll cells in infected cotyledons became hypertrophied and thick-walled (Varma, Singh, and Singh, 1990).

5.5.4.2.4 Diplodia

Heald, Wilcox, and Pool (1909) reported that *Diplodia zeae* is internally seed-borne in maize kernels. The dormant mycelium occurs in the endosperm and embryo. Miller's (1952) report that the fungal mycelium first invades the embryo and subsequently spreads to the endosperm and pericarp needs confirmation.

   Diplodia gossipina, which causes black boll rot in cotton, infects seed and lint, making them smutty (Crowford, 1923; Crosier, 1944; Roncadori, McCarter, and Crawford, 1971). Snow and Sachdev (1977) observed that *D. gossipina* enters cotton bolls through the epidermis, multicellular hairs, and stomata to reach the ovary and fruit cavity.

5.5.4.2.5 Phoma

Phoma lingam, which causes blackleg in oilseed and vegetable crucifers, expresses as crown canker, stem canker, and leaf lesions. The disease is seed-borne and seed transmitted (Petrie and Vanterpool, 1974; Gabrielson, 1983). In cabbage seeds, *P. lingam* is common in the outer epidermis and subepidermal parenchyma and occasionally in other layers of the seed coat. It is frequently found in the peripheral layers of cotyledons and only rarely infects the radicle (Jacobsen and Williams, 1971). *Phoma lingam* infection survives for more than 10 years in vegetable crucifer seed (Gabrielson, 1983).

   Hyphae and pycnidia of *Phoma lycopersici*, which causes stem rot in tomato, are located in spaces left by the absorption of the parenchymatous cells of the middle zone of the seed coat (Fischer, 1954).

   Rastogi, Singh, and Singh (1991) reported that seeds of different cultivars of sorghum infected by *Phoma sorghina* bear black, pinhead-like pycnidia on the surface. Their number correlates directly with the severity of infection. Sorghum seeds are with or without testa. Bold asymptomatic seeds with or without testa were free of infection. Seeds without testa had weaker infection than those with testa. Symptomatic seeds with testa carried thick, dark brown, and septate mycelium in the epicarp and mesocarp, and only occasionally invaded the endocarp in weakly infected seeds. In moderately infected seeds, inter- and intracellular mycelium occurred in all layers of the pericarp and rarely spread in the hilar region, aleurone layer, transfer cells (modified endosperm cells), and peripheral layers of the scutellum. Pycnidia were common on the seed surface and in the pericarp. Abundant hyphae and pycnidia occurred in the pericarp, testa, aleurone layer, and endosperm in heavily infected seeds. The infection in the endosperm proceeded centripetally, and the mycelium formed a network around the cells in floury endosperm. Embryonal infection was moderate to heavy, and abundant mycelium and pycnidia occurred all over (Rastogi, Singh, and Singh, 1991).

5.5.4.2.6 Phomopsis

Phomopsis spp. cause blight, stem canker, leaf spot, and fruit rot in many vegetable, ornamentals, fruit, and other economic crops. In soybean *Phomopsis* spp. complex comprising *Diaporthe phaseolorum* var. *sojae, D. phaseolorum* var. *caulivora*, their *Phomopsis* anamorphs and an unidentified species cause seed decay. Ilyas et al. (1975) reported hyphae of *Phomopsis* sp. in the seed coat and cotyledons of soybean. Subsequently, Singh and Sinclair (1986) observedi the mycelium of *Phomopsis* sp. on the surface and in various layers of the seed coat in naturally infected seeds, i.e., palisade cells, hourglass cells, and parenchyma cells. It also occurred in the hilar region, including the stellate parenchyma and tracheid bar, aleurone layer, other layers of endosperm, and the embryo (cotyledons). Hyphae formed a thick mat in the seed coat parenchyma and endosperm and also colonized the space around the embryo and between the two cotyledons.

Phomopsis heveae, which causes die-back of stem and mature seed pods, is a common seed-borne fungus of *Hevea brasiliansis* (Richardson, 1990; Singh and Singh, 1990). Decoated infected seeds may be symptomless or discolored, light to dark brown. The symptomless infected seeds carry thin and hyaline mycelium in the tegmen, endosperm, and space between the endosperm and embryo. The hyphae are relatively thick, inter- and intracellular in the above tissues in light brown seeds. The dark brown discolored seeds (heavy infection) contain hyphae in all seed components. In the tegmen, hyphae also occur around vascular bundles and in vascular elements, particularly the vessels. Abundant inter- and intracellular mycelium is seen in the perisperm, endosperm, and embryo. The embryo is surrounded by a mycelial mat, and cotyledonary infection is more on the abaxial side than the adaxial side (Varma, Singh, and Singh, 1991).

5.5.4.2.7 Septoria and Stagnospora

Diseases caused by these fungi are worldwide and affect many crops. The most common and most serious disease is leaf and glume blotch in wheat, caused by *Stagnospora nodorum*. It is seed-borne and seed transmitted. Important seed-borne *Septoria* spp. are *S. glycines* in *Glycine max, S. linicola* in *Linum, S. apiicola* in celery and parsley, and *S. lactucae* in *Lactuca*. The pycnidia are present on seeds of celery and parsley (Neergaard, 1979). Dormant mycelium of *S. glycines* and *S. linicola* occurs in the seed coat of soybean and flax seeds, respectively (Lafferty, 1921; Sinclair and Backman, 1989).

Wheat seeds infected by *S. nodorum* form an important source of primary inoculum for leaf and glume blotch disease. The seed-borne infection may survive for 10 or more years, depending upon the storage conditions (Hewett, 1987; Siddiqui and Mathur, 1989; Cunfer, 1991). Ponchet (1966) reported the occurrence of *S. nodorum* mycelium beneath the testa (probably the pericarp) of diseased wheat kernels, but Kietreiber (1961) found that all seed tissues, including the embryo, may be colonized in severely damaged seeds. Agarwal et al. (1985) found that wheat seeds infected by *S. nodorum* may be symptomless (bold) or symptomatic, having a loose cracked pericarp, shriveling, and discoloration. The hyphae of *S. nodorum* are confined mostly to the pericarp in bold and cracked seeds, to the pericarp and aleurone layer in shriveled seeds, and are found in all the parts, including the embryo, in discolored, heavily infected seeds (Figure 5.2A). The internal infection may vary in seeds of the same category in different cultivars. In symptomatic seeds, the pericarp is always infected, but the infection gradually decreases from outside to inside in seeds of each category, even in the same cultivar (Table 5.6). Usually in bold symptomless seeds having a loose pericarp (weakly infected), hyphae are confined to the pericarp, but in shriveled and discolored seeds, hyphae are frequently present in all tissues, e.g., cultivar Svenno, but in cultivar Starke II, no infection was seen in the testa, aleurone layer, endosperm, or embryo in shriveled seeds, and these components were only rarely invaded in discolored seeds (Table 5.6).Varma, Singh, and Singh (1991) determined the distribution of hyphae of an unidentified *Stagnospora* sp. in decoated rubber seeds. The pathogen colonized the tegmen, perisperm, endosperm, and embryo. The fungus caused lysigenous cavities in the endosperm and embryo, containing abundant mycelium.

5.5.4.2.8   Rhizoctonia

Rhizoctonia species are soil inhabitants. They are both soil-borne and seed-borne. *Rhizoctonia* species, particularly *Rhizoctonia bataticola* and *Rhizoctonia solani*, have a very wide host range and are internally seed-borne in a variety of hosts (Table 5.5). The seeds infected with *R. bataticola* (*Sclerotium bataticola*) are mostly symptomatic and occasionally symptomless. The symptomatic seeds have black pinhead-like microsclerotia on the seed surface. Their number is directly correlated with the severity of infection and also with the internal infection and damage caused to seed tissues (Singh and Singh, 1979; Kunwar et al., 1986b; Mathur, 1992; Varma, Singh, and Singh, 1992a,b; Bhatia, Singh, and Singh, 1998).

In seeds of different leguminous crops, the course and extent of infection of *R. bataticola* is basically similar (Table 5.5). In soybean seeds, the fungus is readily observed on the outer and inner surfaces of the seed coat in SEM photomicrophs. Hyphae are seen penetrating through cracks, micropores, and the funiculus (Mathur, 1992). In asymptomatic seeds, fungal hyphae are confined to the seed coat and hilar tracheids. Thick, branched, septate, inter- and intracellular mycelium, and microsclerotia occur in all the layers of the seed coat in weakly, moderately, and heavily infected symptomatic seeds. In heavily infected seeds hyphae traverse horizontally and vertically in cells of seed components and also in spaces between components. Cotyledons are invaded from adaxial as well as abaxial surfaces. In the hilar region abundant mycelium occurs on either side of the counter palisade layer, in the tracheid bar, and stellate parenchyma. The fungus causes lysis of cells, cell vacuolation and necrosis, and formation of sclerotia in the cotyledons, radicle, and plumule in severely infected seeds. TEM micrographs show necrosis of the cell wall, enlargement of protein bodies with pronounced changes in optically dense and stained bodies, which become unrecognizable, digestion of lipid bodies, loss of discreteness in cell organelles, and appearance of ergastic bodies with surface striations, usually seen along the cell wall and plasma membrane (Mathur, 1992).

Basu Chaudhary and Pal (1982) found infection of *M. phaseolina* in all parts of sunn hemp seeds. In seeds of *Cyamopsis tetragonoloba*, having well-developed endosperm, *R. bataticola* infection occurs in all components (Bhatia, Singh, and

TABLE 5.6
Location of Mycelium of *Stagnospora nodorum* in Different Categories of Seeds of Two Wheat Samples

| Seed Categories | Sample No. 1 (Cv. Svenno) | | | | | | | Sample No. 2 (Cv. Starke II) | | | | | | |
| | Pericarp | | | Testa | Aleurone Layer | Endosperm | Embryo | Pericarp | | | Testa | Aleurone Layer | Endosperm | Embryo |
	Epicarp	Hypoderm	Cross Cell Layer					Epicarp	Hypoderm	Cross Cell Layer				
Bold seeds	4	4	1	0	0	0	0	3	2	0	0	0	0	0
Seeds with loose and cracked pericarp	5	4	3	0	0	0	0	5	3	1	0	0	0	0
Shriveled seeds	5	5	3	2	2	2	2	5	4	2	0	0	0	0
Discolored seeds	5	5	5	3	3	3	3	5	5	2	1	1	1	1

Note: Evaluated using microtome sections, five seeds per category.

From Agarwal, K., Singh, T., Singh, D., and Mathur, S.B. 1985. *Phytomorphology* 35: 87–94.

Singh, 1998). In sunflower, infection is common in the pericarp and endosperm, but rare in the embryo (Raut, 1983; Godika, Agarwal, and Singh, 1999).

Singh and Singh (1979) observed that *M. phaseolina* causes usually symptomatic and rarely symptomless infection in sesame seeds. In symptomatic seeds, profuse thick, knotty, dark brown to black, and septate mycelium and also microsclerotia were seen in all components and also between the spaces among the components (Figure 5.16C to F). Weak infection induced cell divisions in the endosperm and embryo, quite characteristic in the palisade cell region of cotyledons (Figure 5.3A, B). Heavy infection resulted in abundant aggregation of mycelium and microsclerotia in seed components and depletion of food contents (Figure 5.16F).

Hedgecock (1904) first reported *Rhizoctonia solani* to be internally seed-borne in beans. Mycelium and sclerotia were present in the seed coat of seeds obtained from pods in contact with soil. Baker (1947), who examined seeds of *Capsicum frutescens* from rotted fruits that were in contact with soil, found *R. solani* as mycelium and sclerotia on the surface of the seed coat and as mycelium in the remnants of the funiculus, seed coat, endosperm, and embryo, particularly at the tip of the radicle. Chitkara, Singh, and Singh (1986b) observed *R. solani* mycelium and microsclerotia in the seed coat and endosperm in seeds of *C. annuum* obtained from fruits borne on plants above the soil surface.

5.6 ENDOPHYTES

Endophytic fungi constitute an interesting group present in conifers to grasses (Alexopoulos, Mims, and Blackwell, 1996). Taxonomically they belong to Hypocreales (Clavicipitaceae), Loculoascomycetes, and Inoperculate Discomycetes (Rhytismatales). The term *endophyte* has been defined as an organism contained or growing entirely within a plant, parasitically or symbiotically (Snell and Dick, 1971; Siegel, Latch, and Johnson, 1987). The endophytic fungi, which are seed-borne and belong to the tribe Balansiae of the Clavicipitaceae (Diehl, 1950; Siegel, Latch, and Johnson, 1987), are considered here. The infection is systemic, and nonhaustorial and intercellular mycelium occur in all parts of the affected plant except roots.

The presence of endophytic mycelium in seed of a grass (*Lolium temulentum*) was first reported by Vogl (1898). Sampson (1933, 1935, 1937) showed the maternal transmission of the endophyte. This has been confirmed on $E_- \times E_+$ pair crosses by Neill (1940, 1941), Siegel et al. (1984a, 1985), and Manuel and Do Valle (1993). The effects of the fungal endophytes on the reproductive cycle of their hosts vary, ranging from complete sterility to complete fertility. There is an inverse correlation of host and fungus sexuality. In parasitic associations, the fungus is sexual (produces both ascospores and conidia) and the host is asexual or with serious derangements in the reproductive cycle. In symbiotic or mutualistic associations, the host is sexual (produces viable seeds) and the fungus is asexual (Clay, 1986). The behavior of the same endophyte may also vary in different hosts (Sampson, 1935).

Table 5.7 gives information on infection of endophytes in flowers and seeds of grasses. The taxonomic identity of the endophyte in many of these cases has not been determined. Latch, Christensen, and Samuels (1984) isolated five endophytes of *Lolium* and *Festuca* in New Zealand and discussed their taxonomy and taxonomic

TABLE 5.7
Fungal Endophytes in Reproductive Structures and Seeds

Pathogen	Host	Part(s) of Seed and Fruit	Important References
Epichloe typhina	Festuca rubra	Developing ovaries, seed-pericarp, endosperm, embryo	Sampson, 1935
E. typhina	Dactylis glomerata	Stromatic sheath around inflorescence, totally or partially suppressed panicles: seeds in the latter abortive or contained mycelium in all parts	Western and Cavett, 1959
E. typhina (Acremonium typhinum)	Agrostis hiemalis	Seed including embryo	White and Chambless, 1991
Unidentified endophyte	Lolium perenne	Nucellus, periphery of aleurone layer	Lloyd, 1959
Unidentified endophyte	Lolium perenne Festuca arundinacea	Inflorescence primordium, floral apices, ovary, ovule, embryo sac, seed-pericarp, aleurone layer, endosperm, embryo including plumule apex	Philipson and Christey, 1986
Pseudocercosporella trichachinicola	Trichachne insularis	Scutellum, coleoptile and embryonic leaves	White and Morrow, 1990
Acremonium coenophialum	Festuca arundinacea	Between seed coat and aleurone layer	Manuel et al., 1994
A. lolii	Lolium perenne	Between seed coat and aleurone layer	Manuel et al., 1994

relationships with previously reported unidentified fungal endophytes. The endophytes identified are *Acremonium lolii* (*A. loliae*) and a *Gliocladium*-like species found in *Lolium perenne*, *A. coenophialum*, and *Phialophora*-like species in *Festuca arundinacea*, and *Epichloe typhina* in *Festuca rubra*.

5.6.1 STROMATIC INFECTION

Many grasses and sedges are infected by endophytes that produce a stromatic sheath around the panicle. *Epichloe typhina* produces a stromatic sheath that almost completely covers the panicle in *Dactylis glomerata* (Cocksfoot grass). Usually the panicle is completely suppressed, but it is rarely partially emerged. The partially emergent panicles have ovules and developing seeds that are invaded by fungal mycelium. At maturity such panicles bear seeds of all kinds from completely aborted to apparently sound seeds that germinate normally (Western and Cavett, 1959).

   Cyperus virens plants infected by the endophyte, *Balansia cyperi,* produce aborted inflorescence covered with fungal stroma. The stromatic tissue bears abundant conidial fructifications and, more rarely, ascostromata. Viviparous plantlets are occasionally produced on the aborted panicles of infected plants. The plantlets are also infected by *B. cyperi* (Clay, 1986). Sampson and Western (1954) have also reported that when *Poa bulbosa*, a viviparous grass, is infected by *E. typhina*, the bulbils are also infected. Clay (1986) considers the induced vivipary in *C. virens* to represent a mechanism of vegetative reproduction wherein host and fungus are dispersed simultaneously by the same propagule.

5.6.2 Nonstromatic Infections

Symbiotic or mutualistic associations between the endophyte and the host usually do not affect the sexual reproduction, and viable seeds are produced. Although fungal sexual stages occur in many of these endophytes, it is their anamorph stage that is usually encountered in host tissues. Sampson (1935) traced *Epichloe typhina* in all parts of flowers — palea, lemma, lodicules, stamens, and young ovaries — of *Festuca rubra*. The fungus invades the ovule and embryo sac (Figure 5.19A), and in seed it occurs in the pericarp, endosperm including aleurone layer, and, rarely, in embryo (Figure 5.19 B to F). Thus in *F. rubra* (red fescue) viable, but often infected, seeds are produced. On germination these seeds produce seedlings with mycelium in the plumule (Figure 5.19G, H). The occurrence of endophyte mycelium in seeds of several other grasses has been demonstrated (Table 5.7).

   Philipson and Christey (1986) have provided an elegant account of the relationship of host and endophyte mycelium during flowering, fruiting, and seed germination in *Lolium perenne* and *Festuca arundinacea*. The endophyte progresses intercellularly from the vegetative apex into the inflorescence primordium, floral apices, ovary, and ovule. The mycelium aggregates outside the embryo sac wall (Figure 5.20A) and subsequently penetrates it. It has been observed in antipodal cells. During early embryogenesis, mycelium occurs on the surface of the embryo and penetrates it at the notched stage. At seed maturity, hyphae are widespread within the embryo, including the plumule apex (Figure 5.20C, D) and in the pericarp, aleurone layer, and the space between them (Figure 5.20B).

5.6.3 Viability of Mycelium in Seed

Endophyte viability in infected seed declines as the seed ages (Latch and Christensen, 1982; Siegel et al., 1984b). Siegel, Latch, and Johnson (1985) believe that most endophyte-infected seeds, which have been stored in warehouses for 2 years, contain little or no viable endophyte. Endophyte viability in infected seeds of tall fescue is lost after 7 to 11 months of storage at 21°C (Siegel et al., 1984b), but at low temperatures and low humidity it is retarded. The perennial rye grass seed infected by the endophyte and stored at 0 to 5°C and near-zero humidity contained living endophyte mycelium for 15 years.

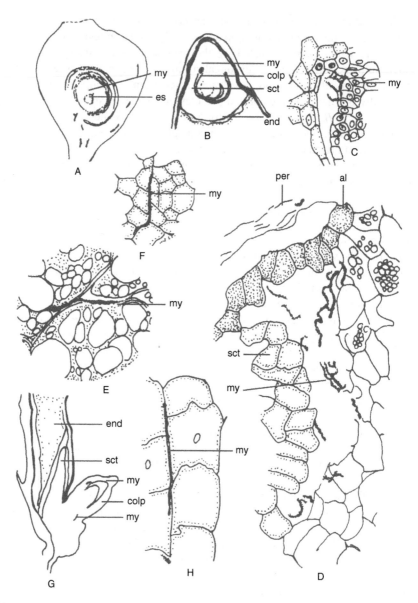

FIGURE 5.19 *Epichloe typhina* (endophyte) in ovule and seed of *Festuca rubra*. A, Ls ovary at the time of pollination showing mycelium on the edge of the embryo sac. B, Part of mature caryopsis in transverse section having mycelium between coleoptile and scutellum. C, Part from B magnified to show mycelium. D, Ls part of caryopsis showing mycelium in the pericarp, endosperm, and embryo. E, F, Mycelium in between the cells of endosperm and embryo, respectively. G, Ls germinating 3-day-old seedling showing mycelium in plumule. H, Part magnified from G to show mycelium. (Abbreviations: al, aleurone layer; colp, coleoptile; end, endosperm; es, embryo sac; my, mycelium; per, pericarp; sct, scutellum.) (From Sampson, K. 1935. *Trans. Br. Mycol. Soc.* 19: 337–343.)

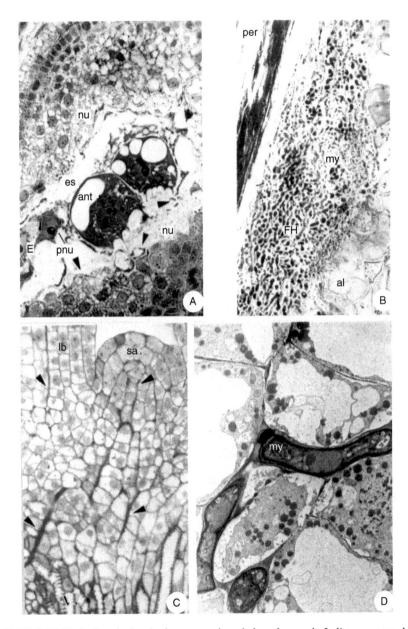

FIGURE 5.20 Endophyte in developing caryopsis and plumule apex in *Lolium perenne*. A, Ls ovule showing part of embryo sac with antipodal cells and endophyte hyphae (arrowheads). B, Ls part of caryopsis having abundant mycelium between pericarp and aleurone layer. C, Ls water-imbibed seed showing endophyte in shoot apex, developing leaf, and mesocotyl (arrowheads). D, TEM photomicrograph of cells of shoot apex with hyphae in between host cells. (Abbreviations: al, aleurone layer; ant, antipodal cells; es, embryo sac; lb, leaf base; my, mycelium; nu, nucellus; per, pericarp; pnu, polar nuclei; sa, shoot apex.) (From Philipson, M.N. and Christey, M.C. 1986. *N.Z. J. Bot.* 24: 125–135. With permission.)

5.7 IMPLICATION OF INTERNAL INFECTION

The importance of seed-borne inoculum is in its ability to survive and affect seed and seedlings at the time of germination. Agarwal and Sinclair (1997) have tabulated the information on the longevity of fungi in seed. It varies for different fungi and host contaminations from 1 year to more than 13 years. However, there is no information on the longevity of a fungus in different hosts or different fungi in seeds of a host. Information on the nature of infection, superficial or deep, and its survival period in seed will be useful for deciding on control strategies.

Naturally infected seeds cause mostly deleterious effects, viz., seed rot, early damping off, seedling blight, spots, and necrosis on cotyledons, leaf spots, spots and streaks on hypocotyl, and primary root rot (Maden et al., 1975; Singh and Singh, 1982; Agarwal et al., 1986). Seeds with heavy internal infection of fungi mostly fail to germinate with the exception of a few specialized biotrophs, e.g., loose smuts and endophytes. *Ascochyta rabiei,* with localized deep infection on the seed surface, affects seedling emergence only slightly. The seedling symptoms develop as spots and rotting on the lower part of the stem and wilting and dryness of the leaflets (Maden et al., 1975). Heavy infection of *Macrophomina phaseolina* causes failure of seed germination and browning and rotting of seedlings in sesame, whereas moderate and weak infections produce diseased seedlings (Singh and Singh, 1982).

Albugo candida survives on plant debris in soil (Butler, 1918; Walker, 1969). The pathogen is shown to be seed-borne in Canada (Petrie, 1975; Verma and Petrie, 1975; Verma et al. 1975) and also in India (Verma and Bhowmik, 1988; Sharma, Agarwal, and Singh, 1997). The seed-borne inoculum remains viable and causes seedling infection (Sharma, 1989). Verma and Bhowmik (1988) and Sharma, Agarwal, and Singh (1994) failed to recover oospores from plant debris buried in the field after 6 months and have argued that the seeds with contaminated and infected oospores provide an important source of perennation and primary infection of white rust disease in India.

The location of pathogens in seed affects seed viability as well as control of seed-borne infection. Fungal pathogens survive for a longer period in seed even under normal storage conditions, and under optimal conditions this is considerably prolonged. The dormant mycelium and fruiting bodies (sclerotia and reproductive propagules) occurring in seed tissues are difficult to inactivate or control by fungicidal treatments. The deeper the location of the inoculum, the more difficult it is to control. Routine treatments and dosages prove ineffective. Treatment, if given, should be tested using growing-on tests. However, an adequate knowledge of the behavior of internal inoculum during storage seems essential before steps to control it are taken. It is also known that the inoculum in seed may lose viability much before the seed viability is seriously affected.

Grain quality is also adversely affected in moderately and heavily infected seeds, particularly due to the depletion of food materials. Infection by mycotoxin-producing fungi may cause health hazards.

5.8 CONCLUDING REMARKS

Histopathology provides information on the exact expanse of mycelium in seed, and the use of categorized infected seeds has revealed a direct correlation between the severity of infection and the expanse of mycelium in seed tissues. Most seed infections of field fungi occur in the developing ovule and seed in the field. The extent of invasion of seed tissues varies with the stage of development of the host, its nature (susceptibility and degree of resistance), and the environment, particularly humidity and temperature, at the time of infection. Early infections in congenial environments usually result in deep internal infections, whereas infections during advanced stages of seed development generally cause superficial invasion. The nutritional status of the fungus, necrotroph or biotroph, seems to have little influence on the invasion of tissues. The effects of invasion on the development of seed and also the degradation of its tissues depend on the harmonic status that exists between the host and the fungus. Necrotrophs cause more damage than biotrophs, and specialized biotrophs, such as loose smuts and endophytes, cause apparently little damage to seed tissues.

In mixed fungal infections of seeds, hyphae of different fungi compete for colonization, showing antagonistic or synergistic behavior.

Dermal coatings, cuticula of different components, lignified or suberized cells, and polyphenols or other similar compounds in seed components appear to act as barriers to fungal invasion; however, the evidence is not wholly conclusive.

Deep internal infection remains viable for various durations. This may play an important role in disease spread over long distances and also in the recurrence of disease from one crop season to another.

REFERENCES

Agarwal, K., Singh, T., Singh, D., and Mathur, S.B. 1985. Studies on glume blotch disease of wheat. I. Location of *Septoria nodorum* in seed. *Phytomorphology* 35: 87–94.

Agarwal, K., Singh, T., Singh, D., and Mathur, S.B. 1986. Studies on glume blotch disease of wheat. II. Transference of seed-borne inoculum of *Septoria nodorum* from seed to seedling. *Phytomorphology* 36: 291–297.

Agarwal, K., Sharma, J., Singh, T., and Singh, D. 1987. Histopathology of *Alternaria tenuis* infected black pointed kernels of wheat. *Bot. Bull. Academia Sinica* 28: 123–130.

Agarwal, N.K. and Jain, B.L. 1978. Histopathological studies of 'Rajma' infected by *Macrophomina phaseolina* (Abstr.). *J. Mycol. Plant Pathol.* 8: 59.

Agarwal, S. and Singh, T. 2000. Effect of extra- and intraembryonal infection of *Macrophomina phaseolina* on disease transmission in okra seeds. *J. Mycol. Plant. Pathol.* 6: 135–139.

Agarwal, V.K. and Sinclair, J.B. 1997. *Principles of Seed Pathology*, 2nd ed. CRC Press, Boca Raton, FL.

Alcock, N.L. 1931. Notes on common diseases sometimes seed-borne. *Trans. Bot. Soc. Edinburgh* 30: 332–337.

Alexopoulos, C.J., Mims, C.W., and Blackwell, M. 1996. *Introductory Mycology*. John Wiley & Sons, New York.

Arya, H.C. and Sharma, R. 1962. On the perpetuation and recurrence of the green ear disease of bajra (*Pennisetum typhoides*) caused by *Sclerospora graminicola* (Sacc.) Schroet. *Indian Phytopathol.* 15: 166–172.

Atkinson, R.G. 1950. Studies on the parasitism and variation of *Alternaria raphani*. *Can. J. Res. Sect. C* 28: 288–317.

Baker, K.F. 1947. Seed transmission of *Rhizoctonia solani* in relation to control of seedling damping off. *Phytopathology* 37: 912–924.

Baker, K.F. 1952. A problem of seedsmen and flower growers — seed-borne-parasites. *Seed World* 70: 38, 40, 44, 46–47.

Baker, K.F. and Davis, L.H. 1950. Heterosporium disease of Nasturtium and its control. *Phytopathology* 40: 553–566.

Bandyopadhyay, R., Mughogho, L.K., Manohar, S.K., and Satyanarayana, M.V. 1990. Stroma development, honey-dew formation and conidial production in *Claviceps sorghi*. *Phytopathology* 80: 812–818.

Basu Chaudhary, K.C. and Mathur, S.B. 1979. Infection of sorghum seeds by *Colletotrichum graminicola*. I. Survey, location in seed and transmission of the pathogen. *Seed Sci. Technol.* 7: 87–92.

Basu Chaudhary, K.C. and Pal, A.K. 1982. Infection of sunn hemp (*Crotalaria juncea*) seeds by *Macrophomina phaseolina*. *Seed Sci. Technol.* 10: 151–154.

Batts, C.C.V. 1955. Observations on the infection of wheat by loose smut (*Ustilago tritici* [Pers.] Rostr.). *Trans. Br. Mycol. Soc.* 38: 465–475.

Batts, C.C.V. and Jeater, A. 1958. The development of loose smut (*Ustilago tritici*) in susceptible varieties of wheat, and some observations on field inspection. *Trans. Br. Mycol. Soc.* 41: 115–125.

Bedi, P.S. and Dhiman, J.S. 1984. Spread of *Neovossia indica* in a wheat ear. *Indian Phytopathol.* 37: 335–337.

Bennum, A. 1972. *Botrytis anthophila* Bondarzew, med særligt henblik på pathologisk anatomi. Ph.D. thesis, Royal Veterinary and Agricultural University, Copenhagen, Denmark.

Bhat, R.S. 1946. Studies in the Ustilaginales. I. The mode of infection of the bajra plant (*Pennisetum typhoides*) by the smut, *Tolyposporium penicillariae*. *J. Indian Bot. Soc.* 25: 163–186.

Bhatia, Anshu 1995. Studies on Important Field and Storage Seed-borne Fungi of Guar (*Cyamopsis tetragonoloba* [L.] Taub.) Ph.D. thesis, University of Rajasthan, Jaipur, India.

Bhatia, A., Singh, T., and Singh, D. 1996a. Infection of *Fusarium oxysporum* in guar (*Cyamopsis tetragonoloba*). *Acta Bot. Indica* 23: 257–259.

Bhatia, A., Singh, T., and Singh, D. 1996b. Histopathology of *Colletotrichum dematium* (Fr.) groves in cluster bean seeds. *Proc. Natl. Acad. Sci. India* 66(B): 289–291.

Bhatia, A., Singh, T., and Singh, D. 1998. Seed-borne infection of *Rhizoctonia bataticola* in guar and its role in disease development. *J. Mycol. Plant Pathol.* 28: 231–235.

Boek, K. 1952. Einige Untersuchungen an samenübertragbaren Krankheitserregern der Kruziferen. Diplomarbeit, University of Hamburg, Hamburg, Germany.

Bondarzew, A.S. 1914. Eine neue Krankheit der Bluten des Rotlees im Zusammenhange mit seiner Fruktifikation. [In Russian.] *Bolezni Rastenij*, St. Petersburg 8: 1–25.

Brefeld, O. 1903. Neue Untersuchugen und Ergebnisse über die natürliche Infektion und Verbreitung der Brandkrankheiten des Getreides. *Klub Landw. Berlin. Nachr.* 466: 4224–4234.

Brown, H.B. 1916. Life history and poisonous properties of *Claviceps purpurea*. *J. Agric. Res.* 7: 401–406.

Butler, E.J. 1918. *Fungi and Diseases in Plants*. Thacker, Spinck and Co., Calcutta, India.

Campbell, W.P. 1958. Infection of barley by *Claviceps purpurea*. *Can. J. Bot.* 36: 615–619.

Cashion, N.L. and Luttrell, E.S. 1988. Host-parasitic relationships in karnal bunt of wheat. *Phytopathology* 78: 75–84.

Chahal, S.S., Thakur, R.P., and Mathur, S.B. 1994. *Seed-Borne Diseases and Seed Health Testing of Pearl Millet*. Danish Government Institute of Seed Pathology for Developing Countries, Copenhagen, Denmark.

Channon, A.G. 1969. Infection of flowers and seed of parsnip by *Itersonilia pastinacae*. *Ann. Appl. Biol.* 64: 281–288.

Cheeran, A. and Raj, J.S.. 1972. A technique for the separation of the whole embryo of the grains for the detection of fungal mycelium. *Agric. Res. J. (Kerala) India* 10: 125–127.

Chikuo, Y. and Sugimoto, T. 1989. Histopathology of sugar beet flowers and seed balls infected with *Colletotrichum dematium* f. *spinaciae*. *Ann. Phytopathol. Soc. Japan* 55: 404–409.

Chitkara, S., Singh, T., and Singh, D. 1986a. *Alternaria tenuis* in chili seeds of Rajasthan. *Biol. Bull. India* 8: 18–23.

Chitkara, S., Singh, T., and Singh, D. 1986b. *Rhizoctonia solani* in chili seeds of Rajasthan. *Indian Phytopathol.* 39: 565–567.

Chitkara, S., Singh, T., and Singh, D. 1990. Histopathology of *Colletotrichum dematium* infected chili seeds. *Acta Bot. Indica* 18: 226–230.

Chowdhury, S. 1946. Mode of transmission of the bunt of rice. *Curr. Sci.* 15: 111.

Christen, A.A. 1982. Demonstration of *Verticillium albo-atrum* within alfalfa seed. *Phytopathology* 72: 412–414.

Chung, H.S. and Lee, C.U. 1983. Detection and transmission of *Pyricularia oryzae* in germinating rice seed. *Seed Sci. Technol.* 11: 625–637.

Chupp, C. 1935. *Macrosporium* and *Colletotrichum* rots of turnip roots. *Phytopathology* 25: 269–274.

Clay, K. 1986. Induced vivipary in the sedge *Cyperus virens* and the transmission of the fungus *Balansia cyperi* (Clavicipitaceae). *Can. J. Bot.* 64: 2984–2988.

Cohen, Y. and Sackston, W.E. 1974. Seed infection and latent infection of sunflowers by *Plasmopara halstedii*. *Can. J. Bot.* 52: 231–238.

Crosier, W.F. 1944. *Diplodia gossypina* and other fungi in cotton seed. *Assoc. Off. Seed Anal. Newsl.* 18: 13–15.

Crowford, R.F. 1923. Fungi isolated from the interior of cotton seed. *Phytopathology* 13: 501–503.

Cunfer, B.M. 1991. Long term viability of *Septoria nodonum* in stored wheat seed. *Cereal Res. Commun.* 19: 347–349.

Dekker, J. 1957. Inwendige ontsmetting van door *Ascochyta pisi* aangetaste erwtezaden met de antibiotica rimocidine en pimaricine, benevens enkele aspecten van het parasitisme van deze schimmel. *Tijdschr. Plant Ziekt.* 63: 65–144.

Dhaliwal, H.S., Randhawa, A.S., Chand, K., and Singh, Dalip. 1983. Primary infection and further development of karnal bunt of wheat. *Indian J. Agric. Sci.* 53: 239–244.

Diehl, W.W. 1950. *Balansia* and the *Balansiae* in America. Agricultural Monograph 4. U.S. Department of Agriculture, Washington, D.C.

Djerbi, M. 1971. Etude ecologique et histologique des actions parasitaires des espèces fusariennes à l'égard du blé. Thesis, Paris.

Doken, M.T. 1989. *Plasmopara halstedii* (Farl.) Berl. et de Toni in sunflower seeds and the role of infected seeds in producing plants with systemic symptoms. *J. Phytopathol.* 124: 23–26.

Domsch, K.H. 1957. Die Raps- und Kohlschwarze. *Z. Pflanzenkr. Pflanzenschutz* 64: 65–79.

Dubey, A.K. 2000. Studies on Seed-Borne Microorganisms of Sesame (*Sesamum indicum* L.). Ph.D. thesis, University of Rajasthan, Jaipur, India.

Dubey, A.K. and Singh, T. 1999. Oospore contamination and infection of *Phytophthora* in sesame seeds. *J. Indian Bot. Soc.* 78: 103–106.

Dwivedi, A.K. 1994. Field and Storage Fungal Diseases of Garlic and Onion. Ph.D. thesis, University of Rajasthan, Jaipur, India.

Ellis, M.B. 1971. *Dematiaceous Hyphomycetes*. Commonwealth Mycological Institute, Kew, Richmond, Surrey, U.K.

Ellis, M.A., Ilyas, M.B., and Sinclair, J.B. 1975. Effect of three fungicides on internally seed-borne fungi and germination of soybean seeds. *Phytopathology* 65: 553–556.

Fazli, I.S.F. and Schroeder, H.W. 1966. Effect of kernel infection of rice by *Helminthosporium oryzae* on yield and quality. *Phytopathology* 56: 1003–1005.

Fischer, D.E. 1954. Stem and fruit rot of outdoor tomatoes. *Rep. Natl. Veg. Res. Sta.* Warwick 1953: 15–16.

Frank, B. 1883. Ueber einige neue und weniger bekannte Pflanzenkrankheiten. *Ber. Deutsch. Bot. Gesell.* 1: 29–34.

Gabrielson, R.L. 1983. Blackleg disease of crucifers caused by *Leptosphaeria maculans (Phoma lingam)* and its control. *Seed Sci. Technol.* 11: 749–780.

Genau, A. 1928. Methoden der künstlichen Infektion der Gerste mit *Helminthosporium gramineum* und Studien über die Anfälligkeit verschiedener Sommergersten diesen Pilz gegenüber. *Kuhn-Arch.* 19: 303–351.

Goates, B.J. 1988. Histology of infection of wheat by *Tilletia indica*, the Karnal bunt pathogen. *Phytopathology* 78: 1434–1441.

Godika, S. 1996. Mycoflora Associated with Seeds of Sunflower and Its Phytopathological Effects. Ph.D. thesis, University of Rajasthan, Jaipur, India.

Godika, S., Agarwal, K., and Singh, T. 1999. Incidence of *Rhizoctonia bataticola* in sunflower seeds grown in Rajasthan. *J. Mycol. Plant Pathol.* 29: 255–256.

Griffith, R.B., Zscheile, F.P., and Oswald, J.W. 1955. The influence of certain environmental features on expression of resistance to bunt in wheat. *Phytopathology* 45: 428–434.

Grummer, G. and Mach, F. 1955. Die Behandlung brennfleckenkranker Bohnen mit Patulin und Streptomycin. *Zentralbl. Bakteriol. Parasitenkd.* Abt. 2, 108: 449–464.

Gupta, J.S. 1962. Pathological anatomy of the floral parts and fruits of coriander affected with *Protomyces macrosporus* Ung. *Agra Univ. J. Res. (Sci.)* 11: 307–320.

Halfon-Meiri, A. 1983. Seed transmission of safflower rust (*Puccinia carthami*) in Israel. *Seed Sci. Technol.* 11: 835–851.

Halfon-Meiri, A., Kunwar, I.K., and Sinclair, J.B. 1987. Histopathology of achenes and seeds of *Ranunculus asiaticus* infected with an *Alternaria* sp. *Seed Sci. Technol.* 15: 197–205.

Harrison, J.G. 1978. Role of seed-borne infection in epidemiology of *Botrytis fabae* on field beans. *Trans. Br. Mycol. Soc.* 70: 35–40.

Heald, P.D., Wilcox, E.M., and Pool, Y.W. 1909. The life history and parasitism of *Diplodia zeae* (Schw.) Lev. *Nebr. Agric. Exp. Sta.* 22nd Ann. Rep. 1–7.

Hedgecock, G.G. 1904. A note on *Rhizoctonia*. *Science* 19: 268.

Hellwig, L.C. 1699. *Kurzes Send-Schreiben wegen des so genanndten Honig Tanes, welcher sich am heurigen Korn sehen lassen, und die grossen Schwarzen Korner (ins gemein Mutter-Korn genanndt) hervor bracht, was davon au halten, wovon es entstanden, und ob es nützlich, oder schadlich sey.* J.C. Bachmann, Langen Saltza.

Heursel, J. 1961. Gozondheidstoestand van Tabakzaad (Health conditions of tobacco seed). *Verh. Rijkssta. Pfziekt. Gent* 12 (Abstr. in *Rev. Appl. Mycol.* 42: 171).

Hewett, P.D. 1987. Pathogen viability on seed in deep freeze storage. *Seed Sci. Technol.* 15: 73–77.

Hildebrand, A.A. and Koch, L.W. 1951. A study of the systemic infection of downy mildew of soybean with special reference to symptomatology, economic significance and control. *Sci. Agric.* 31: 505–518.

Huang, H.C., Hanna, M.R., and Kokke, E.G., 1985. Mechanism of seed contamination by *Verticillium albo-atrum* in alfalfa. *Phytopathology* 75: 482–488.

Hungerford, C.W. 1920. Rust in seed wheat and its relation to seedling infection. *J. Agric. Res.* 19: 257–277.

Ilyas, M.B., Dhingra, O.D., Ellis, M.A., and Sinclair, J.B. 1975. Location of mycelium of *Diaporthe phaseolorum* var *sojae* and *Cercospora kikuchii* in infected soybean seeds. *Plant Dis. Rep.* 59: 17–19.

Jacobsen, B.J. and Williams, P.H. 1971. Histology and control of *Brassica oleracea* seed infection by *Phoma lingam*. *Plant Dis. Rep.* 55: 934–938.

Johnson, H.W. and Lefebvre, C.L. 1942. Downy mildew of soybean seeds. *Plant Dis. Rep.* 26: 49–50.

Jones, B.L., Leeper, J.C., and Frederiksen, R.A. 1972. *Sclerospora sorghi* in corn: its location in carpellate flowers and mature seeds. *Phytopathology* 62: 817–819.

Kaveriappa, K.M. and Safeeulla, K.M. 1975. Investigation on the sorghum downy mildew. *J. Mysore Univ. Sect. B* 26: 201–214.

Kaveriappa, K.M. and Safeeulla, K.M. 1978. Seed-borne nature of *Sclerospora sorghi* in sorghum. *Proc. Indian Acad. Sci.* (Plant Science) 87: 303–308.

Keshavamurthy, G.T. 1990. Studies on seed-borne nature and cultural aspects of *Drechslera setariae* incitant of leaf spot or leaf blotch of navne. Ph.D. thesis, University of Agricultural Sciences, Bangalore, India.

Khanzada, A.K., Rennie, W.J., Mathus, S.B., and Neergaard, P. 1980. Evaluation of two routine embryo test procedures for assessing the incidence of loose smut infection in seed samples of wheat (*Triticum aestivum*). *Seed Sci. Technol.* 8: 363–370.

Kietreiber, M. 1961. Die Erkennung des Septoria-Befalles von Weizenkörnern bei der Saatgutprüfung. *Pflanzenschutzberichte* 26: 129–157.

Klisiewicz, J.M. 1963. Wilt-incitant *Fusarium oxysporum* f. *carthami* present in seed infected safflower. *Phytopathology* 53: 1046–1049.

Klisiewicz, J.M. 1975. Survival and dissemination of *Verticillium* in infected safflower seed. *Phytopathology* 65: 696–698.

Knox-Davies, P.S. 1979. Relationships between *Alternaria brassicicola* and *Brassica* seeds. *Trans. Br. Mycol. Soc.* 73: 235–248.

Kozera, W. 1968. Przebieg pierwszych etapow zakazenia Jeczmienia przez glownie pykowa, *Ustilago nuda* (Jens.) Rostr. *Hodowla Rosl. Aklim. Narienn.* 12: 269–281.

Kumar, V. and Shetty, H.S. 1983. Seed-borne nature and transmission of *Botryodiplodia theobromae* in maize (*Zea mays*). *Seed Sci. Technol.* 11: 781–790.

Kumi, M.A., Wolffhechel, H., Hansen, H.J., and Mathur, S.B. 1996. Seed transmission of *Phytophthora* in cacao. *Seed Sci. Technol.* 24: 593–595.

Kunwar, I.K., Singh, T., and Sinclair, J.B. 1985. Histopathology of mixed infections by *Colletotrichum truncatum* and *Phomopsis* spp. or *Cercospora sojina* in soybean seeds. *Phytopathology* 75: 489–592.

Kunwar, I.K., Manandhar, J.B., and Sinclair, J.B. 1986a. Histopathology of soybean seeds infected with *Alternaria alternata*. *Phytopathology* 76: 543–546.

Kunwar, I.K., Singh, T., Machado, C.C., and Sinclair, J.B. 1986b. Histopathology of soybean seed and seedling infection by *Macrophomina phaseolina*. *Phytopathology* 76: 532–535.

Kuniyasu, K. 1980. Seed transmission of Fusarium wilt of bottle gourd, *Lagenaria siceraria*, used as root stock of watermelon. *Jap. Agric. Res.* 14: 157–162.

Lafferty, H.A. 1921. The 'browning' and 'stem-break' disease of cultivated flax (*Linum usitatissimum*), caused by *Polyspora lini* n. gen. et sp. *Sci. Proc. R. Dubl. Soc.* 16 (N.S.): 248–274.

Latch, G.C.M. and Christensen, M.J. 1982. Ryegrass endophyte, incidence and control. *N.Z.. J. Agric Res.* 25: 443–448.

Latch, G.C.M., Christensen, M.J., and Samuel, G.J. 1984. Five endophytes of *Lolium* and *Festuca* in New Zealand. *Mycotaxon* 15: 311–318.

Leach, L.D. 1931. Downy mildew of beet, caused by *Peronospora schachtii* Fuckel. *Hilgardia* 6: 203–251.

Lee, D.H., Mathur, S.B., and Neergaard, P. 1984. Detection and location of seed-borne inoculum of *Didymella bryoniae* and its transmission in seedling of cucumber and pumpkin. *Phytopathol. Z.* 108: 301–308.

Lehman, S.G. 1953. Systemic infection of soybean by *Peronospora manshurica* as affected by temperature. *Elisha Mitchell Scient. Soc. J.* 69: 83.

Lloyd, A.B. 1959. The endophytic fungus of perennial ryegrass. *N.Z. J. Agric. Res.* 2: 1187–1194.

Loiselle, R. and Shands, R.G. 1960. Influence of maternal tissue on loose smut infection of hybrid barley kernels. *Can. J. Bot.* 38: 741–746.

Luttrell, E.S. 1977. The disease cycle and fungus-host relationships in dalligrass ergot. *Phytopathology* 67: 1461–1468.

Maddox, F. 1896. Eastfield experiments, smut and bunt. *Agr. Gaz. J. Counc. Agr. Tasmania* 4: 92–95.

Maden, S., Singh, D., Mathur, S.B., and Neergaard, P. 1975. Detection and location of seed-borne inoculum of *Ascochyta rabiei* and its transmission in chickpea (*Cicer arietinum*). *Seed Sci. Technol.* 3: 667–681.

Malik, M.M.S. and Batts, C.C.V. 1960a. The infection of barley by loose smut (*Ustilago nuda*) (Jens.) Rostr. *Trans. Br. Mycol. Soc.* 43: 112–125.

Malik, M.M.S. and Batts, C.C.V. 1960b. The development of loose smut (*Ustilago nuda*) in the barley plant, with observations on spore formation in nature and in culture. *Trans. Br. Mycol. Soc.* 43: 126–131.

Manuel, A.M. and Do Valle, R. 1993. Transmission and survival of *Acremonium* and the implications for grass breeding. *Agric. Ecosystems Environ.* 44: 193–213.

Manuel, A.M., Do Valle, R., Gwinn, K.D., Siegel, M.R., and Bush, L.P. 1994. Transmission of *Acremonium* spp. infection. In *Proceedings of the International Conference on Harmful and Beneficial Microorganisms in Grassland, Pastures and Turf*, Paderborn, Germany, IOBC/WPRS Bulletin 17: 111–121.

Mathre, D.E. 1982. *Compendium of Barley Diseases*. American Phytopathological Society, St. Paul, MN.

Mathur, R. 1992. Pathological and Physiological Studies on Some Seed Disorders in Soybean. Ph.D. thesis, University of Rajasthan, Jaipur, India.

Mathur, S.B. and Cunfer, B.M., Eds. 1993. *Seed-Borne Diseases and Seed Health Testing of Wheat*. Danish Government Institute of Seed Pathology for Developing Countries, Jordbrugsforlager, Frederiksberg, Denmark.

Mathur, S.K., Mathur, S.B., and Neergaard, P. 1975. Detection of seed-borne fungi in sorghum and location of *Fusarium moniliforme* in the seed. *Seed Sci. Technol.* 3: 683–690.

Maude, R.B. 1996. *Seed-Borne Diseases and Their Control*. CAB International, Wallingford, U.K.

Maude, R.B. and Presly, A.H. 1977. Neck rot (*Botrytis allii*) of bulb onions. I. Seed-borne infection and its relationship to the disease in the onion crop. *Ann. Appl. Biol.* 86: 163–180.

Maude, R.B. and Humpherson-Jones, F.M. 1980. Studies on the seed-borne phases of dark leaf spot (*Alternaria brassicicola*) and grey leaf spot (*Alternaria brassicae*) of brassicas. *Ann. Appl. Biol.* 95: 311–319.

McKenzie, T.R. and Wyllie, T.D. 1971. The effect of temperature and lesions size on the sporulation of *Peronospora manshurica*. *Phytopathol. Z.* 71: 321–326.

Mead, H.W. 1942. Host–parasite relationships in a seed-borne disease of barley caused by *Helminthosporium sativum* Pammel, King and Bakke. *Can. J. Res. Sec. C* 20: 501–523.

Meffert, M.-E. 1950. Ein Beitrag zur Biologie und Morphologie der Erreger der Parasitaren Blattdürre des Mohns. *Z. Parasitenkd.* 14: 442–498.

Melhus, I.E., 1931. The presence of mycelium and oospores of certain downy mildews in the seeds of their hosts. *Iowa St. Coll. J. Sci.* 5: 185–188.

Miller, J.H. 1952. The presence of internal mycelium in corn grains in relation to external symptoms of corn ear rot. *Phytopathology* 42: 286.

Mitter, V. and Siddiqui, M.R. 1995. Histopathological evidence for infection process in smut of pearl millet. *J. Mycol. Plant Pathol.* 25: 43 (Abstr.).

Muralidhara Rao, B., Shetty, H.S., and Safeeulla, K.M. 1984. Production of *Peronosclerospora sorghi* oospores in maize seeds and further studies on the seed-borne nature of the fungus. *Indian Phytopathol.* 37: 278–283.

Muralidhara Rao, B., Prakash, H.S., Shetty, H.S., and Safeeulla, K.M. 1985. Downy mildew inoculum in maize seeds: Techniques to detect seed-borne inoculum of *Peronosclerospora sorghi* in maize. *Seed Sci. Technol.* 13: 593–600.

Neergaard, E. 1989. Histological investigation of flower parts of cucumber infected by *Didymella bryoniae*. *Can. J. Plant Pathol.* 11: 28–38.

Neergaard, P. 1979. *Seed Pathology*. Vols. 1 and 2. Macmillan Press, London.

Neill, J.C. 1940. The endophyte of ryegrass (*Lolium perenne*). *N.Z. J. Sci. Technol.* 21A: 280–291.

Neill, J.C. 1941. The endophytes of *Lolium* and *Festuca*. *N.Z. J. Sci. Technol.* 23A: 185–193.

Netzer, D. and Kenneth, R.G. 1969. Resistance and transmission of *Alternaria dauci* (Kuhn) Groves and Skolko in the semi-arid conditions of Israel. *Ann. Appl. Biol.* 63: 289–291.

Nisikado, Y. and Nakayama, T. 1943. Notes on the pathological anatomy of rice grains affected by *Helminthosporium oryzae*. *Ber. Ohara Inst. Landw. Forsch.* 9: 208–213.

Novotelnova, N.S. 1963. Nature of parasitism of the pathogen of downy mildew of sunflower at seed infection. [In Russian.] *Bot. J. Acad. Sci. USSR* 48: 845–860.

Ohms, R.E. and Bever, W.M. 1956. Effect of time of inoculation of winter wheat with *Ustilago tritici* on the percentage of embryos infected and on the abundance of hyphae. *Phytopathology* 46: 157–158.

Parmar, S.M.S. 1981. Studies on Some *Alternaria* Species Infecting Plants of the Family Compositae. Ph.D. thesis, University of Rajasthan, Jaipur, India.

Parnis, E.M. and Sackston, W.E. 1979. Invasion of lupin seed by *Verticillium alboatrum*. *Can. J. Bot.* 57: 597–601.

Paul, M. 1987. Studies on the Pathological Disorders in Seeds of Paddy from Tribal Areas of Rajasthan. Ph.D. thesis, University of Rajasthan, Jaipur, India.

Pavgi, M.S. and Mukhopadhyay, A.N. 1972. Development of coriander fruit infected by *Protomyces macrosporus* Unger. *Cytologia* 37: 619–627.

Pedersen, P.N. 1956. Infection of barley by loose smut, *Ustilago nuda* (Jens.) Rostr. *Friesia* 5: 341–348.

Petrie, G.A. 1975. Prevalence of oospores of *Albugo cruciferarum* in *Brissica* seed samples from western Canada, 1967–1973. *Can. Plant Dis. Surv.* 55: 19–24.

Petrie, G.A. and Vanterpool, T.C. 1974. Infestation of crucifer seed in western Canada by the blackleg fungus *Leptosphaeria maculans. Can. Plant Dis. Surv.* 54: 119–123.

Pirson, H. 1960. Prüfung verschiedener Winterweizensorten auf Anfälligkeit gegen *Septoria nodorum* Berk mit Hilfe von kunstlichen Infektionen. *Phytopathol. Z.* 37: 330–342.

Philipson, M.N. and Christey, M.C. 1986. The relationship of host and endophyte during flowering, seed formation and germination of *Lolium perenne. N.Z. J. Bot.* 24: 125–135.

Platenkamp, R. 1975. Investigation on the infection pathway of *Drechslera graminea* in germinating barley. Royal Veterinary and Agricultural University, Copenhagen, Denmark. Yearbook 1976: 49–64.

Ponchet, J. 1966. *Etude des Communaute's mycopericarpiques de caryopsis de ble.* Institut National de la Recherche Agronomique, Paris.

Popp, W. 1959. A new approach to the embryo test for predicting loose smut of wheat in adult plant. *Phytopathology* 49: 75–77.

Porter, D.M., Wright, F.S., Taber, R.A., and Smith, D.H. 1991. Colonization of peanut seed by *Cylindrocladium crotalariae. Phytopathology* 81: 896–900.

Prabhu, M.S.C., Safeeulla, K.M., Venkatasubbaiah, P., and Shetty, H.S. 1984. Detection of seed-borne inoculum of *Peronosclerospora sorghi* in sorghum. *Phytopathol. Z.* 111: 174–178.

Prakash, H.S., Shetty, H.S. and Safeeulla, K.M. 1980. Histology of carpel infection of *Claviceps fusiformis* in pearl millet. *Proc. Indian Natl. Sci. Acad.* 49B: 459–465.

Prasad, K.V.V., Khare, M.N. and Jain, A.C. 1985. Site of infection and further development of *Colletotrichum graminicola* (Ces.) Wilson in naturally infected sorghum grains. *Seed Sci. Technol.* 13: 37–40.

Purakusumah, H. 1965. Seed-borne primary infection in downy mildew. *Sclerospora maydis* (Ricib.) Butler. *Nature* 207: 1312–1313.

Raghuvendra, S. and Safeeulla, K.M. 1979. Histopathological studies on ragi (*Eleusine coracana* L. Gaertnr.) infected by *Sclerophthora macrospora* (Sacc.). *Proc. Indian Acad. Sci.* (Plant Science) 88: 19–24.

Ranganathaiah, K.G. 1994. Seed-borne diseases of minor millets. In *Vistas in Seed Biology,* Singh, T. and Trivedi, P.C., Eds. Printwell, Jaipur, India. Vol. 1, 251–261.

Rangnathaiah, K.G. and Mathur, S.B. 1978. Seed health testing of *Eleusine coracona* with special reference to *Drechslera nodulosa* and *Pyricularia grisea. Seed Sci. Technol.* 6: 943–951.

Rao, C.G.P. 1972. Anatomical studies on abnormal growth caused by *Protomyces macrosporus* on *Coriandrum sativum. Indian Phytopathol.* 25: 483–496.

Rastogi, A. 1995. Studies on Seed-borne Fungi and Diseases in Cumin and Fenugreek Grown in Rajasthan. Ph.D. thesis, University of Rajasthan, Jaipur, India.

Rastogi, A., Singh, T., and Singh, D. 1998a. Contamination, infection and transmission of *Alternaria alternata* in cumin. *J. Indian Bot. Soc.* 77: 151–154.

Rastogi, A., Singh, T., and Singh, D. 1998b. Colonization and phytopathological effects of *Alternaria alternata* in seeds of fenugreek (*Trigonella foenum-graecum* L.). *J. Phytol. Res.* 11: 137–139.

Rastogi, A., Singh, T., and Singh, D. 1998c. Incidence, location and transmission of natural seed-borne inoculum of *Cercospora traversiana* in fenugreek. *J. Mycol. Plant Pathol.* 28: 218–222.

Rastogi, R. 1984. Studies on Seed-Borne Mycoflora and Some Important Diseases of Sorghum (*Sorghum vulgare* Pres.) with Special Reference to Rajasthan. Ph.D. thesis, University of Rajasthan, Jaipur, India.

Rastogi, R., Singh, T., and Singh, D. 1990. Infection of *Curvularia lunata* in sorghum seeds. *J. Indian Bot. Soc.* 69: 71–73.

Rastogi, R., Singh, T., and Singh, D. 1991. Histopathology of *Phoma sorghina* infected sorghum seeds. *J. Indian Bot. Soc.* 70/71: 329–333.

Rathore, R.S., Siradhana, B.S., and Mathur, K. 1987. Transmission and location of *Peronosclerospora heteropogoni* in maize seeds. *Seed Sci. Technol.* 15: 101–107.

Raut, J.G. 1983. Transmission of seed-borne *Macrophomina phaeolina* in sunflower. *Seed Sci. Technol.* 11: 807–814.

Remmant, R. 1637. *A Discourse or Historie of Bees. Whereunto is added the causes, and cure of blasted wheat. And some remedies for blasted Hops and Rie, and Fruit. Together with the causes of smutty wheat: all of which are very useful for this later age.* Thomas Slater, London.

Richardson, M.J. 1990. *An Annotated List of Seed-Borne Diseases.* 4th ed. International Seed Testing Association, Zurich, Switzerland.

Roncadori, R.W., McCarter, S.M., and Crawford, J.L. 1971. Influence of fungi on cotton seed deterioration prior to harvest. *Phytopathology* 61: 1326–1328.

Roongruangsree, U-Tai, Olson, L.W., and Lange, L. 1988. The seed-borne inoculum of *Peronospora manshurica*, causal agent of soybean downy mildew. *J. Phytopathol.* 123: 233–243.

Roy, K.W. and Abney, T.S. 1988. Colonization of pods and infection of seeds by *Phomopsis longicolla* in susceptible and resistant soybean lines inoculated in the greenhouse. *Can. J. Plant Pathol.* 10: 317–320.

Roy, S. 1984. Studies on Ergot Disease of Bajra (*Pennisetum typhoides* (Burm.) Stapf and Hubb.) in Rajasthan. Ph.D. thesis, University of Rajasthan, Jaipur, India.

Rudolph, B.A. 1944. The unimportance of tomato seed in the dissemination of *Verticillium* wilt in California. *Phytopathology* 34: 622–630.

Ruttle, M.L. 1934. Studies on barley smuts and on loose smut of wheat. *N.Y. Agr. Exp. Sta. Tech. Bull.* 221: 1–39.

Sackston, W.E. and Martens, J.W. 1959. Dissemination of *Verticillium albo-atrum* on seed of sunflower *(Helianthus annuus).* *Can. J. Bot.* 37: 759–768.

Safeeulla, K.M. 1976. *Biology and Control of the Downy Mildew of Pearl Millet, Sorghum and Finger Millet.* Wesley Press, Mysore, India.

Sampson, K. 1933. The systemic infection of grasses by *Epichloe typhina* (Pers.) Tul. *Trans. Br. Mycol. Soc.* 18: 30–47.

Sampson, K. 1935. The presence and absence of an endophytic fungus in *Lolium temulentum* and *Lolium perenne.* *Trans. Br. Mycol. Soc.* 19: 337–343.

Sampson, K. 1937. Further observations on the systemic infection of *Lolium.* *Trans. Br. Mycol. Soc.* 21: 84–97.

Sampson, K. and Western, J.H. 1954. *Diseases of British Grasses and Herbage Legumes,* 2nd ed. Cambridge University Press, Cambridge, U.K.

Savile, D.B.O. 1973. Rusts that pass import inspection. *Can. Plant Dis. Surv.* 53: 105–106.

Schmiedeknecht, M. 1958. Morphologische Untersuchungen zur Frage der Rassenbildung bei *Helminthosporium papaveris* Saw. *Arch. Mikrobiol.* 28: 404–416.

Schneider, R.W., Dhingra, O.D., Nicholson, J.F., and Sinclair, J.B. 1974. *Colletotrichum truncatum* borne within the seedcoat of soybean. *Phytopathology* 64: 154–155.

Schuster, M.L. 1956. Investigations on the foot and root phase of safflower rust. *Phytopathology* 46: 591–595.

Schuster, M.L. and Christiansen, D.W. 1952. A foot and root disease of safflower caused by *Puccinia carthami* Cda. *Phytopathology* 42: 211–212.

Sehgal, S.P. and Prasad, N. 1966. Studies on the perennation and survival of *Sesamum Phytophthora. Indian Phytopathol.* 19: 173–177.

Semangoen, H. 1970. Studies on downy mildew of maize in Indonesia, with special reference in the perennation of the fungus. *Indian Phytopathol.* 23: 307–320.

Sharma, J. 1989. Studies on Seed-Borne Mycoflora and Some Important Seed-Borne Diseases of Rape and Mustard Grown in Rajasthan. Ph.D. thesis, University of Rajasthan, Jaipur, India.

Sharma, J. 1992. Mycoflora of Soybean Seeds and Their Pathological Effects. Ph.D. thesis, University of Rajasthan, Jaipur, India.

Sharma, J. Agarwal, K., and Singh, D. 1994. Seed-borne infection of *Albugo candida* in mustard and its role in disease development. *Seed Pathol. News,* Copenhagen 23: 3.

Sharma, J., Agarwal, K., and Singh, D. 1997. Infection of *Albugo candida* (Pers. ex Lev.) Kunze in mustard seeds. *J. Phytol. Res.* 10: 25–29.

Sharma, K. 1999. Studies on Seed and Seedling Diseases of Green Gram (Mung Bean) Grown in Rajasthan. Ph.D. thesis, University of Rajasthan, Jaipur, India.

Sharma, M. 1996. Studies on Seed-borne Mycoflora of Pigeonpea Grown in Rajasthan. Ph.D. thesis, University of Rajasthan, Jaipur, India.

Sharma, Neelam, 1992. Seed-borne disease of *Eruca sativa* Mill and *Linum usitatissimum* L. Grown in Rajasthan. Ph.D. thesis, University of Rajasthan, Jaipur, India.

Sharma, N., Singh, T., and Singh, D. 1993. Incidence and infection of *Alternaria tenuis* Nees in Eruca seeds of Rajasthan. *J. Indian Bot. Soc.* 72: 159–161.

Shetty, H.S., Khanzada, A.K., Mathur, S.B., and Neergaard, P. 1978. Procedures for detecting seed-borne inoculum of *Sclerospora graminicola* in pearl millet (*Pennisetum typhoides*). *Seed Sci. Technol.* 6: 935–941.

Shetty, H.S., Mathur, S.B., and Neergaard, P. 1980. *Sclerospora graminicola* in pearl millet seeds and its transmission. *Trans. Br. Mycol. Soc.* 74: 127–134.

Shetty, H.S., Mathur, S.B., and Neergaard, P. 1982. *Drechslera setariae* in Indian pearl millet seeds, its seed-borne nature, transmission and significance. *Trans. Br. Mycol. Soc.* 78: 170–173.

Shinohara, M. 1972. Anatomical studies on barley loose smut (*Ustilago nuda* [Jens.] Rostrup). I. Path of embryo infection in the developing caryopsis. II. Host-parasite morphological features at the point of the entry and in the tissue of the developing caryopsis. *Bull. Coll. Agric. Vet. Med. Nihon Univ.* 20: 84–113.

Siddiqui, M.R. and Mathur, S.B. 1989. Survival of *Septoria nodorum* Berk. in wheat seed stored at 5°C. *FAO/IBPGR Plant Genetic Resources Newsletter* 75/76: 7–8 (CN III).

Siegel, M.R., Latch, G.C.M. and Johnson, M.C. 1985. *Acremonium* fungal endophytes of tall fescue and perennial ryegrass: significance and control. *Plant Dis.* 69: 179–185.

Siegel, M.R., Latch, G.C.M., and Johnson, M.C. 1987. Fungal endophytes of grasses. *Ann. Rev. Phytopathol.* 25: 293–315.

Siegel, M.R., Johnson, M.C., Varney, D.R., Nesmith, W.C., Buchner, R.C., Bush, L.P., Burrus, P.B., Jones, T.A., and Boling, J.A. 1984a. A fungal endophyte in tall fescue: incidence and dissemination. *Phytopathology* 74: 932–937.

Siegel, M.R., Varney, D.R., Johnson, M.C., Nesmith, W.C., Buchner, R.C., Bush, L.P., Burrus, P.B., and Hardison, J.R. 1984b. A fungal endophyte of tall fescue I: evaluation of control methods. *Phytopathology* 74: 937–941.

Silow, R.A. 1933. A systemic disease of red clover caused by *Botrytis anthophila* Bond. *Trans. Br. Mycol. Soc.* 18: 239–248.

Simmonds, P.M. 1946. Detection of the loose smut fungi in embryos of barley and wheat. *Sci. Agric.* 26: 51–58.

Sinclair, J.B. and Backman, P.A. Eds. 1989. *Compendium of Soybean Diseases*, 3rd ed. American Phytopathological Society, St. Paul, MN.

Singh, B.K. 1991. Studies on Seed-Bborne Mycoflora of Some Umbelliferous Spices with Special Reference to Rajasthan. Ph.D. thesis, University of Rajasthan, Jaipur, India.

Singh, B.K., Singh, T., and Singh, D. 2001. Symptoms and histopathology of flowers and fruits of coriander induced by *Protomyces macrosporous* Unger. In *Some Aspects of Research in Applied Botany,* Chauhan, S.V.S. and Singh, K.P., Eds. Printwell, Jaipur, India, pp. 227–286.

Singh, D. 1983. Histopathology of some seed-borne infections: a review of recent investigations. *Seed Sci. Technol.* 11: 651–663.

Singh, D. and Singh, T. 1977. Location of *Fusarium monaliforme* in kernels of maize and disease transmission. *J. Mycol. Plant Pathol.* 7: 32–38.

Singh, D., Mathur, S.B., and Neergaard, P. 1977. Histopathology of sunflower seeds by *Alternaria tenuis. Seed Sci. Technol.* 5: 579–586.

Singh, D., Mathur, S.B., and Neergaard, P. 1980. Histopathological studies of *Alternaria sesamicola* penetration in sesame seed. *Seed Sci. Technol.* 8: 85–93.

Singh, D. and Singh, K.G. 1990. Occurrence of fungi in rubber seeds of Malaysia. *Indian J. Natl. Rubber Res.* 3: 64–65.

Singh, K. and Singh, D. 1989. *Nigrospora* in corn kernels from tribals in Rajasthan. *Biol. Bull. India* 11: 68–73.

Singh, K., Singh, T., and Singh, D. 1985. Colonization of corn seeds by *Fusarium monaliforme* and *Trichothecium roseum. Ann. Biol.* 1: 232–234.

Singh, K., Singh, T., and Singh, D. 1986a. Histopathology of *Drechslera maydis* infected maize kernels from tribal areas of Rajasthan. *Indian Phytopathol.* 39: 432–434.

Singh, K., Singh, T., and Singh, D. 1986b. *Botryodiplodia theobromae* Pat. in maize kernels from tribal areas of Rajasthan. *Ann. Biol.* 2: 25–31.

Singh, K., Khare, M.N., and Mathur, S.B. 1993. *Ascochyta fabae* f. sp. *lentis* in seeds of lentil, its location and detection. *Acta Phytopathol. Entomol. Hung.* 28: 2–4.

Singh, R.S. 1991. *Plant Disease*, 6th ed. Oxford and IBH Publishing Co., New Delhi, India.

Singh, R.S., Joshi, M.M., and Chaube, H.S. 1968. Further evidence of the seed-borne nature of corn downy mildews and their possible control with chemicals. *Plant Dis. Rep.* 52: 446–449.

Singh, T. and Singh, D. 1979. Anatomy of penetration of *Macrophomina phaseolina* in seeds of sesame. In *Recent Research in Plant Science*, Bir, S.S., Ed. Kalyani Publishers, New Delhi, India.

Singh, T. and Singh, D. 1982. Transmission of seed-borne inoculum of *Macrophoma phaseolina* from seed to plant. *Proc. Indian Acad. Sci.* (Plant Science) 91: 357–370.

Singh, T. and Sinclair, J.B. 1985. Histopathology of *Cercospora sojina* in soybean seeds. *Phytopathology* 75: 185–189.

Singh, T. and Sinclair, J.B. 1986. Further studies on the colonization of soybean seeds by *Cercospora kikuchi*i and *Phomopsis* spp. *Seed Sci. Technol.* 14: 71–77.

Sinha, O.K. and Khare, M.N. 1977. Site of infection and further development of *Macrophomina phaseolina* and *Fusarium equiseti* in naturally infected cowpea seeds. *Seed Sci. Technol.* 5: 721–725.

Skoropad, W.P. 1959. Seed and seedling infection of barley by *Rhynchosporium secalis. Phytopathology* 49: 623–626.

Snell, W.H. and Dick, E.A. 1971. *A Dictionary of Mycology.* Harvard University Press, Cambridge, MA.

Snow, J.P. and Sachdev, M.G. 1977. Scanning electron microscopy of cotton ball invasion by *Diplodia gossipina. Phytopathology* 67: 589–591.

Soteros, J.J. 1979. Detection of *Alternaria radicina* and *A. dauci* from imported carrot seed in New Zealand. *N.Z. J. Agric. Res.* 22: 185–190.

Strandberg, J.O. 1983. Infection and colonization of inflorescences and mericarps of carrot by *Alternaria dauci. Plant Dis.* 67: 1351–1353.

Subramanian, C.V. 1971. *Hyphomycetes. An Account of Indian Species, except Cercosporae.* Indian Council of Agricultural Research, New Delhi, India.

Subramanian, C.V. and Jain, B.L. 1966. A revision of some graminicolous *Helminthosporia. Curr. Sci.* 35: 352–353.

Subramanya, S., Safeeulla, K.M., and Shetty, H.S. 1981. Carpel infection and establishment of downy mildew mycelium in pearl millet seed, *Proc. Indian Acad. Sci.* (Plant Science) 90: 99–106.

Sumner, D.R. 1966. Histology of corn kernels and seedlings infected with *Fusarium moniliforme* and *Cephalosporium* sp. *Phytopathology* 56: 903 (Abstr.)

Suryanarayana, D. 1962. Infectivity of oospore material of *Sclerospora graminicola* (Sacc.) Schroet, the bajra green ear pathogen. *Indian Phytopathol.* 15: 247–249.

Suzuki, H. 1930. Experimental studies on the possibility of primary infection of *Piricularia oryzae* and *Ophiobolus miyabeanus* internal of rice seed. *Ann. Phytopathol. Soc. Japan* 2: 245–275. [In Japanese with English summary.]

Suzuki, H. 1934. Studies on an infection type of rice disease analogous to the flower infection. *Ann. Phytopathol. Soc. Japan* 3: 1–14.

Tarp, G. 1978. *Alternaria zinniae* on *Zinnia elegans* seed pathological studies. *Kgl. Vet. Landbohøjsk. Årsskr.* (Royal Veterinary and Agricultural University Yearbook, Copenhagen, Denmark): 13–20.

Thakkar, R. 1988. Studies on Some Seed-Borne Infections of Barley (*Hordeum vulgare*) Grown in Rajasthan. Ph.D. thesis, University of Rajasthan, Jaipur, India.

Thakkar, R., Aggarwal, A., Singh, T., and Singh, D. 1991. Histopathology of *Drechslera graminea* infected barley seeds. *Acta Bot. Indica* 19: 150–156.

Thakur, R.P., Rao, V.P., and Williams, R.J. 1984. The morphology and disease cycle of ergot caused by *Claviceps fusiformis. Phytopathology* 74: 201–205.

Thirumalachar, M.J. and Mundkur, B.B. 1947. Morphology and the mode of transmission of the ragi smut. *Phytopathology* 37: 481–486.

Tiffani, L.H. 1951. Delayed sporulation of *Colletotrichum* on soybean. *Phytopathology* 41: 975–985.

Tollenaar, H. and Bleiholder, H. 1971. Distribucion del micelio de *Sclerotinia sclerotiorum* en semilla de Girasol (*Helianthus annuus*). *Agricultura tecnica, Santiago de Chile* 31: 44–47.

Tull, J. 1733. *Horse-hoeing husbandry: or an essay on the principles of tillage and vegetation.* London, pp. 65–67.

Tulsane, L.R. 1853. Memoirs sur l'ergot des glumaces. *Ann. Sci. Nat.* 20: 5–56.

Ullstrup, A.J. 1952. Observations on crazy top of corn. *Phytopathology* 42: 675–680.

Van der Spek, J. 1965. *Botrytis cinerea* als parasiet van vlas. Laboratorium v. Fytopathologie, Meded. 213, Wageningen, the Netherlands.

Van der Spek, J. 1972. Internal carriage of *Verticillium dahliae* by seeds and its consequences. *Meded. Fak. Landb. Wet. Gent* 37: 567–573.

Vanderwalle, R. 1942. Note sur la Biologie d'*Ustilago nuda tritici* Schaf. *Bull. Inst. Agric. Sta. Recherches de Gembious* 11: 103–113.

Varma, R. 1990. Seed-borne Mycoflora and Diseases of Moth bean (*Vigna aconitifolia*) and Cowpea (*Vigna unguiculata*) Grown in Rajasthan. Ph.D. thesis, University of Rajasthan, Jaipur, India.

Varma, R. and Singh, T. 1991. *Trichothecium roseum* (Pers.) Link Ex Fr. in cowpea seeds. *Geobios New Reports* 10: 156–157.

Varma, R., Singh, T., and Singh, D. 1990. Colonization of *Botryodiplodia theobromae* Pat. in rubber seeds. *Indian J. Nat. Rubber Res.* 3: 66–68.

Varma, R., Singh, T., and Singh, D. 1991. Histopathology of rubber seeds infected with *Phomopsis heveae* Boedign. and *Stagnospora*. *J. Indian Bot. Soc.* 71: 303–306.

Varma, R., Singh, T., and Singh, D. 1992a. Incidence and colonization of *Rhizoctonia bataticola* in cowpea seeds. *Acta Bot. Indica* 20: 104–107.

Varma, R., Singh, T., and Singh, D. 1992b. *Rhizoctonia bataticola* in mothbean seeds of Rajasthan. *J. Indian Bot. Soc.* 71: 1–2.

Varma, R., Singh, T., and Singh, D. 1992c. Seed-borne infection of *Colletotrichum dematium* in *Vigna aconitifolia* (Jaco.) Marechal. *Proc. Natl. Acad. Sci. India* 62B: 63–65.

Vaughan, D.A., Kunwar, I.K., Sinclair, J.B., and Bernard, R.L. 1988. Routes of entry of *Alternaria* sp. into soybean seed coats. *Seed Sci. Technol.* 16: 725–731.

Velicheti, R.K. and Sinclair, J.B. 1991. Histopathology of soybean seeds colonized by *Fusarium oxysporum*. *Seed Sci. Technol.* 19: 445–450.

Verma, P.R. and Petrie, G.A. 1975. Germination of oospores of *Albugo candida*. *Can. J. Bot.* 53: 836–842.

Verma, P.R., Harding, H., Petrie, G.A., and Williams, P.H. 1975. Infection and temporal development of mycelium of *Albugo candida* in cotyledons of four *Brassica* species. *Can. J. Bot.* 53: 1016–1020.

Verma, U. and Bhowmik, T.P. 1988. Occurrence of *Albugo candida* (Purse ex Lev.) Kunze — its germination and role as the primary source of inoculum for the white rust disease of rapeseed and mustard. *Int. J. Trop. Plant Dis.* 6: 365–369.

Vishnuawat, K., Agarwal, V.K., and Singh, R.S. 1985. Relationship of discoloration of mustard (*Brassica campestris* var. sarson) seed infection by *Alternaria brassicae* and its longevity in seed. *Seed Res.* 13: 53–56.

Vogl, A.E. 1898. Mehl und die anderen Mehlprodukte der Cerealien und Leguminosen. *Nahrungsm. Unters. Hyg. Warenk.* 12: 25–29.

Vogt, E. 1923. Ein Beitrag zur Kenntnis von *Helminthosporium gramineum* Rbh. *Arb. biol. Reichsanst. Ld-u. Forstw.* 11: 387–397.

Wallen, V.R. 1964. Host-parasite relations and environmental influences in seed-borne diseases. *Symposia of the Society for General Microbiology* 14: 187–212.

Walker, J.C. 1969. *Plant Pathology*, 3rd ed. McGraw-Hill, New York.

Weihing, J.L. 1956. False smut of ballograss. *Nebr. Agric. Exp. Stn. Res. Bull.* 180: 1–26.

Weniger, W. 1923. Pathological morphology of durum wheat grains affected with black point. *Phytopathology* 13: 48–49.

Weniger, W. 1925. Black point disease caused by *Helminthosporium sativum* Pam. King and Bak. *Plant Dis. Rep.* Suppl. 40: 136.

Weston, W.H. 1920. Philippine downy mildew of maize. *J. Agric. Res.* 19: 97–122.

Western, J.H. and Cavett, J.J. 1959. The choke disease of cocksfoot (*Dactylis glomerata*) caused by *Epichloe typhina* (Fr.) Tul. *Trans. Br. Mycol. Soc.* 42: 298–307.

White, J.F. and Chambless, D.A. 1991. Endophyte–host associations in forage grasses. XV. Clustering of stromata-bearing individuals of *Agrostis hiemalis* infected by *Epichloe typhina*. *Am. J. Bot.* 78: 527–533.

White, J.F. and Morrow, A.C. 1990. Endophyte-host associqations in forage grasses. XII. A fungal endophyte of *Trichachne insularis* belonging to *Pseudocercosporella. Mycologia* 82: 218–226.

Wilson, M., Noble, M. and Gray, E.G., 1945. The blind seed disease of rye-grass and its causal fungus. *Trans. Roy. Soc. Edin.* 61: 327–340.

Yadav, V. 1984. Studies on Seed-Borne Inoculum of *Drechslera* spp. in Rajasthan Grown Wheat. Ph.D. thesis, University of Rajasthan, Jaipur, India.

Zad, S.J. 1989. Transmission of soybean downy mildew by seed. *Meded. Fac. Landbouwwet. Rijksuniv (Gent)* 54: 561–566.

Zaumeyer, W.J. and Thomas, H.R. 1957. A monographic study of bean diseases and methods for their control. *U.S. Dep. Agric. Tech. Bull.* 868 (rev. ed.), pp. 255.

Zimmer, D.E. 1963. Spore stages and life cycle of *Puccinia carthamii. Phytopathology* 53: 316–319.

Zimmer, R.C., McKeen, W.E., and Campbell, C.G. 1992. Location of oospores in buckwheat seed and probable roles of oospores and conidia of *Peronospora ducometi* in the disease cycle on buckwheat. *J. Phytopathol.* 135: 217–223.

Zscheile, F.P. and Anken, M. 1956. Limited development of *Tilletia* chlamydospores within wheat kernels. *Phytopathology* 46: 183–186.

6 Seed Infection by Bacteria

Bacteria are prokaryotes that have a rigid cell wall, cell membrane, and often one or more flagella. A large number of bacteria are saprophytes, and plant pathogenic bacteria are basically facultative saprophytes. The nonfilamentous phytopathogenic bacteria generally belong to *Acidovorax, Agrobacterium, Burkholderia, Erwinia, Pantoea, Pseudomonas, Ralstonia, Xanthomonas,* and coryneform plant pathogens. These genera are usually seed-borne. The coryneform phytopathogenic species that were placed earlier in *Corynebacterium* have been transferred to *Arthrobacter, Clavibacter, Curtobacterium, Rathayibacter,* and *Rhodococcus.* Some of the *Pseudomonas* species are now placed in *Acidovorax* and *Burkholderia,* and *Erwinia* has been placed in *Pantoea* (Agrios, 1988; Young et al., 1996).

The bacteria may cause seed infestation, i.e., carried on the surface of the seed or seed infection, that occurs in the seed coat and other parts of the seed. Both types are known to be seed-transmitted, causing failure in seed germination and/or disease symptoms in seedlings and plants. The effects of the seed infecting bacteria on host tissues can be determined in histological preparations. Some examples of these infections are included here.

6.1 PENETRATION

Unlike fungi, bacteria lack mechanisms for forcing their way physically through protective barriers, such as cuticle, epidermis, and bark. Before invasion can take place, the bacteria must establish themselves on the plant surface, i.e., they should be able to find a proper niche. There are numerous examples of growth of bacteria on plant foliage. Bacteria multiply with rapidity, and their significance as pathogens may primarily depend on the fact that they can produce large numbers of cells in a short time. The penetration and spread of bacteria in plant tissues have been studied in natural infections and in infections produced after artificial inoculations (Zaumeyer, 1930, 1932; Thiers and Blank, 1951; Wiles and Walker, 1951; Pine, Grogan, and Hewitt,, 1955; Layne, 1967; Getz, Fulbright, and Stephens, 1983; Kritzman and Zutra, 1983; Rudolph, 1993).

The developing ovules and seeds occupy a specific position in angiosperms, as discussed in detail in Chapter 2. In order to reach the ovule or seed on a plant the infection must find passages for invasion of the ovary or developing fruit. This situation was discussed in Chapter 4 with respect to fungi but the details are also applicable to bacteria. Seeds after harvest and threshing are directly exposed to the environments in storage and in soil.

6.1.1 Invasion of Plant Parts

Bacterial diseases of seedling and plant may be established from seed-borne infection (Neergaard, 1979) or may take place through soil, air, water, insects, and nematodes. The entry of phytopathogenic bacteria may be passive and occur through natural openings (stomata, lenticels, hydathodes, or nectariferous surfaces [nectarthodes]), through wounds including scars left by dropping of hairs, cracks, or by means of enzymes. Entry through noncutinized surface of hairs, stigma, and anthers has also been suggested. The enzymatic penetration of the surface depends on macerating and digesting enzymes that degrade the barrier presented by the complex plant cell wall. Enzymatic degradation of plant tissue is important in pathogenesis, and enzymes involved in soft rotting by bacteria are polygalacturonase, pectate lyases, cellulase, proteases, and nucleases (Klement, Rudolph, and Sands, 1990). Only a few phytopathogenic bacteria can degrade cellulose at a rate comparable to that of many cellulolytic fungi (Kelman, 1979). Although bacterial degradation of cell wall components, including lignin, has been reported in wood following extensive exposure to high moisture, conclusive evidence for lignolytic capabilities of the specific bacteria involved has not been demonstrated (Liese, 1970). Furthermore, no bacterial plant pathogens that degrade lignin are known (Kirk and Connors, 1977). This is important since the seed coat in the majority of true seeds and the pericarp in one-seeded indehiscent fruits possess lignified cells.

The stomata on the leaf and stem are the common passagex for entry of bacteria. Zaumeyer (1930) found that *Xanthomonas axonopodis* pv. *phaseoli* (*Bacterium phaseoli*) enters the leaves, stems, and pods of the bean plant through the stomata (Figure 6.1A to E). The entry of bacteria through the stomata seems to be common and has been reported for *X. axonopodis* pv. *malvacearum* in cotton (Thiers and Blank, 1951); *Pseudomonas syringae* pv. *pisi* in pea (Skoric, 1927); *P. s.* pv. *lachrymans* in cucumber (Wiles and Walker, 1951); *Burkholderia plantari, B. glumae,* and *Pantoea agglomerans* (*Erwinia herbicola*) in the lemma and palea of rice grains (Azegami, Tabei, and Fukuda, 1988; Tabei et al., 1988, 1989); *Clavibacter michiganensis* subsp. *michiganensis* in tomato (Layne, 1967); and *Curtobacterium flaccumfaciens* pv. *flaccumfaciens* in beans (Schuster and Sayre, 1967). Fukuda, Azegami, and Tabei (1990) found that *P. s.* pv. *syringae* (*P. s.* pv. *japonica*) invaded the leaf blade, leaf sheath, lemma, and palea through the stomata in barley and wheat. Azegami, Tabei, and Fukuda (1988) also observed that *B. plantari* and *B. glumae* entered young rice seedlings through the stomata present in the surfaces of coleoptiles and leaf sheaths. Tabei (1967) found that nonpathogenic bacteria and *Xanthomonas oryzae* pv. *oryzae* (*X. c.* pv. *oryzae)* entered through the stomata in rice, and pointed out that this seems to occur easily since some stomata in coleoptile and leaf sheaths are always open. Ramos and Valin (1987), while examining the role of stomatal opening and frequency on infection of *Lycopersicon* species by *Xanthomonas vesicatoria*, noted positive correlation between stomatal frequency on adaxial and abaxial leaf surfaces and the number of spots and lesions produced on artificial inoculation. Bacterial spots also were considerably reduced when the stomatal closure was physiologically induced or chemically suppressed by abscissic acid or phenylmercuric acetate before inoculation with *X. vesicatoria*.

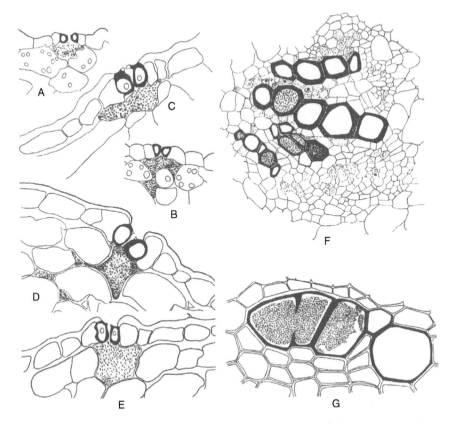

FIGURE 6.1 Stomatal penetration and vascular invasion by *Xanthomonas axonopodis* pv. *phaseoli* in bean. A to E, Stomatal penetration showing bacteria in the substomatal cavity and spreading from there into intercellular spaces. A, B, Leaf. C, D, Stem. E, Pod. F, Cs of midrib region of leaf showing bacteria in the xylem vessel. G, Cs stem showing invasion of metaxylem vessels by bacteria. (From Zaumeyer, W.J. 1930. *U.S. Dept. Agric. Bull.* 180: 1–36.)

In addition to the stomata, *C. m.* subsp. *michiganensis* enters tomato leaves through trichomes (Layne, 1967). Getz, Fulbright, and Stephens (1983) also observed that *P. syringae* pv. *tomato* entered tomato ovaries during the anthesis through eglandular and glandular hairs (Figure 6.2A to C). After the separation of trichomes, the open trichome bases also served as sites for infection (Figure 6.2D).

Meier (1934) studied early infection of *Xanthomonas c.* pv. *campestris* in cabbage and cauliflower and showed that the bacterium enters through the hydathodes. The fire blight bacterium, *Erwinia amylovora*, invaded flowers of pear and apple through a noncutinized surface of stigma and anthers, hydathodes on sepals, stomata on the style, and sepals and nectarthodes on hypanthium (Hildebrand and MacDaniels, 1935). After artificial inoculation of pear and apple flowers by *E. amylovora*, Pierstorff (1931) and Rosen (1935) found that the most common site of infection was through the nectariferous surface of hypanthium. Thomson (1986) observed that *E. amylovora* occurs predominantly on the stigmas of flowers in *Pyrus*, *Malus*,

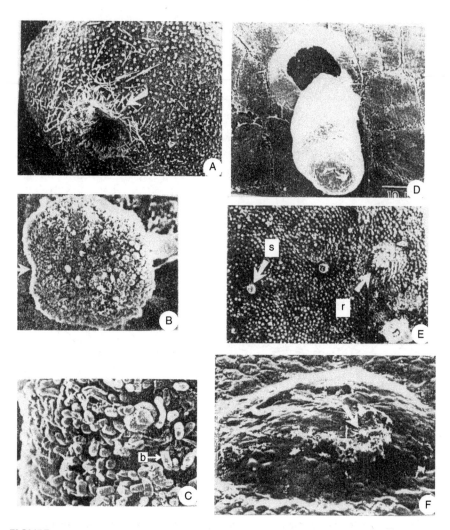

FIGURE 6.2 SEM photomicrographs of tomato ovaries inoculated with *Pseudomonas syringae* pv. *tomato* showing infection sites. A, Tomato ovary at anthesis covered with nonglandular and glandular hairs (arrow). B, C, Capitate glandular hair (arrow) with bacteria on the head. Bacteria seen in C. D, Natural opening caused by detachment of the trichome on ovary surface. E, Ovary surface 11 days after inoculation showing swollen (arrow) and raised ruptured (arrow) areas with bacteria. F, Ruptured bacterial lesion magnified showing bacteria (arrow) extruded from the crack. (Abbreviations: b, bacteria; r, raised area on ovary surface; s, swollen area.) (From Getz, S., Fulbright, D.W., and Stephens, C.T. 1983. *Phytopathology* 73: 39–43. With permission.)

Pyracantha, Crataegus, and *Cotoneaster* (Figure 6.3A to C). The bacterium survived better on the stigma than on the hypanthium, but the bacterial populations moved to the hypanthium to cause infection.

Smith (1922) and Ark (1944) have shown that the pollen of *Juglans regia* may be contaminated by *Xanthomonas juglandis,* causing walnut blight, and Ark found

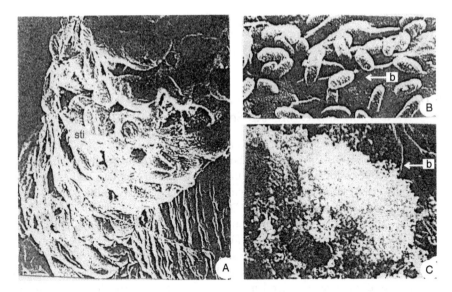

FIGURE 6.3 SEM photomicrographs of stigma of flowers colonized by *Erwinia amylovora*. A, Stigma of pear flower pistil with bacterial cells present on the stigmatic surface. B, Enlargement of bacterial cells on stigmatic surface of pear flower pistil. C, Mass of bacterial cells on stigmatic surface of apple flower. (Abbreviations: b, bacteria; sti, stigma.) (From Thompson, S.V. 1986. *Phytopathology* 76: 476–482. With permission.)

that most of the flowers that were experimentally pollinated by contaminated pollen developed infected nuts. Ivanoff (1933) also reported that maize pollen may be infected or contaminated by *Pantoea stewartii* subsp. *stewartii* (*Erwinia stewartii*), and this may cause infection of maize florets through the stigma; however, no histopathological studies of these two systems have been made.

Wounds or openings caused by any injury, whether mechanical or by insects or nematodes, form easy passages for entry of bacteria (Zaumeyer, 1932; Skoric, 1927; Thiers and Blank, 1951; Nelson and Dickey, 1970). Artificial inoculations (Figure 6.2E, F) pricking the plant surface also demonstrate this (Kelman, 1953; Offutt and Baldridge, 1956; Dickey and Nelson, 1970; Getz, Fulbright, and Stephens, 1983). *Ralstonia solanacearum* enters through insect wounds and lenticels in peanut plants (Miller and Harvey, 1932). Azegami, Tabei, and Fukuda (1988) observed that during seedling growth, ruptures occur in the surfaces of coleoptile and roots at the points of emergence of leaves or secondary roots. *B. plantari* enters through such wounds as well as stomata. Vakili (1967) has shown the importance of wounds in bacterial spot disease of tomato caused by *X. vesicatoria*.

Available information on the role of plant–parasitic nematodes in the initiation of bacterial diseases was summarized by Pitcher in 1963. According to Pitcher, all nematodes puncture plant cells and can act as inoculants, but the micropuncture made by a nematode stylet is not the type of wound most likely to favor the entry of bacteria. The endoparasitic nematodes, which enter the host bodily and, perhaps, cause local necrosis, seem better suited for entry and establishment of bacteria. *Anguina tritici* for *Rathayibacter tritici* in wheat and *Aphelenchoides ritzema-bosi* for *Rhodococcus facians* (*Corynebacterium fascians*) in strawberry and cauliflower are reported to be essential as vectors of these pathogens (Hunger, 1901; Carne, 1926; Vasudeva and Hingorani, 1952; Crosse and Pitcher, 1952; Pitcher and Crosse, 1958).

6.1.2 SPREAD IN PLANT AND COURSE OF ENTRY INTO OVARY AND FRUIT AND OVULE AND SEED

After entering the plant tissue, the bacteria multiply in a substomatal cavity (Figure 6.1A, C, E) or space beneath the site of entry. Within the host, the bacteria follow two courses for further spread: they invade xylem elements and move vertically after coming in contact with vascular elements and they multiply and spread in intercellular spaces (Figure 6.1B, D). The latter course usually results in a restricted spread. Vascular bacteria can move over long distances, and they can also spread laterally after dissolving the walls of the vessels (Smith, 1911; Yu, 1933; Pine, Grogan, and Hewitt, 1955) or after escaping through openings (pits) in the walls of vessel elements (Ivanoff, 1933; Wolf and Nelson, 1969). Nelson and Dickey (1966) have also suggested lateral movement of bacteria through plasmodesmatal connections in the pit membranes of vessels. Association of bacteria with phloem is observed, but it probably plays no significant role in their spread (Pine, Grogan, and Hewitt, 1955).

Several seed infecting bacteria are systemic vascular pathogens (Zaumeyer, 1930; Pine, Grogan, and Hewitt, 1955; Cormack and Moffat, 1956; Mukerjee and Singh, 1983; Du Plessis, 1990). Pine, Grogan, and Hewitt (1955) have given a detailed account of the spread of *C. m.* subsp. *michiganensis*, a bacterial canker pathogen, in young tomato plants and provided evidence of longitudinal and lateral movement of the organism in xylem elements (Figure 6.4A to D). Bacterial pockets were formed in phloem and pith, but they gave no indication of vertical movement. During earlier studies, *C. m.* subsp. *michiganensis* was described primarily as a phloem parasite (Smith, 1914, 1920). However, Zaumeyer (1930) has also observed invasion of xylem elements in leaf (Figure 6.1F) and stem (Figure 6.1G) by *X. a.* pv. *phaseoli*, and its subsequent spread throughout the organism. *X. oryzae* pv. *oryzae* has a similar course in rice (Mukerjee and Singh, 1983).

The systemic vascular infection often enters the reproductive shoot and causes ovary and fruit infection (Zaumeyer, 1930; Cormack and Moffatt, 1956; Mukerjee and Singh, 1983; Du Plessis, 1990). Zaumeyer (1930) traced the passage of *X. a.* pv. *phaseoli* from plant to pods in beans and noted that the bacteria in the xylem vessels of the stem travels up and through the vascular elements of the pedicel and passes into the vascular supply of the pod. Mukerjee and Singh (1983) observed that *X. o.* pv. *oryzae* reaches the rice husk through vessels of the infected rachilla.

Skoric (1927) reported that *Pseudomonas syringae* pv. *pisi* is largely a parenchyma invader and only occasionally may enter the vessels under field conditions. The bacterium penetrates pods through wounds and spreads in intercellular spaces, forming abundant slime on the inner side of the pod. After artificial inoculation by *P. s.* pv. *lachrymans* of the stem 1.5 m from the fruit, Kritzman and Zutra (1983) observed systemic invasion of the cucumber fruits. No histopathological studies were

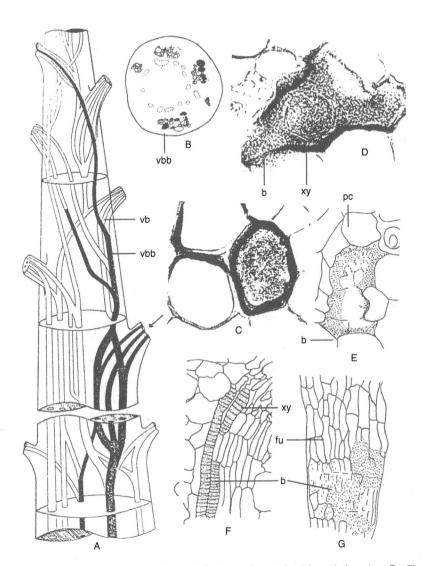

FIGURE 6.4 Path of movement of bacteria in plant after artificial inoculation. A to D, *Clavibacter michiganensis* ssp. *michiganensis* in young tomato plant. A, Diagram from plant collected after 5 days of inoculation. Arrow indicates the point of inoculation and solid black areas, the primary vascular bundles that carry bacteria. B, Cross section from the main axis of the inoculated plant. Solid black areas represent bacteria in and around xylem. C, Bacteria in vessel. D, Bacterial pocket formed by escape of bacterial cells from a vessel into the surrounding intercellular spaces and xylem parenchyma. E to G, *Pseudomonas syringae* pv. *lachrymans* in naturally infected cucumber fruit tissues. E, F, Bacteria in and between parenchyma cells and within xylem in the mesocarp, respectively. G, Bacteria invading the parenchyma cells of the funiculus. (Abbreviations: b, bacteria; fu, funiculus; pc, parenchyma cells; vb, vascular bundle; vbb, vascular bundle with bacteria; xy, xylem elements.) (A to D, from Pine, T.S., Grogan, R.G., and Hewitt, Wm. B. 1955. *Phytopathology* 45: 267–271; E to G, from Wiles, A.B. and Walker, J.C. 1951. *Phytopathology* 41: 1059–1064. With permission.)

made by them. However, earlier observations by Wiles and Walker (1951) of the same pathogen showed that the bacterium advanced intercellularly as well as within xylem elements of the fruit mesocarp (Figure 6.4E, F) after invasion through the stomata. Zaumeyer (1932) has also noted the movement of *P. s.* pv. *phaseolicola* through the xylem and parenchyma tissues of bean plants.

The above information reveals that the ovary and fruit infection may take place from the mother plant via vascular elements or through intercellular spaces in the parenchyma. Local infection of any part of the fruit including the pedicel may also become vascular if it gains entry into xylem elements, or it may spread in intercellular spaces, or both. Whether vascular, or from intercellular spaces, the infection can reach the ovule either through the funiculus, which connects ovule and seed to the fruit at the placenta, or it may enter the ovary and fruit cavity after breaking the wall. In the latter condition, bacterium may spread on the ovule surface in part or completely, and it may enter the ovule through the micropyle and ovule surface directly or through the stomata, if present, as in *Gossypium* and *Hibiscus*.

X. axonopodis pv. *phaseoli* in beans enters the ovule and seed through the vascular elements and also the parenchyma of the funiculus or from the bacterial mass occurring in pod cavity. In cases of severe funicular infection, its tissues degrade and the ovule and seed fail to develop. If weak to moderate infection takes place at a late stage of seed development, the bacterium spreads in the seed coat and around the cotyledons. Bacterial mass in the pod cavity surrounding or occurring in the vicinity of the ovule and seed may easily invade through the micropyle (Zaumeyer, 1930). In 1932 Zaumeyer found that *C. f.* pv. *flaccumfaciens* also enters the ovule through the vascular elements in the funiculus or through the micropyle. After the bacterium has escaped from the vessels in the funiculus and the dorsal suture and the parenchyma of the pod wall, it may gain entry into the micropyle.

Skoric (1927) found invasion of pea seeds by *P. s.* pv. *pisi* through the parenchymatous cells of the funiculus and also through the micropyle (Figure 6.8A, B; see p. 186). In the pea pod, abundant bacterial mass also occurs on the inner side of infected pods. Although Skoric (1927) reports that the micropylar infection occurs from the funiculus, its invasion from the bacterial mass present in the pod cavity cannot be ruled out. Seed infection in cucumber by *P. s.* pv. *lachrymans* occurs through bacteria growing intra- and intercellularly (Figure 6.4G) in the funiculus (Wiles and Walker, 1951). In the mesocarp of the fruit it is detected in xylem elements and also in parenchyma (Figure 6.4E, F). However, Wiles and Walker (1951) could not detect the bacterium in seed. But Nauman (1963) observed that *P. s.* pv. *lachrymans* penetrates through the funiculus and micropyle and spreads in the seed coat, endosperm, and embryo. *X. c.* pv. *campestris* also reaches cabbage seed through the funiculus as shown by Cook, Larson, and Walker (1952) in artificially inoculated plants. The invasion is systemic and spreads through the vascular system in pedicels and siliculas. The bacterial mass was observed throughout the funiculus up to its junction with the seed and it disrupted the xylem in severely affected areas.

In artificially inoculated tomato plants with *Clavibacter michiganensis* subsp. *sepedonicus*, Larson (1944) found that the pathogen entered the fruit through the vessels and/or tracheids in the fleshy placenta to the funiculus of the seed. Fukuda,

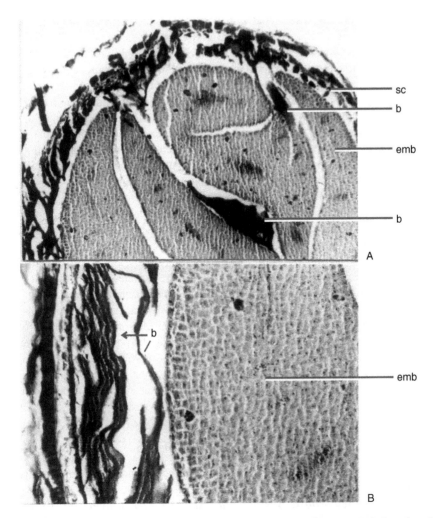

sc

b

emb

b

A

b

emb

B

FIGURE 6.5 Location of *Xanthomonas campestris* pv. *campestris* in naturally infected seeds of *Brassica campestris*. A, Ls part of seed showing bacterial masses in seed coat and in the spaces between endosperm and embryo, and in between the cotyledons. B, A magnified view of part of seed showing aggregation of bacteria in seed coat, endosperm, and the space outside the embryo. (Abbreviations: b, bacteria; emb, embryo; sc, seed coat.) (From Sharma, J., Agarwal, K., and Singh, D. 1992. *Seed Res.* 20: 128–133.)

Azegami, and Tabei (1990) reported that *P. s.* pv. *syringae* invaded the kernel in the wheat and barley caryopsis through the funiculus.

6.1.3 Invasion of Threshed and Disseminated Seeds

After harvest and threshing or dissemination seeds find an entirely different environment from the environment during their development inside the fruit. Such seeds are directly exposed to the surroundings and have a well-formed cuticle; wax deposits, when present; a seed coat with protective thick-walled and lignified layers; a hilum; micropyles, closed or open to various extents; raphe, when present,; and pits or micropores and cracks on the surface. Since seeds are stored under dry conditions, no free moisture is available. The soil environments of seeds are highly variable and range from dry to water-logged conditions.

Field bacteria in general are not expected to cause internal infection in dry seeds during storage; however, seed contamination is possible during threshing. Under favorable conditions of humidity and temperature, in true seeds, the avenues for infection are (1) the seed surface through the cuticle, natural openings such as the stomata as in cotton, micropores, cracks, or through injuries caused during threshing; (2) the hilum, which is covered by cuticle, but usually has fissures; (3) micropyles, particularly the open type; and (4) accessory structures, e.g., hairs, wings, aril, and caruncle.

One-seeded fruits such as caryopsis, cypsils, achenes, cremocarp, and mericarp may also become infected or contaminated during the postharvest period. Contamination is common during threshing as well as during storage and in soil, but infection will depend on favorable environmental conditions. Avenues for infection in such seeds could be the pericarp and adhering bract surface, separation scar (analogous to hilum), remnants or scars caused by the style and stigma, and persistent accessory structures. Information on penetration and establishment of bacterial infection of seeds (true seeds as well as one-seeded fruits) during the postharvest period either in storage or in soil is meager. Grogan and Kimble (1967) have reported that in halo-blight of beans caused by *P. s.* pv. *phaseolicola* the hilum and cracks in the seed coat during threshing or from wetting provide sites at which the bacterium gains entry.

6.2 HISTOPATHOLOGY OF INFECTED SEEDS

The important bacterial species that cause seed infection and have been investigated histologically are listed in Table 6.1. These belong to the genera *Xanthomonas, Pseudomonas, Acidovorax, Burkholderia, Rathayibacter, Clavibacter, Curtobacterium,* and *Pantoea.* The nomenclature followed is as given in Young et al.'s (1996) article "Names of Plant Pathogenic Bacteria, 1854–1995."

6.2.1 *Xanthomonas*

Xanthomonas species cause some of the serious bacterial diseases in crucifers, beans, cotton, and rice (Table 6.1). *Xanthomonas c.* pv. *campestris* is a serious pathogen that causes black rot in crucifers. Cook, Larson, and Walker (1952) reported systemic invasion by this bacterium through the vascular system of pedicels and pods (silicula). Further invasion through the xylem of the funiculus was recorded and infection of the seed coat was suspected. Walker (1950) observed that *X. c.* pv. *campestris* reaches the seed coat through the vascular system in the funiculus. Sharma, Agarwal, and Singh (1992) have described the location of *X. c.* pv. *campestris* in rape and mustard seeds. The infected seeds were asymptomatic and symptomatic and were categorized into bold-symptomless, bold-discolored, and) shriveled-discolored. The

TABLE 6.1
Seed Infecting Bacterial Pathogens

Bacteria	Host (Disease)	Seed Part	Important References
Xanthomonas			
Xanthomonas axonopodis pv. *phaseoli* (Smith) Dye (*X. phaseoli* (Smith) Dowson; *Bacterium phaseoli* (Smith) Smith)	*Phaseolus, Lablab* (bacterial blight)	Seed coat, embryo	Burkholder, 1921; Zaumeyer, 1930; Weller and Saettlar, 1980
X. a. pv. *phaseoli* var. *fuscans* Vauterin et al.	*Phaseolus* (fuscus blight)	Seed coat, hilum	Weller and Saettler, 1980
X. a. pv. *malvacearum* (Smith) Dye (*X. malvacearum* (Smith) Dowson)	*Gossypium* (bacterial blight, angular leaf spot)	Seed coat, embryo	Tennyson, 1936; Brinkerhoff and Hunter, 1963
X. a. pv. cajani (Kulkarni et al.) Vauterin et al. (=*X. cajani* Kulkarni et al.)	*Cajanus* (leaf spot, canker)	Seed coat, embryo	Sharma, 1996, Sharma et al., 2001
X. campestris pv. *campestris* (Pammel) Dowson (*X. campestris*) (Pammel) Dowson)	*Brassica* (black rot)	Seed coat, endosperm, embryo	Cook et al.,1952; Sharma et al., 1992
X. oryzae pv. *oryzae* (Ishiyama) Swings et al. (*X. oryzae* (Uyeda and Ishiyama) Dowson; *X. c.* pv. *oryzae* (Ishiyama) Dye)	*Oryza* (bacterial leaf blight)	Glumes, endosperm	Fang et al., 1956; Srivastava and Rao, 1964; Mukerjee and Singh, 1983
X. oryzae pv. *oryzicola* (Fang et al.) Swings et al. (*X. c.* pv. *oryzicola* (Fang et al.) Dye)	*Oryza* (bacterial leaf streak)	Glumes, endosperm surface	Shekhawat et al., 1969
Pseudomonas			
Pseudomonas savastanoi pv. *phaseolicola* (Burkholder) Garden et al.	*Phaseolus* (bacterial halo blight)	Seed coat, surface of cotyledons	Taylor et al., 1979
P. s. pv. *glycinea* (Coerper) Gardan et al. (*P. glycinea* Coerper)	*Glycine* (bacterial blight)	Seed coat, embryo	Parashar and Leben, 1972

(continued)

TABLE 6.1 (CONTINUED)
Seed Infecting Bacterial Pathogens

Bacteria	Host (Disease)	Seed Part	Important References
Pseudomonas (continued)			
P. syringae pv. *lachrymans* (Smith and Bryan) Young et al. (*P. lachrymans* (Smith and Bryan) Carsner)	*Cucumis* (angular leaf spot)	Seed coat, embryo	Wiles and Walker, 1951; Nauman, 1963
P. s. pv. *pisi* (Sackett) Young, Dye and Wilkie (*P. pisi* Sackett)	*Pisum* (bacterial blight)	Seed coat	Skoric, 1927
P. s. pv. *helianthi* (Kawamura) Young et al.	*Helianthus*	Pericarp, seed coat, endosperm, embryo	Godika, 1995
P. s. pv. *syringae* van Hall	*Hordeum, Triticum* (bacterial black node)	Palea, lemma	Fukuda et al., 1990
Acidovorax			
Acidovorax avenae subsp. *citrulli* Willems et al. (*P. pseudoalcaligenes* ssp. *citrulli* Schaad et al.)	*Citrullus* (bacterial fruit blotch)	Seed coat, embryo	Rane and Latin, 1992
Burkholderia			
Burkholderia glumae (Kurita and Tabei) Urakami et al. (*P. glumae* Kurita and Tabei)	*Oryza* (bacterial seedling rot)	Lemma, palea, surface of endosperm, embryo	Tabei et al., 1989
Rathayibacter			
Rathayibacter tritici (ex Hutchinson) Zgurskaya et al. (*Corynebacterium tritici* (Hutchinson) Burkholder)	*Triticum* (yellow shine disease, yellow ear rot, tundu disease)	Spikes completely or partially affected	Cheo, 1946; Sabet 1954a,b; Swarup and Singh, 1962
R. iranicus (ex Scharif) Zgurskaya et al. (*Corynebacterium iranicum* Scharif)	*Triticum* (yellow ear rot)	Spikes affected	Scharif, 1961

(continued)

TABLE 6.1 (CONTINUED)
Seed Infecting Bacterial Pathogens

Bacteria	Host (Disease)	Seed Part	Important References
	Rathayibacter (continued)		
R. rathayi (Smith) Zgurskaya et al. (*Corynebacterium rathayi* (Smith) Dowson)	*Dactylis, Festuca* (yellow slime disease, Rathay's disease)	Spikes affected	Neergaard, 1979
	Lolium (seed galls)	Seeds affected	Stynes et al., 1979; Bird et al., 1980
	Clavibacter		
Clavibacter michiganensis subsp. *insidiosus* (McCulloch) Davis et al. (*Corynebacterium insidiosum* (McCulloch) Jensen)	*Medicago* (bacterial wilt)	Seed coat, aleurone layer	Cormack and Moffatt, 1956
C. m. subsp. *michiganensis* (Smith) Davis et al. (*Corynebacterium michiganense* (Smith) Jensen)	*Lycopersicon, Capsicum* (bacterial canker)	Seed coat	Bryan, 1930; Grogan and Kendrick, 1953; Patino-Mendez, 1964
C. m. subsp. *nebraskensis* (Vidaver and Mandel) Davis et al. (*Corynebacterium nebraskense*) Vidaver and Mandel)	*Zea* (Goss's bacterial wilt and leaf blight)	Chalaza, vicinity of embryo	Schuster, 1972; Biddle et al., 1990
C. m. subsp. *tessellarius* (Carlson and Vidaver) Davis et al. (*Corynebacterium michiganense* ssp. *tessellarius* Carlson and Vidaver)	*Triticum* (bacterial mosaic)	Seed coat, endosperm, near embryo	McBeath and Adelman, 1986

(continued)

TABLE 6.1 (CONTINUED)
Seed Infecting Bacterial Pathogens

Bacteria	Host (Disease)	Seed Part	Important References
Curtobacterium			
C. flaccumfaciens pv. *flaccumfaciens* (Hedges) Collins and Jones (*Corynebacterium flaccumfaciens* pv. *flaccumfaciens* (Hedges) Dowson)	*Phaseolus, Glycine* (bacterial wilt)	Raphe, seed coat	Burkholder, 1930; Zaumeyer, 1932; Schuster and Sayre, 1967; Schuster and Smith, 1983
Pantoea			
Pantoea stewartii subsp. *stewartii* (Smith) Mergaert et al. (*Erwinia stewartii* (Smith) Dye)	*Zea* (bacterial wilt, Stewart's disease)	Chalaza, endosperm	Rand and Cash, 1921; Ivanoff, 1933
Pantoea agglomerans (Beijerinck) Gavini et al. (*E. herbicola* (Lohnis) Dye)	*Oryza* (bacterial palea browning)	Lemma, palea	Tabei et al., 1988

discoloration varied from small water-soaked to slimy brown spots or general browning of seeds. Occasionally brownish sticky ooze occurred on the seed surface of symptomatic seeds. Histopathology of infected seeds revealed direct correlation between the severity of infection and internal invasion of tissues. In infected bold seeds, the bacterium usually occurred in the seed coat and rarely in the endosperm. But in bold-discolored and shriveled-discolored seeds, the bacterium was found in the seed coat, endosperm, and embryo (Figure 6.5A, B). The distortion of seed tissues was greater in shriveled seeds than in bold-discolored seeds. The embryo was relatively small and thin in the former, and the infection was extra- as well as intraembryal. Symptomatic infected seeds on germination produced seedlings with brown-black necrotic spots on the cotyledons.

Mihail, Taylor, and Versluses (1993) detected *X. campestris* pv. *campestris* in siliqua and the surface of seeds of *Crambe abyssinica*. Zaumeyer (1930) found invasion of *X. a.* pv. *phaseoli* in bean seed through the xylem in the funiculus and/or micropyle. Infection occurred in intercellular spaces in the seed coat and also in the space around the embryo. No infection was seen in the cells of the palisade and hourglass layers of seed coat. It is only at the time of seed germination that the bacteria enter into the cotyledons.

Sharma et al. (2001) have described the location of *Xanthamonas a.* pv. *cajani* in seeds of pigeon pea. Seed infection varied from weak to severe and severely infected seeds were small and shriveled (Figure 6.6). In asymptomatic infected seeds, bacteria occurred in the palisade, hourglass cells, and parenchyma of the seed coat.

FIGURE 6.6 (Color figure follows p. 146.) Pigeon pea (*Cajanus cajan*) seeds healthy (upper row) and infected with *Xanthomonas axonopodis* pv. *cajani*. Seeds moderately shriveled-discolored (middle row) and heavily shriveled-discolored (lower row) with water-soaked translucent areas. (From Sharma, M. et al. 2001. *J. Mycol. Plant Pathol.* 31: 216–219.)

The embryo remained free of infection. Moderately shriveled and discolored seeds carried infection in all the layers of seed coat and superficial layers of the cotyledons (Figure 6.7A, D). Bacteria occurred inter- and intracellularly and were present in the stellate parenchyma and persistent part of funiculus (Figure 6.7C). Affected cotyledonary cells also showed necrosis. Heavily infected seeds had infection in all parts including parts of the embryo (Figure 6.7B). Bacterial masses occurred in the seed coat (Figure 6.7E), stellate parenchyma, and remnants of the funiculus. Necrosis and formation of the lytic cavities, occupied by bacterial cells, were seen. Spaces in seed, between the seed coat and embryo, between the cotyledons, and around the shoot-root axis also contained bacteria. Bacteria also occurred intracellularly in cotyledonary cells (Figure 6.7F) and caused depletion of cell contents including starch grains, which were small and only of the simple type (simple and compound starch grains in unaffected cotyledons).

   X. axonopodis pv. *malvacearum*, a causal organism of bacterial blight, angular leaf spot, or black arm of cotton, occurs externally and internally in seed. Internal infection was detected in the micropylar as well as chalazal halves of the seed coats, and rarely in the embryo (Tennyson, 1936; Brinkerhoff and Hunter 1963). Brinker-hoff and Hunter (1963) tried to determine the course of entry of the bacterium into the seed. They soaked ginned fuzzy seeds of several varieties of cotton in aqueous dye (light green) and aqueous suspension of the bacterium. The dye readily entered through the blunt chalazal end, and only after the embryo swelled, entered through

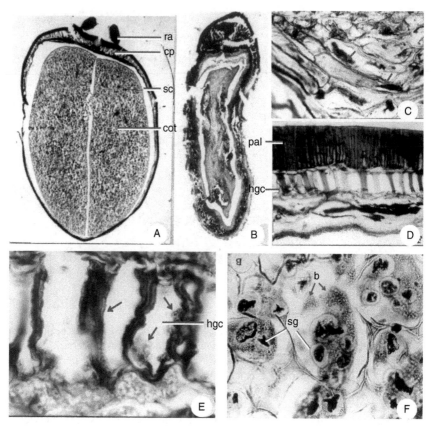

FIGURE 6.7 Histology of normal and *Xanthomonas axonopodis* pv. *cajani*-infected pigeon pea seeds. A, Ts seed through hilar region. B, Ts shriveled and discolored seed showing presence of bacterial masses in different parts of seed. C, Section of part of funiculus having bacterial masses in cells. D, Ts part of seed coat with bacterial cells in hourglass cells and inner parenchymatous cells. E, Ts part of seed coat from heavily infected seed. Hourglass cells are distorted and possess bacterial masses. F, Inter- and intracellular colonies of bacteria in cells of cotyledon. Starch grains are surrounded by bacteria and show corrosion. (Abbreviations: b, bacteria; cot, cotyledon; cp, counter palisade; hgc, hourglass cells; pal, palisade cells; ra, rim aril; sc, seed coat; sg, starch grains.) (A, B, E, F, From Sharma, M. et al. 2001. *J. Mycol. Plant Pathol.* 31: 216–219; C, D, From Sharma, M. 1996. Ph.D. thesis, University of Rajasthan, Jaipur, India.)

the micropylar end. Similarly, the causal bacterium entered the fuzzy seed through the chalazal end. It may be mentioned that the stomata are aggregated in the surface of the chalaza.

Crossan and Morehart (1964) isolated *X. vesicatoria* from vascular and cortical tissues of pedicel, ovary and seed of *Capsicum annuum*. SEM studies of artificially inoculated stalks and mature fruits of Golden King plum with *X. arboricola* pv. *pruni* (*X. c.* pv. *pruni*) have shown that the vascular elements of the stalk, mesocarp, stony endocarp, and seed are filled with masses of the bacterium (Du Plessis, 1990).

X. oryzae pv. *oryzae* and *X. o.* pv. *oryzicola* (*X. campestris* pv. *oryzicola*) occur beneath glumes and rarely in the endosperm of rice kernels (Fang, Liu, and Chu, 1956; Srivastava and Rao, 1964; Shekhawat and Srivastava, 1972a,b; Mukerjee and Singh, 1983). While making detailed histological studies, Mukerjee and Singh (1983) found that the bacterial colonization was common in the innermost parenchyma layers and rarely in the sclerenchyma and xylem parenchyma of husk (lemma and palea). In the upper part of the peduncle, bacteria occurred in the cortical tissue, intercellular spaces, xylem vessels, and xylem parenchyma. Abundant bacteria were seen in the endosperm adjacent to the scutellum, but none was seen in the aleurone layer and testa. No bacterial cells occurred in the embryo, but after 48 hours of soaking, bacterial masses were seen in the space between the coleoptile and the first leaf.

6.2.2 PSEUDOMONAS, ACIDOVORAX, AND BURKHOLDERIA

The more important *Pseudomonas* species, for which evidence of natural seed infection is available, are listed in Table 6.1. The classicl example is bacterial halo blight of beans caused by *Pseudomonas savastanoi* pv. *phaseolicola*. The halo blight organism, after entering through raphe or the micropyle, is mainly located in the parenchyma layers of the seed coat. It may be present on the surface of the cotyledons and in severely infected seeds in the cells of the embryo (Zaumeyer, 1932; Taylor, Dudley, and Presley, 1979). Using hand-harvested pods with lesions, Grogan and Kimble (1967) found that 81% of pods had positive transmission from one or more seeds in seed transmission studies under controlled conditions. A total of 51% of the seeds from beneath or adjacent to a lesion transmitted the disease, whereas only 18% of the seeds situated at least one seed away from a lesion did so. Such seeds were examined under a dissecting microscope, and the authors concluded that the contaminating inoculum could be internal as well as external.

Skoric (1927) observed that *P. s.* pv. *pisi* enters pea seeds through the funiculus and micropyle (Figure 6.8A, B) and is usually distributed in the seed coat. The bacteria are inter- and intracellular in the parenchymatous cells and cause their lysis (Figure 6.8C). The bacterial cavities are small or large and full of slime and bacterial cells.

Wiles and Walker (1951) demonstrated histologically the presence of *P. s.* pv. *lachrymans* in the placental tissue and the funiculus in cucumber. Seeds from diseased fruits on germination produced seedlings with bacterial lesions on the surface of cotyledons. Nauman (1963) found *P. s.* pv. *lachrymans* in the embryo, confined to the outermost layers of the radicle. Kritzman and Zutra (1983) isolated *P. s.* pv. *lachrymans* from 60% of seeds of the infected cucumber fruits and concluded that 16% of seeds carried the pathogen internally. Fukuda, Azegami, and Tabei (1990) have reported that *P. s.* pv. *syringae*, which causes bacterial black node of barley and wheat, after invading the grains through the stomata in the lemma and palea then infects the caryopsis. The pathogen multiplies in the intercellular spaces of the parenchyma of the caryopsis and funiculus.

The fruit blotch pathogen of watermelon, *Acidovorax avenae* subsp. *citrulli* (*Pseudomonas pseudoalcaligenes* subsp. *citrulli),* was isolated from seed coats and

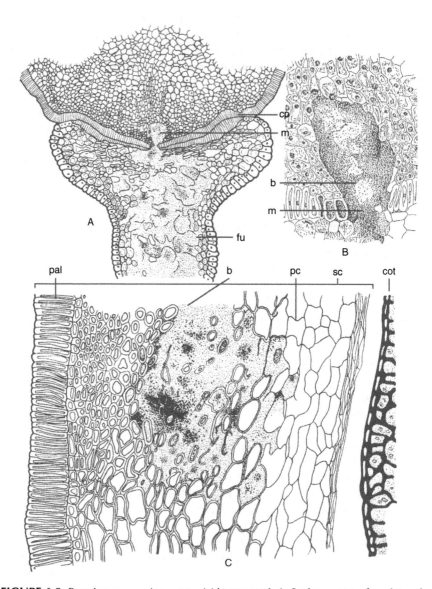

FIGURE 6.8 *Pseudomonas syringae* pv. *pisi* in pea seed. A, Ls lower part of seed together with the part of funiculus. Note bacterial mass in the funiculus and its entry through the micropyle. B, A part from A magnified showing bacteria in micropyle and the cavity in seed. C, Cross section of seed coat with bacteria in the parenchyma cells. (Abbreviations: b, bacteria; cot, cotyledon; cp, counter palisade; fu, funiculus; m, micropyle; pal, palisade layer; pc, parenchyma cells; sc, seed coat.) (From Skoric, V. 1927. *Phytopathology* 17: 611–627. With permission.)

embryos of naturally infected as well as artificially inoculated symptomatic fruits (Wall and Santos, 1988; Wall, 1989; Rane and Latin, 1992).

Burkholderia glumae, the causal bacterium of rice grain rot, enters lemma and palea through stomata, multiplies in intercellular spaces in the parenchyma, and reaches the surface of the endosperm and embryo. The bacterium never invaded the endosperm and embryo (Tabei et al., 1989). The sites of occurrence of *B. glumae* in naturally infected and artificially inoculated grains were the same.

6.2.3 RATHAYIBACTER AND CLAVIBACTER

Several phytopathogenic bacteria previously placed in the genus *Corynebacterium* have been transferred to the genera *Rathayibacter* and *Clavibacter* (Table 6.1). *Rathayibacter tritici*, which causes yellow slime disease or yellow ear rot in wheat, has drawn considerable attention (Cheo, 1946; Vasudeva and Hingorani, 1952; Sabet, 1954a,b; Gupta and Swarup, 1968). Neergaard (1979) mentioned that *Rathayibacter tritici* (*Conynebacterium tritici*) and *R. rathayi* (*Conynebacterium rathayi*) are probably vascular pathogens, but none of the histopathological investigations conform to this possibility. The bacterium is transmitted by *Anguina tritici*. Sabet (1954a) determined that infection takes place in the soil, and aerial infection does not occur under field conditions even in plants infested with the nematode. Using sterilized and unsterilized soil, sterilized and unsterilized nematode larvae, and nematode galls, Swarup and Singh (1962) observed that the bacterial disease symptoms appeared only when the nematode galls as such were used for inoculation. They concluded that possibly the bacterium was carried on the gall surface.

Sabet (1954a) found that *R. tritici* spreads on the surface of the affected plants and only in leaf sheath and terminal internode (ear-stalk); it may invade internally, occupying intercellular spaces. Early spike infection is from the inner surface of the sheath. Bacteria spread and fill in all gaps between the organs of spikelet, causing disintegration of tissues and results in yellow slime. If the bacterium is sparsely distributed on the leaf sheath, rachis, and parts of florets, slight or no disintegration of the internal tissues occurs. In such spikes either no grains or distorted ungerminable or germinable grains are formed. The germinable grains may give rise to either healthy or affected seedlings (Sabet, 1954a).

In plants raised from artificially inoculated seeds, Sabet (1954b) observed that the bacterium spread on the surface of affected plants and caused complete or partial inflorescence infection. In partially affected inflorescence, the stamens were affected, one or more of the anther's four pollen chambers compressed and tissues disintegrated, and slimy mass was seen on the surface. The bacterium either covered virtually the whole surface of the styles and ovary (carpel) or was restricted to certain areas on the ovary. The infection might reach the ovary cavity, and depending on the amount of slime, the developing ovule was affected. Severe infection prevented pollination and further development of grains. However, in less affected florets, grains may develop. The affected grains had mostly bacterial crust in and around the groove. Bacterial cells also occurred between the pericarp and seed coat, endosperm below the groove, and rarely in between the pericarp and embryo.

Rathayibacter iranicus produces profuse honey-yellow ooze in wheat spikelets in Iran (Scharif, 1961). The ooze develops from ovaries resulting in abortion of grains. No histopathological studies have been made on this pathogen, but its general behavior is comparable to that of *R. tritici.*

Rathayibacter rathayi is also transferred by a nematode, *Anguina agrostis,* and causes yellow slime disease or Rathay's disease in *Dactylis glomerata* and *Festuca rubra* var. *fallax* (Neergaard, 1979). In these hosts, the bacterium produces profuse yellow slime, which may cover the uppermost leaves, stem, and parts of inflorescence. Contaminated seeds have a varnish-like crust of bacteria.

Rathayibacter rathayi and *A. agrostis* complex cause gall formation in the place of seeds in rye grass (*Lolium rigidum*). The presence of nematodes is considered essential for the gall initiation, but mature galls are colonized predominantly by either nematodes or bacteria. The two types of galls are almost of the same shape and size, but differ in color and contents (Stynes et al., 1979; Bird, Stynes, and Thompson, 1980). The bacterial galls are bright yellow in comparison to dark brown for galls containing nematodes. The wall of bacterial galls is thin, and the cavity is packed with bacterial cells (Figure 6.9 A to C).

C. michiganensis subsp. *insidiosum,* which causes bacterial wilt of alfalfa, occurs in the vascular system of the flowering rachis (peduncle) and pedicel, and in vascular and parenchymatous tissues in seed pods. Masses of bacterium were present in the vicinity of aborted seeds. The bacterial cells were seen in the intercellular spaces below the malpighian cells in immature seeds, but these were rare in mature seeds (Cormack and Moffatt, 1961).

C. michiganensis subsp. *michiganensis,* causing bacterial canker of tomato, is a well-known vascular invader (Pine, Grogan, and Hewitt, 1955). Detailed histopathology of infected seed has not been worked out. However, the available information reveals that *C. m.* subsp. *michiganensis* sometimes forms large bacterial pockets in the chalazal region and can be seen in the inner cells of the seed coat. It does not invade the endosperm and embryo (Patino-Mendez, 1964). Similarly *C. m.* subsp. *nebraskensis,* causing Goss's bacterial wilt and leaf blight of corn, could be seen in the chalazal region, the area between the scutellum and endosperm, and in the vicinity of the embryo of the heavily infected seeds (Schuster, 1972). Biddle, McGee, and Braun (1990), after leaf inoculation of a susceptible corn inbred, detected the bacterium in seeds, ear shanks, and stalks. In seeds the pathogen occurred internally and externally.

McBeath and Adelman (1986) detected *C. m.* subsp. *tessellarius* in wheat using a scanning electron microscope. The clusters of bacterial cells were seen in the seed coat (pericarp), endosperm, and interface near the embryo in infected grains.

6.2.4 CURTOBACTERIUM

Curtobacterium flaccumfaciens pv. *flaccumfaciens,* causing bacterial wilt of beans, can be carried in bean seeds either externally or internally (Zaumeyer, 1932). According to Zaumeyer, *C. f.* pv. *flaccumfaciens* is primarily a vascular pathogen and tends to become systemic. The bacterium is located principally in the seed coat (Zaumeyer,

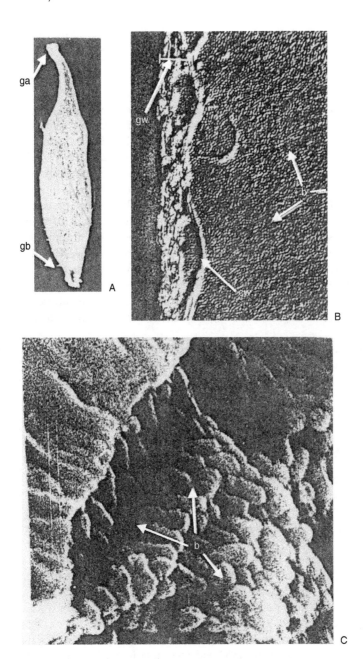

FIGURE 6.9 Seed gall of *Lolium rigidum* caused by *Rathayibacter rathayi*. A, Gall colonized by bacteria. B, Interference contrast photograph of a section through bacterial gall showing the wall and bacteria filling the cavity. C, A part from B magnified showing the closely packed bacteria. (Abbreviations: b, bacteria; cw, cell wall; ga, gall apex; gb, gall base; gw, gall wall.) (From Bird, A.F., Stynes, B.A., and Thomson, W.W. 1980. *Phytopathology* 70: 1104–1109. With permission.)

1932). Embryos in infected seeds are occasionally surrounded by bacterial masses (Schuster and Smith, 1983).

6.2.5 PANTOEA

The genus, *Erwinia*, has two major groups of plant pathogenic species, viz., the pectolytic bacteria that cause soft rot (often placed in *Pectobacterium*) and bacteria that cause necrosis and wilt have been reorganized into *Enterobacter, Erwinia,* and *Pantoea* Gavini. The two species *Enterobacter stewartii* and *E. herbicola*, known to cause seed infection in corn and rice, have been transferred to *Pantoea*.

Ivanoff (1933) has reported the location of *Pantoea stewartii* subsp. *stewartii* in the tissues of corn kernel. The bacterium occurs in the chalazal tissues, between the chalazal tissue and endosperm and in endosperm. In the chalazal region, it occurs in spiral vessels, sometimes completely plugging them, and after breaking through the vessel walls, forms cavities in the surrounding parenchyma. After disrupting the inner epidermis, the bacterium invades the space between the chalazal tissue and aleurone layer and the inner layers of the endosperm. Earlier Rand and Cash (1921) isolated *P. stewartii* subsp. *stewartii* from the endosperm of infected corn seeds. Pepper (1967) also concluded that *P. stewartii* subsp. *stewartii* is internally located in seed.

Pantoea agglomerans, a causal agent of bacterial palea browning of rice, infects the lemna and palea through the stomata (Tabei, Azegami, and Fukuda, 1988). It multiplies in intercellular spaces, and browning occurred only of the parenchyma toward the inner surface in the palea. The browning of palea is suspected to be a defense reaction.

6.3 SURVIVAL IN SEED

The seed-borne bacteria do not form spores. Some bacterial pathogens die before the seed loses its viability, while many others survive even beyond the time of seed germinability. Skoric (1927) found that *P. s.* pv. *pisi* survives in pea seed for at least 10 months. *X. o.* pv. *oryzae* in rice seeds is also short-lived, only 2 to 6 months (Kauffman and Reddy, 1975; Trimurthy and Devadath, 1984). *X. o.* pv. *oryzicola* and *X. a.* pv. *phaseoli* survive in rice and bean seeds, respectively, from one season to another (Zaumeyer, 1930; Shekhawat and Srivastava, 1972a,b). Zaumeyer and Thomas (1957) reported that *C. f.* pv. *flaccumfaciens* was viable after 5 to 24 years in bean seeds. Schuster and Sayre (1967) isolated virulent *C. f.* pv. *flaccumfaciens* var. *aurantiacum*, *X. a.* pv. *phaseoli* and its pigmented var. *fuscans* from 15-year-old bean seeds. Basu and Wallen (1966) found *X. a.* pv. *phaseoli* viable in bean seeds after 3 years at 20 to 35°C and nonviable in another sample of 2-year-old seeds, revealing the variability in its viability. *Acidovorax avenae* subsp. *avenae* was found to survive in rice seed lots stored for 8 years at 5°C (Shakya, Vinther, and Mathur, 1985).

Pantoea stewartii subsp. *stewartii* survived and was pathogenic when isolated from 1-year-old maize kernels. This bacterium has another very interesting mechanism of survival and transmission of the disease. *P. stewartii* subsp. *stewartii* survives in the intestinal tract of corn flea beetle (*Chaetocnema pullicaria*). Robert (1955) found that 10 to 70% of the beetles emerging from hibernation carried the bacterial wilt organism, and up to 75% of the beetles feeding on corn in midsummer may act as carriers.

Vegetative cells of bacteria are not subject to dormancy; still, many are tolerant to desiccation and survive relatively long periods under dry conditions as mentioned in the preceding paragraphs. Do these bacteria have any means of protection in the seed? They occur in varied conditions in seeds as contaminants of the seed surface in the form of loose cells or embedded in the bacterial slime or ooze, in seed tissues in intercellular spaces, including the vascular elements (xylem). They may occur in seed galls along with the nematode, e.g., *R. tritici* or exclusively, *R. rathayi*. *R. rathayi* cells are closely packed in a regular fashion along the wall of the gall and also within the gall (Bird, Stynes, and Thompson, 1980). Bird, Stynes, and Thompson (1980) regard this as an anhydrobiotic state, although they have not carried out studies on the survival of the bacterium. Chand (1967) observed that *R. tritici* survived in soil debris under laboratory conditions for about 7 months only, whereas Mathur and Ahmad (1964) had reported that the bacterium remained viable for at least 5 years in the gall of *A. tritici*.

Bacterial ooze or slime has been considered protective to bacterial cells (Rosen, 1929, 1938; Hildebrand, 1939). *Erwinia amylovora* is believed to be very susceptible to desiccation, but survives a long time in dry exudate (Rosen, 1929, 1938). Hildebrand (1939) recovered virulent bacterial cells of *E. amylovora* from dry exudate after 15 and 25 months, but the organism survived only 13 days in moist exudate. Leach et al. (1957) found an appreciable number of viable *X. a.* pv. *phaseoli* cells in exudate for as long as 1325 days under different conditions.

Bacteria present in seed tissues may have better chances of survival than those present on the seed surface. However, environmental conditions, inherent seed factors (structural features), and the inherent characteristics of the pathogen affect survival as well as transmission. Schuster and Coyne (1974) have reviewed the literature on this aspect. It may, however, be indicated that complete information on the above aspects is not available even for one disease.

6.4 CONCLUDING REMARKS

Good information is available on initial penetration, multiplication, and spread of bacteria in plant tissues and penetration of ovule and seed (Zaumeyer, 1930, 1932; Skoric, 1927; Cook, Larson, and Walker, 1952; Wiles and Walker, 1951; Pine, Grogan, and Hewitt, 1955; Getz, Fulbright, and Stephens, 1983; Mukerjee and Singh, 1983; Tabei, 1967; Tabei et al., 1988, 1989; Azegami, Tabei, and Fukuda, 1988; Fukuda, Azegami, and Tabei, 1990). There is, however, very little experimental evidence on penetration and association or establishment of bacterial infection in seeds during the postharvest period. The former may become systemic with long distance movement through tracheids and vessels or it may be localized, multiplying and spreading in the intercellular spaces. The course of short distance systemic infection as shown by Kritzman and Zutra (1983) needs support from histological studies.

Presence of *C. m.* subsp. *tessellarius* in sieve element of wheat needs confirmation because earlier observations by Smith (1914, 1920) on *C. m.* subsp. *michiganensis* that described the bacterium as primarily a phloem parasite have proved erroneous (Pine, Grogan, and Hewitt, 1955). In their excellent review on histopathology of plants infected with vascular bacterial pathogens, Nelson and Dickey (1970) have clearly shown that primary infection of tissues by such pathogens is in the xylem elements, and subsequently the infection may spread to surrounding tissues, mostly in the parenchyma cells.

Studies on the location of bacteria are meager and mostly inconclusive. Recent observations by Sharma, Agarwal, and Singh (1992) on seed infection of rapeseed and mustard and *X. c.* pv. *campestris* and by Sharma et al. (2001) on *X. a.* pv. *cajani* (*X. c.* pv. *cajani*) in pigeon pea seeds have clearly demonstrated that affected seeds vary in the degree of the severity of infection, and the location and effects on seed tissues are directly correlated with it. The spread of infection varies in asymptomatic and symptomatic seeds. Severely infected seeds carry infection in all seed tissues. Similar detailed accounts are needed for full insight into the location of bacteria in seeds in other cases.

REFERENCES

Agrios, G.N. 1988. *Plant Pathology,* 3rd ed. Academic Press, San Diego.
Ark, P.A. 1944. Pollen as a source of walnut bacterial blight infection. *Phytopathology* 34: 330–334.
Azegami, K., Tabei, H., and Fukuda, T. 1988. Entrance into rice grains of *Pseudomonas plantarii*, the causal agent of seedling blight of rice. *Ann. Phytopathol. Soc. Japan* 54: 633–636.
Basu, P.K. and Wallen, U.R. 1966. Influence of temperature on the viability, virulence, and physiologic characteristics of *Xanthomonas phaseoli*. *Can. J. Bot.* 44: 1239–1245.
Biddle, J.A., McGee, D.C., and Braun, E.J. 1990. Seed transmission of *Clavibacter nebraskense* in corn. *Plant Dis.* 74: 908–911.
Bird, A.F., Stynes, B.A., and Thomson, W.W. 1980. A comparison of nematode and bacteria-colonized galls induced by *Anguina agrostis* in *Lolium rigidum*. *Phytopathology* 70: 1104–1109.
Brinkerhoff, L.A. and Hunter, R.E. 1963. Internally infected seed as a source of inoculum for the primary cycle of bacterial blight of cotton. *Phytopathology* 53: 1397–1401.
Bryan, M.K. 1930. Studies on bacterial canker of tomato. *J. Agric. Res.* 41: 825–851.
Burkholder, W.H. 1921. The bacterial blight of the bean: a systemic disease. *Phytopathology* 11: 61–69.
Burkholder, W.H. 1930. The bacterial disease of the bean. A comparative study. *New York (Cornell) Agric. Exp. Sta. Mem.* 127, 93 pp.
Carne, W.M. 1926. Earcockle (*Tylenchus tritici*) and a bacterial disease (*Pseudomonas tritici*) of wheat. *J. Dept. Agric. Western Australia* 3 (Ser. 2): 508–512.
Chand, J.N. 1967. Longevity of *Corynebacterium tritici* (Hutchinson) Burk. causing 'tundu' disease of wheat in Haryana. *Sci. and Cult.* 33: 539.
Cheo, C.C. 1946. A note on the relation of nematodes (*Tylenchus tritici*) to the development of the bacterial disease of wheat caused by *Bacterium tritici*. *Ann. Appl. Biol.* 33: 446–449.

Cook, A.A., Larson, R.H., and Walker, J.C. 1952. Relation of the black rot pathogen to cabbage seed. *Phytopathology* 42: 316–320.

Cormack, M.W. and Moffatt, J.E. 1956. Occurrence of the bacterial wilt organism in alfalfa seed. *Phytopathology* 46: 407–409.

Crossan, D.F. and Morehart, A.L. 1964. Isolation of *Xanthomonas vasicatoria* from tissues of *Capsicum annuum. Phytopathology* 54: 356–359.

Crosse, J.E. and Pitcher, R.S. 1952. Studies in the relationship of eelworms and bacteria in certain plant diseases. II. The etiology of strawberry cauliflower disease. *Ann. Appl. Biol.* 39: 475–484.

Dicky, R.S. and Nelson, P.E. 1970. *Pseudomonas caryophylli* in carnation. IV. Unidentified bacteria isolated from carnation. *Phytopathology* 60: 647–653.

Du Plessis, H.J. 1990. Systemic invasion of plum seed and fruit by *Xanthomonas campestris* pv. *pruni* through stalks. *J. Phytopathol.* 13: 37–45.

Fang, C.T., Liu, C.F., and Chu, C.L. 1956. A preliminary study on the disease cycle of the bacterial leaf blight of rice. *Acta Phytopathol. Sinica* 2: 179–185.

Fukuda, T., Azegami, K., and Tabei, H. 1990. Histological studies on bacterial black node of barley and wheat caused by *Pseudomonas syringae* pv. *japonica. Ann. Phytopathol. Soc. Japan.* 56: 252–256.

Getz, S., Fulbright, D.W., and Stephens, C.T. 1983. Scanning electron microscopy of infection sites and lesion development on tomato fruit infected with *Pseudomonas syringae* pv. *tomato. Phytopathology* 73: 39–43.

Godika, S. 1995. Microorganisms of Sunflower Seeds and Their Pathological Effects. Ph.D. thesis, University of Rajasthan, Jaipur, India.

Grogan, R.G. and Kendrick, J.B. 1953. Seed transmission, mode of overwintering and spread of bacterial canker of tomato caused by *Corynebacterium michiganense. Phytopathology* 43: 473.

Grogan, R.G. and Kimble, K.A. 1967. The role of seed contamination in the transmission of *Pseudomonas phaseolicola* in *Phaseolus vulgaris. Phytopathology* 57: 36–42.

Gupta, P. and Swarup, G. 1968. On the ear-cockle and yellow ear rot diseases of wheat. I. Symptoms and histopathology. *Indian Phytopathol.* 21: 318–323.

Hildebrand, E.M. 1939. Studies on fire-blight ooze. *Phytopathology* 29: 142–155.

Hildebrand, E.M. and MacDaniels, L.H. 1935. Modes of entry of *Erwinia amylovora* into the flowers of principal pome fruits (Abst.). *Phytopathology* 25: 20.

Hunger, F.W.T. 1901. Een bacterie-ziekte den tomaat. *S. Lands Plantentuin Meded.* 48: 4–57.

Ivanoff, S.S. 1933. Stewart's wilt disease of corn, with emphasis on the life history of *Phytomonas stewarti* in relation to pathogenesis. *J. Agric. Res.* 47: 749–770.

Kauffman, H.E. and Reddy, A.P.K. 1975. Seed transmission studies of *Xanthomonas oryzae* in rice. *Phytopathology* 65: 663–666.

Kelman, A. 1953. The bacterial wilt caused by *Pseudomonas solanacearum. North Carolina Agric. Exp. Sta. Tech. Bull.* 99: 1–194.

Kelman, A. 1979. How bacteria induce disease. In *Plant Disease: An Advanced Treatise.* Horsfall, J.G. and Cowling, E.B., Eds. Academic Press, New York. Vol. 4, 181–202.

Kirk, T.K. and Connors, W.J. 1977. Advances in understanding the microbiological degradation of lignin. In *Recent Advances in Phytochemistry. The Structure, Biosynthesis and Degradation of Wood.* Loewius, F.A. and Runeckles, V.C., Eds. Plenum, New York. Vol. 2, 369–394.

Klement, Z., Rudolph, K., and Sands, D.C. 1990. *Methods in Phytobacteriology.* Akadémiai Kiadó, Budapest.

Kritzman, G. and Zutra, D. 1983. Systemic movement of *Pseudomonas syringae* pv. *lachrymans* in the stem, leaves, fruits and seeds in cucumber. *Can. J. Plant Pathol.* 5: 273–278.

Larson, R.H. 1944. The ring rot bacterium in relation to tomato and eggplant. *J. Agric. Res.* 69: 309–325.

Layne, R.E.C. 1967. Foliar trichomes and their importance as infection sites for *Corynebacterium michiganense* on tomato. *Phytopathology* 57: 981–985.

Leach, J.G., Lily, V.G., Wilson, H.A., and Purvis, M.R., Jr. 1957. Bacterial polysaccharides: the nature and function of the exudate produced by *Xanthomonas phaseoli*. *Phytopathology* 47: 113–120.

Liese, W. 1970. Ultrastructural aspects of woody tissue disintegration. *Ann. Rev. Phytopathol.* 8: 231–258.

Mathur, R.S. and Ahmad, Z.U. 1964. Longevity of *Corynebacterium tritici* causing tundu disease of wheat. *Proc. Natl. Acad. Sci. India Sect. B* 34: 335–336.

McBeath, J.H. and Adelman, M. 1986. Detection of *Corynebacterium michiganense* ssp. *tessellarius* in seeds of wheat plants (Abstr.). *Phytopathology* 76: 1099.

Meier, D. 1934. A cytological study of the early infection stages of black rot of cabbage. *Bull. Torrey Bot. Club* 61: 173–190.

Mihail, J.D., Taylor, S.J., and Versluses, P.E. 1993. Bacterial blight of *Cambe akyssimica* in Missouri caused by *Xanthomonas campestris*. *Plant Dis.* 77: 569–574.

Miller, J.H. and Harvey, H.W. 1932. Peanut wilt in Georgia. *Phytopathology* 22: 371–383.

Mukerjee, P. and Singh, R.A. 1983. Histopathological studies on infected rice seed with *Xanthomonas campestris* pv. *oryzae* and mode of its passage from seed to seedling. *Seed Res.* 11: 32–41.

Naumann, K. 1963. Über das Anftreten von Bacterien in Gurkensamen aus Fruchten, die durch *Pseudomonas lachrymans* infiziert waren. *Phytopathol. Z.* 48: 258–271.

Neergaard, P. 1979. *Seed Pathology.* Vols. 1 and 2. Macmillan Press, London.

Nelson, P.E. and Dickey, R.S. 1966. *Pseudomonas caryophilli* in carnation. II. Histopathological studies of infected plants. *Phytopathology* 56: 154–163.

Nelson, P.E. and Dickey, R.S. 1970. Histopathology of plants infected with vascular bacterial pathogens. *Ann. Rev. Phytopathol.* 8: 259–280.

Offutt, M.S. and Baldridge, J.D. 1956. Inoculation studies related to breeding for resistance to bacterial wilt in lespedeza. *Mo. Agr. Exp. Sta. Res. Bull.* 600: 1–47.

Parasher, R.D. and Leben, C. 1972. Detection of *Pseudomonas glycinea* in soybean seed lots. *Phytopathology* 62: 1075–1077.

Patino-Mendez, G. 1964. Studies on the pathogenicity of *Corynebacterium michiganense* (E.F. Sm.) Jensen and its transmission in tomato seed. Ph.D. thesis, University of California, Davis.

Pepper, E.H. 1967. Stewart's bacterial wilt of corn. *Am. Phytopathol. Soc. Monogr.* 4: 1–56.

Pierstorff, A.L. 1931. Studies on the fire blight organism, *Bacillus amylivora. N.Y. (Cornell) Agric. Exp. Stn. Mem.* 136: 1–53.

Pine, T.S., Grogan, R.G., and Hewitt, Wm. B. 1955. Pathological anatomy of bacterial canker of young tomato plants. *Phytopathology* 45: 267–271.

Pitcher, R.S. 1963. Role of plant-parasitic nematodes in bacterial diseases. *Phytopathology* 53: 33–39.

Pitcher, R.S. and Crosse, J.E. 1958. Studies in the relationship of eelworms and bacteria in certain plant diseases. III. Further analysis of the strawberry cauliflower disease complex. *Nematologica* 3: 244–256.

Ramos, L.J. and Valin, R.B. 1987. Role of stomatal opening and frequency of infection of *Lycopersicon* spp. by *Xanthomonas campestris* pv. *vesicatoria*. *Phytopathology* 77: 1311–1317.

Rand, F.V. and Cash, L.C. 1921. Stewart's disease of corn. *J. Agric. Res.* 21: 263–264.

Rane, K.K. and Latin, X.L. 1992. Bacterial fruit blotch of watermelon: association of the pathogen with seed. *Plant Dis.* 76: 509–512.

Robert, A.L. 1955. Bacterial wilt and Stewart's leaf blight of corn. *U.S. Dept. Agric. Farmers Bull.* 2092:1–13.

Rosen, H.R. 1929. The life history of the fire blight pathogen, *Bacillus amylovorus*, as related to the means of over-wintering and dissemination. *Ark. Agric. Exp. Sta. Bull.* 244: 1–96.

Rosen, H.R. 1935. The mode of penetration of pear and apple by the fire blight pathogen. *Science* 81: 26.

Rosen, H.R. 1938. Life span and morphology of the fire blight bacteria as influenced by relative humidity, temperature and nutrition. *J. Agric. Res.* 56: 239–258.

Rudolph, K. 1993. Infection of the plant by *Xanthomonas*. In *Xanthomonas*. Swings, J.G. and Civerolo, E.L., Eds. Chapman & Hall, London, 193–264.

Sabet, K. 1954a. On the sources and mode of infection with the yellow slime disease of wheat. *Bull. Fac. Agric. Cairo Univ.* 42: 15.

Sabet, K. 1954b. Pathological relationships between host and parasite in the yellow slime disease of wheat. *Bull. Fac. Agric. Cairo Univ.* 43: 10.

Scharif, G. 1961. *Corynebacterium iranicum* sp. nov. on wheat (*Triticum vulgare* L.) in Iran, and a comparative study of it with *C. tritici* and *C. rathayi*. *Ent. Phytopathol. Appl. Tehran* 19: 1–24.

Schuster, M.L. 1972. Leaf freckles and wilt, a new corn disease. In *Proc. Acad. Corn Sorghum Res. Conf.* 27. American Seed Trade Association, Washington, D.C., pp. 176–191.

Schuster, M.L. and Coyne, D.P. 1974. Survival mechanisms of phytopathogenic bacteria. *Ann. Rev. Phytopathol.* 12: 199–221.

Schuster, M.L. and Sayre, R.M. 1967. A coryneform bacterium induces purple-colored seed and leaf hypertrophy of *Phaseolus vulgaris* and other leguminosae. *Phytopathology* 57: 1064–1066.

Schuster, M.L. and Smith, C.C. 1983. Seed transmission and pathology of *Corynebacterium flaccumfaciens* in beans (*Phaseolus vulgaris*). *Seed Sci. Technol.* 11: 867–875.

Shakya, D.D., Vinther, P., and Mathur, S.B. 1985. World wide distribution of a bacterial stripe pathogen of rice identified as *Pseudomonas avenae*. *Phytopathol. Z.* 114: 256–259.

Sharma, J., Agarwal, K., and Singh, D. 1992. Detection of *Xanthomonas campestris* pv. *campestris* (Pammel) Dowson infection in rape and mustard seeds. *Seed Res.* 20: 128–133.

Sharma, M. 1996. Studies on Seed-Borne Mycoflora of Pigeonpea Grown in Rajasthan. Ph.D. thesis, University of Rajasthan, Jaipur, India.

Sharma, M., Kumar, D., Agarwal, K., Singh, T., and Singh, D. 2001. Colonization of pigeonpea seed by *Xanthomonas campestris* pv. *cajani*. *J. Mycol. Plant Pathol.* 31: 216–219.

Shekhawat, G.S. and Srivastava, D.N. 1972a. Mode of seed infection and transmission of bacterial leaf of rice. *Ann. Phytopathol. Soc. Japan* 38: 4–6.

Shekhawat, G.S. and Srivastava, D.N. 1972b. Epidemiology of bacterial leaf streak of rice. *Ann. Phytopathol. Soc. Japan* 38: 7–14.

Shekhawat, G.S., Srivastava, D.N., and Rao, Y.P. 1969. Seed infections and transmission of bacterial leaf streak of rice. *Plant Dis. Rep.* 53: 115–116.

Skoric, V. 1927. Bacterial blight of pea: overwintering, dissemination and pathological histology. *Phytopathology* 17: 611–627.

Smith, E.F. 1911. Wilt of cucurbits. In *Bacteria in Relation to Plant Diseases*. Carnegie Institution, Washington, D.C. Vol. 2, pp. 209–299.

Smith, E.F. 1914. The Grand Rapids tomato disease. In *Bacteria in Relation to Plant Diseases*. Carnegie Institution, Washington, D.C. Vol. 3, pp. 161–165.

Smith, E.F. 1920. Bacterial canker of tomato. In *Introduction to Bacterial Diseases of Plants*. Saunders, Philadelphia, pp. 202–222.

Smith, C.O. 1922. Some studies relating to infection and resistance to walnut blight, *Pseudomonas juglandis*. *Phytopathology* 12: 106.

Srivastava, D.N. and Rao, Y.P. 1964. Seed transmission and epidemiology of the bacterial blight disease of rice in Northern India. *Indian Phytopathol.* 17: 77–78.

Stynes, B.A., Petterson, D.S., Lloyd, J., Payne, A.L., and Laningan, G.W. 1979. The production of a toxin in annual rye grass, *Lolium rigidum*, infected with a nematode, *Anguina* sp. and *Corynebacterium rathayi*. *Aust. J. Agric. Res.* 30: 201–209.

Swarup, G. and Singh, N.J. 1962. A note on the nematode — bacterium complex in tundu disease of wheat. *Indian Phytopathol.* 15: 294–295.

Tabei, H. 1967. Anatomical studies of rice plants affected with bacterial leaf blight with special reference to stomatal infection at the coleoptile and the foliage leaf sheath of rice seedling. *Ann. Phytopathol. Soc. Japan* 33: 12–16.

Tabei, H., Azegami, K., and Fukuda, T. 1988. Infection site of rice grain with *Erwinia herbicola*, the causal agent of bacterial palea browning of rice. *Ann. Phytopathol. Soc. Japan* 54: 637–639.

Tabei, H., Azegami, K., Fukuda, T., and Goto, T. 1989. Stomatal infection of rice grains with *Pseudomonas glumae*, the causal agent the bacterial grain rot of rice. *Ann. Phytopathol. Soc. Japan* 55: 224–228.

Taylor, J.D., Dudley, C.L., and Presley, L. 1979. Studies of halo-blight seed infection and disease transmission. *Ann. Appl. Biol.* 93: 267–277.

Tennyson, G. 1936. Invasion of cotton seed by *Bacterium malvacearum*. *Phytopathology* 26: 1083–1085.

Thiers, H.D. and Blank, L.M. 1951. A histological study of bacterial blight of cotton. *Phytopathology* 41: 499–510.

Thomson, S.V. 1986. The role of stigma in fire blight infections. *Phytopathology* 76: 476–482.

Trimurthy, V.S. and Devadath, S. 1984. Role of seed in survival and transmission of *Xanthomonas campestris* pv. *oryzae*. causing bacterial leaf blight of rice. *Phytopathol. Z.* 110: 15–19.

Vakili, N.G. 1967. Importance of wound in bacterial spot (*Xanthomonas vesicatoria*) of tomatoes in the field. *Phytopathology* 57: 1099–1103.

Vasudeva, R.S. and Hingorani, M.K. 1952. Bacterial disease of wheat caused by *Corynebacterium tritici* (Hutchinson) Bergey et al. *Phytopathology* 42: 291–293.

Wall, G.C. 1989. Control of watermelon fruit blotch by seed heat treatment. (Abstr.) *Phytopathology* 769: 1191.

Wall, G.C. and Santos, V.M. 1988. A new bacterial disease on watermelon in the Mariana Islands (Abstr.) *Phytopathology* 78: 1605.

Walker, J.C. 1950. The mode of seed infection by the cabbage black-rot organism. *Phytopathology* 40: 30.

Weller, D.M. and Saettler, A.W. 1980. Evaluation of seedborne *Xanthomonas phaseoli* and *X. phaseoli* var. *fuscans* as primary inocula bean blights. *Phytopathology* 70: 148–152.

Wiles, A.B. and Walker, J.C. 1951. The relation of *Pseudomonas lachymans* to cucumber fruits and seeds. *Phytopathology* 41: 1059–1064.

Wolf, E.T. and Nelson, P.E. 1969. An anatomical study of carnation stems infected with the carnation strain of *Erwinia chrysanthemi*. *Phytopathology* 59: 1802–1808.

Young, J.M., Saddler, G.S., Takikawa, Y., De Boer, S.H., Vauterin, L., Garden, L., Gvozdyak, R.I., and Stead, D.E. 1996. Names of plant pathogenic bacteria 1864–1995. *Rev. Plant Pathol.* 75: 721–763.

Yu, T.F. 1933. Pathological and physiological effects of *Bacillus tracheiphilus* E.F. Smith on species of Cucurbitaceae. *Nanking Univ. Coll. Agric. Forest Bull.* 5: 1–72.

Zaumeyer, W.J. 1930. The bacterial blight of beans caused by *Bacterium phaseoli. U.S. Dept. Agric. Bull.* 180: 1–36.

Zaumeyer, W.J. 1932. Comparative pathological histology of three bacterial diseases of bean. *J. Agric. Res.* 44: 605–632.

Zaumeyer, W.J. and Thomas, H.R. 1957. A monographic study of bean diseases and methods for their control. *U.S. Dept. Agric. Tech. Bull.* 868 (rev. ed.): 1–255.

7 Seed Infection by Viruses

Viruses are submicroscopic entities composed of a nucleic acid, ribonucleic acid (RNA), or deoxyribonucleic acid (DNA), surrounded by a protective protein or lipoprotein coat. They are acellular, but each virus has a characteristic shape, which may be a rigid rod, flexuous thread, spherical (polyhedral), or bacilliform (bullet-shaped). One fourth of all known viruses cause diseases in plants. Plant viruses number about 2000, and new viruses are described regularly (Agrios, 1988). Approximately 18 to 20% of the described plant viruses are seed-transmitted (Mathews, 1991; Johansen, Edwards, and Hampton, 1994). Stace-Smith and Hamilton (1988) believe that one third of plant viruses will eventually prove to be seed-borne.

Mathews (1991), Mink (1993), and Agarwal and Sinclair (1997) consider that most seed-transmitted viruses are carried within the embryo. The seed-borne and seed-transmitted viruses are listed in several publications (Bennett, 1969; Phatak, 1974; Bos, 1977, Mink, 1993; Shukla, Ward, and Bunk, 1994; Agarwal and Sinclair, 1997). Table 7.1 provides a list of seed-borne and seed-transmitted viruses, excluding cryptic viruses, using the nomemclature of the updated International Committee on Taxonomy of Viruses (ICTV) lists of 1995 and 1999. The cryptic viruses are listed separately in Table 7.2. The occurrence of viruses in seed is shown in seed transmission, infectivity tests, and serological tests using whole seed or seed components. Histopathological investigations using TEM are few, but in recent years enzyme-linked immunosorbent assay (ELISA), immunosorbent electron microscopy (ISEM), and molecular techniques have been used more frequently. Viroids, composed of naked single-stranded, low-molecular-weight, and circular RNA, were initially treated under viruses. They are also seed-borne and seed-transmitted. Table 7.3 lists the most common seed-borne viroids. Histopathological information on seed infections by viroids is lacking at present.

7.1 INFECTION AND MULTIPLICATION

Viruses are unique as they multiply intracellularly. They enter cells through wounds made mechanically, by vectors (insects, aphids, thrips, whiteflies, leafhoppers, beetles, mites, nematodes, and fungi), systemically through seeds, and by infected pollen grains. Ectodesmata have also been implicated as possible routes of infection for plant viruses (Brants, 1964; Thomas and Fulton, 1968). Outside the cell, viruses do not divide and are not known to produce any specialized reproductive structures. The viral parasitism is at the genetic level, using the internal cellular machinery during its replication. They multiply by inducing the host cell to form more virus particles, and in doing so, they disturb the normal cellular processes, upset the cell metabolism, and prove injurious to functions and life of the host cell. Since viruses

TABLE 7.1
Seed-Borne Viruses Excluding Crypto-Viruses (Nomenclature as in the Reports of the International Committee on Taxonomy of Viruses)

Virus Name	Acronym	Genus	Important Host (Genus)
Alfalfa mosaic virus	AMV	*Alfamovirus*	*Medicago, Melilotus, Glycine, Nicandra, Solanum, Capsicum*
Apple stem grooving virus	ASGV	*Capillovirus*	*Malus*
Arabis mosaic virus	ArMV	*Nepovirus*	*Chenopodium, Capsella, Beta, Glycine*
Arracacha virus B	AVB	*Nepovirus*	*Solanum*
Artichoke yellow ringspot virus	AYRSV	*Nepovirus*	*Cyanara*
Asparagus virus 2	AV-2	*Ilarvirus*	*Asparagus*
Barley stripe mosaic virus	BSMV	*Hordeivirus*	*Hordeum, Triticum, Avena several other grains*
Bean common mosaic necrosis virus (serotype A of BCMV)	BCMNV	*Potyvirus*	*Phaseolus*
Bean common mosaic virus	BCMV	*Potyvirus*	*Phaseolus, Vigna*
Bean pod mottle virus	BPMV	*Comovirus*	*Glycine*
Bean yellow mosaic virus	BYMV	*Potyvirus*	*Lupinus, Vicia, Pisum, Melilotus*
Blackgram mottle virus	BMoV	*Carmovirus*	*Vigna*
Blueberry leaf mottle virus	BLMV	*Nepovirus*	*Chenopodium, Vitis*
Broad bean mottle virus	BBMV	*Bromovirus*	*Vicia, Cicer, Vigna, Phaseolus, Pisum*
Broad bean stain virus	BBSV	*Comovirus*	*Vicia, Vigna, Pisum, Lens*
Broad bean true mosaic virus	BBTMV	*Comovirus*	*Vicia*
Broad bean wilt virus 1	BBWV-1	*Fabavirus*	*Vicia*
Broad bean wilt virus 2	BBWV-2	*Fabavirus*	*Vicia*
Brome mosaic virus	BMV	*Bromovirus*	*Triticum*
Cassia yellow spot virus	CasYSV	*Potyvirus*	*Cassia*
Cherry leaf roll virus	CLRV	*Nepovirus*	*Prunus, Rhus, Sambucus, Juglans, Glycine, Phaseolus*
Cherry rasp leaf virus	CRLV	*Nepovirus*	*Prunus, Chenopodium, Taraxacum*
Chicory yellow mottle virus	ChYMV	*Nepovirus*	*Cichorium*
Clover yellow mosaic virus	ClYMV	*Potexvirus*	*Trifolium*
Cocoa mosaic virus	CoMV	*Nepovirus*	*Glycine, Phaseolus*
Cowpea aphid-borne mosaic virus (South African Passiflora virus)	CABMV	*Potyvirus*	*Phaseolus, Vigna*
Cowpea green vein banding virus	CGVBV	*Potyvirus*	*Vigna*
Cowpea mild mottle virus	CPMMV	*Carlavirus*	*Vigna, Glycine, Phaseolus*
Cowpea mosaic virus	CPMV	*Comovirus*	*Vigna*
Cowpea mottle virus	CPMoV	*Carmovirus*	*Vigna, Phaseolus*

(continued)

TABLE 7.1 (CONTINUED)
Seed-Borne Viruses Excluding Crypto-Viruses (Nomenclature as in the Reports of the International Committee on Taxonomy of Viruses)

Virus Name	Acronym	Genus	Important Host (Genus)
Cowpea severe mosaic virus	CPSMV	*Comovirus*	*Vigna*
Crimson clover latent virus	CCLV	*Nepovirus*	*Trifolium*
Cucumber green mottle mosaic virus	CGMMV	*Tobamovirus*	*Cucumis, Citrullus, Lagenaris*
Cucumber mosaic virus	CMV	*Cucumovirus*	*Cucumis, Cucurbita, Luffa, Echinocystis, Arachis, Glycine, Lupinus, Phaseolus, Lycopersicon, Solanum*
Cycas necrotic stunt virus	CNSV	*Nepovirus*	*Cycas*
Desmodium mosaic virus	DesMV	*Potyvirus*	*Desmodium*
Dulcamara mottle virus	DuMV		*Solanum*
Eggplant mosaic virus (Andean potato latent virus)	EMV	*Tymovirus*	*Solanum, Nicotiana, Petunia*
Elm mottle virus	EMoV	*Ilarvirus*	*Ulmus*
Grapevine Bulgarian latent virus	GBLV	*Nepovirus*	*Vitis, Chenopodium*
Grapevine fanleaf virus	GFLV	*Nepovirus*	*Vitis, Chenopodium*
Guar symptomless virus	GSLV	*Potyvirus*	*Cyamopsis*
Hippeastrum mosaic virus	HiMV	*Potyvirus*	*Hippeastrum*
Hop mosaic virus	HpMV	*Carlavirus*	*Humulus*
Humulus japonicus latent virus	HJLV	*Ilarvirus*	*Humulus*
Hydrangea mosaic virus	HdMV	*Ilarvirus*	*Chenopodium*
Indian peanut clump virus	IPCV	*Pecluvirus*	*Arachis*
Lettuce mosaic virus	LMV	*Potyvirus*	*Lactuca, Senecio*
Lucerne Australian latent virus	LALV	*Nepovirus*	*Medicago, Chenopodium*
Lychnis ringspot virus	LRSV	*Hordeivirus*	*Lychnis, Beta, Capsella, Silene, Stellaria*
Maize dwarf mosaic virus	MDMV	*Potyvirus*	*Zea*
Melon necrotic spot virus	MNSV	*Carmovirus*	*Cucumis*
Mulberry ringspot virus	MRSV	*Nepovirus*	*Glycine*
Mungbean mosaic virus	MbMV	*Potyvirus*	*Vigna*
Pea early-browning virus	PEBV	*Tobravirus*	*Pisum, Vicia*
Pea enation mosaic virus	PEMV	*Enamovirus*	*Pisum, Lathyrus*
Pea mild mosaic virus	PMiMV	*Comovirus*	*Pisum*
Pea seed-borne mosaic virus	PSbMV	*Potyvirus*	*Pisum, Lathyrus, Lens, Vicia, Vigna*
Peach rosette mosaic virus	PRMV	*Nepovirus*	*Chenopodium, Taraxacum, Vitis*
Peanut clump virus	PCV	*Pecluvirus*	*Arachis, Setaria*
Peanut mottle virus	PeMoV	*Potyvirus*	*Arachis, Glycine, Vigna, Voandzeia*
Peanut stunt virus	PSV	*Cucumovirus*	*Arachis, Glycine*
Pepper mild mottle virus	PMMoV	*Tobamovirus*	*Capsicum*

(continued)

TABLE 7.1 (CONTINUED)
Seed-Borne Viruses Excluding Crypto-Viruses (Nomenclature as in the Reports of the International Committee on Taxonomy of Viruses)

Virus Name	Acronym	Genus	Important Host (Genus)
Plum pox virus	PPV	*Potyvirus*	*Prunus*
Potato virus T	PVT	*Trichovirus*	*Solanum, Datura, Nicandra*
Potato virus U	PVU	*Nepovirus*	*Nicotiana, Chenopodium*
Prune dwarf virus	PDV	*Ilarvirus*	*Prunus*
Prunus necrotic ringspot virus	PNRSV	*Ilarvirus*	*Prunus, Cucurbita*
Raspberry bushy dwarf virus	RBDV	*Idaeovirus*	*Fragaria, Malus, Chenopodium, Rubus*
Raspberry ringspot virus	RpRSV	*Nepovirus*	*Fragaria, Rubus, Beta, Glycine, Petunia, Stellaria*
Red clover mottle virus	RCMV	*Comovirus*	*Trifolium*
Red clover vein mosaic virus	RCVMV	*Carlavirus*	*Trifolium, Vicia*
Rubus Chinese seed-borne virus	RCSV	*Nepovirus*	*Rubus*
Satsuma dwarf virus	SDV	*Nepovirus*	*Phaseolus*
Southern bean mosaic virus	SBMV	*Sobemovirus*	*Phaseolus*
Southern cowpea mosaic virus	SCPMV	*Sobemovirus*	*Vigna*
Sowbane mosaic virus	SoMV	*Sobemovirus*	*Atriplex, Chenopodium*
Soybean mosaic virus	SMV	*Potyvirus*	*Glycine, Lupinus, Phaseolus*
Spinach latent virus	SpLV	*Ilarvirus*	*Spinacia, Chenopodium, Celosia, Nicotiana*
Squash mosaic virus	SqMV	*Comovirus*	*Citrullus, Cucumis, Cucurbita*
Strawberry latent ringspot virus	SLRSV	*Nepovirus*	*Rubus, Amaranthus, Apium, Pastinacia, Petroselinum, Lamium, Mentha, Solanum, Stellaria, Senecio*
Subterranean clover mottle virus	SCMoV	*Sobemovirus*	*Trifolium*
Sugarcane mosaic virus	SCMV	*Potyvirus*	*Zea*
Sunflower mosaic virus	SuMV	*Potyvirus*	*Helianthus*
Sunn-hemp mosaic virus	SHMV	*Tobamovirus*	*Crotalaria*
Telfairia mosaic virus	TeMV	*Potyvirus*	*Telfairia*
Tobacco mosaic virus	TMV	*Tobamovirus*	*Lycopersicon, Capsicum, Malus, Prunus*
Tobacco rattle virus	TRV	*Tobravirus*	*Capsella, Lamium, Myosutus, Papaver, Solanum, Nicandra, Petunia*
Tobacco ringspot virus	TRSV	*Nepovirus*	*Nicotiana, Cicer, Glycine, Vigna, Lactuca*
Tobacco streak virus	TSV	*Ilarvirus*	*Nicotiana, Nicandra, Lycopersicon, Datura, Glycine, Cicer, Vigna, Phaseolus, Raphanus, Chenopodium, Asparagus*

(continued)

TABLE 7.1 (CONTINUED)
Seed-Borne Viruses Excluding Crypto-Viruses (Nomenclature as in the Reports of the International Committee on Taxonomy of Viruses)

Virus Name	Acronym	Genus	Important Host (Genus)
Tomato aspermy virus	TAV	*Cucumovirus*	*Stellaria*
Tomato black ring virus	TBRV	*Nepovirus*	*Lycopersicon, Datura, Nicotiana, Rubus, Cajanus, Glycine, Phaseolus, Vigna, Capsella*
Tomato bushy stunt virus	TBSV	*Tombusvirus*	*Malus*
Tomato mosaic virus	ToMV	*Tobamovirus*	*Lycopersicon, Physalis*
Tomato ringspot virus	ToRSV	*Nepovirus*	*Lycopersicon, Nicotiana, Rubus, Fragaria, Glycine*
Turnip yellow mosaic virus	TYMV	*Tymovirus*	*Brassica, Camelina, Alliaria*
Urdbean leaf crinkle virus	UBLCV		*Vigna, Vicia*
Wheat streak mosaic virus	WSMV	*Tritimovirus*	*Triticum*
White clover mosaic virus	WClMV	*Potexvirus*	*Trifolium*
Zucchini yellow mosaic virus	ZYMV	*Potyvirus*	*Cucurbita, Ranunculus*

do not show the phenomenon of growth and are also not motile, they depend on the transport systems available in plants for their movement from the site of entry to other tissues. But if this is to accompany the continued multiplication, it must take place mainly through cell-to-cell movement and vascular elements, particularly phloem sieve tubes. Interconnected air spaces and xylem vessels or tracheids may not support multiplication of viruses.

Experimental evidence that viruses can move over long distances in the phloem have been provided by (1) killing or ringing a section of the stem, preventing or substantially delaying the movement (Helms and Wardlaw, 1976); (2) presence of virus particles in sieve elements (Esau, 1967; Esau, Cronshaw, and Hoefert, 1967; Halk and McGuire, 1973); and (3) inoculation of viruses into young leaves, revealing that the first mesophyll cells to show signs of infection are always next to the phloem elements. Once the virus enters the phloem, movement is very rapid (Bennett, 1940; Helms and Wardlaw, 1976). The rapid rates in phloem cells may be due to the very directional nature of protoplasmic streaming in these cells. Phloem-limited viruses that are transmitted by insect vectors through injection directly into sieve elements (SE) must get out of SE into companion cells and/or phloem parenchyma, with which sieve elements have plasmodesmatal connections, for replication (Lucas and Gilbertson, 1994).

The virus particles have also been observed in young xylem cells in leaf veins and they probably move in xylem (Schneider and Worley, 1959). Carroll and Mayhew (1976a) observed barley stripe mosaic virus (BSMV) particles in phloem as well as xylem in the vascular supply of the anther in barley.

TABLE 7.2
Seed-Borne Cryptic Viruses (Nomenclature as in the Reports of the International Committee on Taxonomy of Viruses)

Virus Name	Acronym	Genus	Host Genus
Alfalfa cryptic virus 1	ACV-1	*Alphacryptovirus*	*Medicago*
Alfalfa cryptic virus 2	ACV-2	*Betacryptovirus*	*Medicago*
Beet cryptic virus 1	BCV-1	*Alphacryptovirus*	*Beta*
Beet cryptic virus 2	BCV-2	*Alphacryptovirus*	*Beta*
Beet cryptic virus 3	BCV-3	*Alphacryptovirus*	*Beta*
Carnation cryptic virus 1	CCV-1	*Alphacryptovirus*	*Dianthus*
Carnation cryptic virus 2	CCV-2	*Alphacryptovirus*	*Dianthus*
Cucumber cryptic virus	CuCV	*Alphacryptovirus*	*Cucumis*
Hop trefoil cryptic virus 1	HTCV-1	*Alphacryptovirus*	*Humulus*
Hop trefoil cryptic virus 2	HTCV-2	*Betacryptovirus*	*Humulus*
Hop trefoil cryptic virus 3	HTCV-3	*Alphacryptovirus*	*Humulus*
Poinsettia cryptic virus	PnCV	*Alphacryptovirus*	*Poinsettia*
Red clover cryptic virus 2	RCCV-2	*Betacryptovirus*	*Trifolium*
Red pepper cryptic virus 1	RPCV-1	*Alphacryptovirus*	*Trifolium*
Red pepper cryptic virus 2	RPCV-2	*Alphacryptovirus*	*Trifolium*
Ryegrass cryptic virus	RGCV	*Alphacryptovirus*	*Lolium*
Vicia cryptic virus	VCV	*Alphacryptovirus*	*Vicia*
White clover crytpic virus 1	WCCV-1	*Alphacryptovirus*	*Trifolium*
White clover cryptic virus 2	WCCV-2	*Betacryptovirus*	*Trifolium*
White clover cryptic virus 3	WCCV-3	*Betacryptovirus*	*Trifolium*

TABLE 7.3
Seed-Borne Viroids (Nomenclature as in Updated ICTV Reports)

Viroid	Acronym	Genus	Host
Apple scar skin	ASS Vd	*Apseaviroid*	*Malus*
Avocado sunblotch	ASBVd	*Avsunviroid*	*Persea*
Chrysanthemum stunt	CSVd	*Pospiviroid*	*Chrysanthemum, Lycopersicon*
Citrus exocortis	CEVd	*Pospiviroid*	*Citrus, Lycopersicon*
Coconut cadang-cadang	CCC Vd	*Cocadviroid*	*Cocos*
Coleus blumei	CbVd	*Coleviroid*	*Coleus*
Hop stunt	HSVd	*Hostuviroid*	*Lycopersicon*
Potato spindle tuber	PST Vd	*Pospiviroid*	*Lycopersicon, Scopolia, Solanum*

The tracheary elements have half-bordered or simple pit pairs with the contiguous parenchyma cells (Esau, 1953; Fahn, 1974). However, Maule (1991) is of the opinion that virus movement within the vascular tissues may be a passive process requiring neither virus replication nor gene expression. At present there is no adequate evidence that xylem plays any important role in systemic movement of viruses.

Unlike conventional viruses, not all cryptoviruses are transmitted through vectors, grafting, or mechanical processes. They seem unable to move from cell to cell. Propagation and distribution of cryptic viruses occur only through cell multiplication (Boccardo et al., 1987; Mink, 1993). Pollination experiments among carrier and noncarrier parents have shown that cryptic viruses are transmitted through pollen or ovule or both. When both parents are carriers, all progeny carry the virus. When both parents are free of the virus, the progeny remains free.

7.2 CELLULAR CONTACTS, ISOLATION, AND TRANSPORT SYSTEMS IN OVULE AND SEED

The recent information, based on ultrastructure, has yielded valuable details about cell contacts and barriers in the ovule and developing seed. A brief summary is given because this situation has been largely misunderstood, resulting in several erroneous conclusions. Neergaard (1979) has written that "The virus may not invade the ovule or embryo. The ovule is isolated by lack of plasmodesmatal connections with the surrounding tissues...." Mathews (1991) considers that a few viruses that are confined to vascular elements may be unable to enter the ovule, which has no vascular connection with the parent.

The ovule, progenitor of seed, is connected through the funiculus to the ovary wall at the placenta. Each ovule receives a vascular supply originating from the ventral carpellary trace which, in turn, is in continuation of the pedicel and stem vascular supply. Initially, the ovular supply is of procambial cells, but the xylem and phloem elements differentiate during later stages. The cells of the integument, nucellus, and chalaza in the ovule have plasmodesmatal connections. However, the cuticles of integument and that of the nucellus act as barriers between them. The nucellus is the site of differentiation, and development of the female gametophyte and cell contacts and separation in these tissues are as follows:

1. Megasporocyte, functional megaspore, two- and four-nucleate embryo sacs have plasmodesmatal connections with the surrounding nucellar cells.
2. By the time the mature embryo sac is formed, the plasmodesmatal connections with the surrounding cells are severed.
3. In mature female gametophyte, the synergids and egg have no cell wall in the chalazal one third part, which is delimited by plasmalemma. The central cell is also without the cell wall adjacent to the egg apparatus.
4. After fertilization, the zygote develops the cell wall all around.
5. The proembryo has plasmodesmatal connections among its cells, but it is isolated (no plasmodesmata) from the surrounding endosperm.
6. Wall ingrowths occur in the basal cells of the proembryo, particularly those of suspensor, and are prominent at the globular or precordate stage of the embryo.
7. The embryo proper in early stages is covered by the cuticle, which is absent over the suspensor cells.

8. At the cordate stage of embryo, the suspensor usually shows weakening and obliteration, and simultaneously the cuticle around the embryo proper disappears.

9. The developing endosperm also develops wall ingrowths along the embryo sac boundary as seen in endosperm of *Vicia faba* (Johansson and Walles, 1993).

10. In ovule and developing seed, the sieve elements of ovular supply have numerous plasmodesmatal connections with companion cells, and also some with the parenchyma cells. The companion cells in turn are connected through plasmodesmata to the parenchyma cells (Thorne, 1980, 1981; Offler and Patrick, 1984; Offler, Nerlich, and Patrich, 1989; Singh, 1998).

The above information shows that there are two kinds of transport systems operating in ovule and seed: (1) cells with plasmodesmatal connections have symplastic transport, and (2) those with wall ingrowths (transfer cell structure) carry on apoplastic transport. Johansson and Walles (1993) have concluded that cells with wall ingrowths are common at sites at the junction of different generations (old maternal and the new sporophyte) in ovules.

7.3 VIRUS MOVEMENT

7.3.1 INFECTED PLANT

The information on the movement of viruses in vegetative parts, particularly leaf and stem, of the infected plant has been summarized in several excellent review articles (Broadbent, 1976; Maule, 1991; Mink, 1993; Lucas and Gilbertson, 1994; Johansen, Edwards, and Hampton, 1994). A brief summary of various aspects related to virus movement in these parts is given with the view that similar systems probably operate in the ovule and seed.

1. The conventional viruses that are seed-borne and seed-transmitted, after entering the leaf epidermis or mesophyll cells through wounds or by vectors, multiply and spread to neighboring cells through cell-to-cell movement. When this infection comes in contact with the vascular tissues, it enters the phloem sieve elements. In the sieve tubes, the virus moves over long distances. Long-distance spread through the phloem is also important for phloem-limited viruses.

2. Virus particles have been seen in plasmodesmatal channels and sieve pores (Esau, Cronshaw, and Hoefert, 1967; Kitajima and Lauritus, 1969). Plasmodesmata vary considerably in diameter, 20 to 200 mm, in young tobacco leaves. The properties of plasmodesmata may be altered as a result of virus infection. The altered plasmodesmata have consistently larger openings of uniform diameter throughout their length (Kitajima and Lauritus, 1969).

3. The movement of virus through plasmodesmata may take place on account
    of the capacity of the virus to alter the structure of plasmodesmata or may
    be facilitated by the *spread* and *movement proteins* coded by the virus
    (Meshi et al., 1987; Wolf et al., 1989).
4. Long-distance movement of virus occurs through conducting tissues, par-
    ticularly the phloem. The existence of functional plasmodesmata between
    mesophyll cells and phloem has been observed, providing a symplastic
    pathway for the movement of viruses. The open ends of the vascular
    elements in the leaf mesophyll may also provide sites for the phloem
    loading with virus. The virus moves rapidly over long distances as evident
    from the classic experiments of Samuel (1934) on tobacco mosaic virus
    (TMV) in tomato. Samuel inoculated one terminal leaflet and then fol-
    lowed the spread of the virus in the infected plant with the help of
    infectivity tests and found a systematic spread of the pathogen downward
    as well as upward. Schippers (1963) examined the spread of BCMV in
    bean plants. He inoculated the middle leaflet of the first compound leaf
    on the main stem and thereafter determined its presence in flower buds
    and ovaries at different nodes (Figure 7.1). Five days after inoculation,
    the virus was detected for the first time in buds and ovaries from nodes
    four and five and from axillary shoots at nodes one and three. After 6
    days of inoculation, the virus also could be detected in floral buds and
    ovaries at nodes zero and two (Figure 7.1). This showed upward as well
    as downward transport of virus in the plant.
5. An important question is the form in which the virus moves from cell to
    cell systematically in the plant. Three options are possible: (1) as virus
    particle; (2) as virus nucleic acid; and (3) as virus-specific nucleoprotein.
    It is also possible that the virus may move in more than one form. The
    TMV infection spreads from cell to cell in the absence of protein coat in
    tomato (Siegel, Zaitlin, and Sehgal, 1962; Dawson, Bubrick, and
    Granthan, 1988) but for long-distance transport, virus particles are
    probably essential (Saito, Yamanaka, and Okada, 1990). BSMV causes
    infection in the absence of the protein coat and may become systemic
    (Maule, 1991).

7.3.2 OVULE AND SEED

Neergaard (1979) and Mathews (1991) believe that the viruses may not enter ovules
on account of the lack of plasmodesmatal connections with the surrounding tissues.
Neergaard (1979) further concluded that (1) the virus may not be capable of establishing a compatible relationship with the gametes or the embryo sac of the host;
(2) the virus is lethal to the gametes, thus causing sterility and preventing production
of infected seed;and (3) the virus is not capable of infecting male and female gametes
or the young embryo, either due to lack of virulence to these stages of the host or
due to their resistance. Such conclusions seem to have been made mainly on the
basis of detrimental effects of virus infection on the floral, fruit, and seed parts.
Caldwell (1962) has observed that tomato aspermy virus is not seed-transmitted in

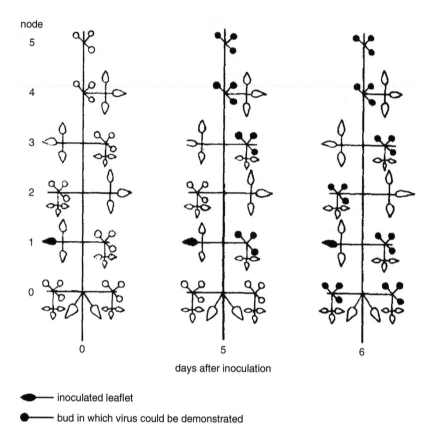

node

FIGURE 7.1 Bean common mosaic virus (BCMV) transmission in *Phaseolus vulgaris*. Transport of virus to flower buds and their ovaries situated at the basal nodes of axillary shoots and at nodes four and five of the main stem, after 5 and 6 days of inoculation of the middle leaflet of the second fully unfolded compound leaf. (From Schippers, B. 1963. *Acta Bot. Neerl.* 12: 433–497.)

tomato plants because it causes sterility of pollen and ovule. Virus infection affects production and quality of pollen in several other cases (Yang and Hamilton, 1974; Mink, 1993). However, the recent histopathological studies using electron microscopy have shown that the invasion, as well as the effects of virus infection, depend on the virus or its strain, host cultivar, and the environment.

Three alternatives for virus invasion of the ovule and seed are recognized. Where virus invades the shoot and floral meristem or causes early invasion of fertile floral appendages, the virus may occur in anthers and pollen grains and in ovule and female gametophyte at all stages of development. Bennett (1969) calls it *ovule infection* from the mother plant. Mathews (1991) believes that virus transmission through the female gametophyte is probably more efficient for most seed-borne viruses than transmission through pollen. The following have noted the presence of virus particles in anthers and ovules and male and female gametophytes during their development: Yang and Hamilton (1974) for tobacco ring spot virus (TRSV) in soybean;

Wilcoxson, Johnson, and Frosheiser (1975) for alfalfa mosaic virus (AMV) in alfalfa; Carroll and Mayhew (1976a,b) and Mayhew and Carroll (1974) using BSMV in barley; and Hunter and Bowyer (1993) for LMV in lettuce.

A second course of invasion of the ovule may take place through infected pollen grains (Mandahar, 1981; Mandahar and Gill, 1984; Mink, 1993). Mandahar (1981) listed 37 viruses as pollen-transmitted, but according to Mink (1993) the current list includes only 9 viruses. Pollen grains may be contaminated by virus, or the virus may occur inside in the cytoplasm and in sperm. BSMV particles in barley pollen grains occur in the cytoplasm as well as the nucleus of sperm cells (Carroll, 1974; Carroll and Mayhew, 1976a). Brlansky, Carroll, and Zaske (1986) have shown that BSMV-infected barley pollen grains are viable, germinate, and on germination the virus particles pass into the pollen tube and reach the embryo sac via the stigma and style. The cytoplasmic contents of the pollen tube together with the sperms are discharged in the degenerating synergid in the embryo sac.

A third course for viruses to reach the ovule and embryo is through the funiculus, its vascular supply, or parenchyma cells. Once the infection has reached the ovule, it may spread through symplastic pathways in cells of the integument and the nucellus. Using pea seed-borne mosaic virus (PSbMV) in pea, Wang and Maule (1992, 1993, 1994) traced this course of virus entry. They found abundant virus particles in and around carpellary vascular bundles of unfertilized carpels (Figure 7.2A, B), but none in the ovule, including the egg cell, synergids, and antipodal cells (i.e., the embryo sac). After fertilization, virus was detected in and around the vascular supply of the ovule (Figure 7.2C to E). Subsequently, it spread in the cells of integument.

7.4 LOCALIZATION IN REPRODUCTIVE SHOOT, OVULE, AND SEED

Table 7.4 lists all those cases that convincingly show the presence of virus in seed components. Histopathological determinations using ultra-thin sections and electron microscopy are few. The detection of seed-borne viruses in mature dry seeds is difficult because of the very laborious procedures for preparing material for electron microscopy transmission and also because of the inactivation of viruses in mature seeds in many cases. The most comprehensive investigations are on BSMV infection in barley by Carroll and co-workers and on PSbMV infection in pea (Wang and Maule, 1992, 1994; Maule and Wang, 1996).

There have been no studies using histopathological techniques of viruses of the cryptovirus group, which are transmitted with high efficiency through pollen and seed (Boccardo et al., 1987; Mathews, 1991).

7.4.1 Barley Stripe Mosaic Virus (BSMV) and Similar Viruses

BSMV is seed-borne and seed-transmitted. It has seed-transmitted as well as non-seed-passage (NSP) strains, and its response in different cultivars also varies. Figure 7.3 is a schematic representation of the infection cycle of a seed-transmitted

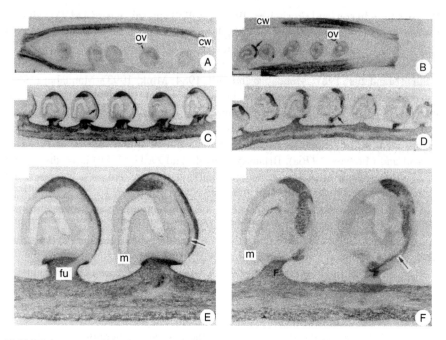

FIGURE 7.2 Distribution of pea seed-borne mosaic virus (PSbMV) RNA in the ovary tissue before and after fertilization (dark areas represent virus RNA concentration also marked by arrow). A, B, Ls through unfertilized ovaries of cultivars Progreta and Vedette, respectively, showing viral RNA restricted mostly to the carpel tissue. C, D, Ls fertilized ovaries showing the ingress of the virus into the ovule along the vascular tissue (arrow). E, F, magnified views of ovules from C, D to show the ingress of the virus. (Abbreviations: cw, carpel wall; fu, funiculus; m, micropyle; ov, ovule.) (From Wang, D. and Maule, A.J. 1994. *Plant Cell* 6: 777–787. With permission.)

strain (MI-1) in a susceptible cultivar *Atlas*. The infection reaches the primordia for fertile appendages — stamen and pistil — quite early. In the anther, cells of wall layers, connective, anther sac-male archesporium, and its derivatives (i.e., microspore mother cells, dyads, and microspores in tetrads), pollen grains contain virus particles in the cytoplasm individually, in clusters, and attached to the microtubules. The virus occurs in vegetative cells and in sperm cells (Figure 7.4A to C), and in the latter in the cytoplasm as well as the nucleus (Carroll, 1974; Carroll and Mayhew, 1976a). The pollen grains are viable and after pollination, they germinate, and pollen tubes enter the stigma and style, ovule, and embryo sac, and discharge their contents in the cytoplasm of the degenerating synergid (Figure 7.5A to D). BSMV particles occur in pollen tubes at all stages and also in the contents discharged in the embryo sac (Brlansky, Carroll, and Zaske, 1986).

In the ovule, the BSMV particles have been observed in the integument, nucellus, female archesporium (Figure 7.6A, B), megaspore mother cell, megaspores, two-, four-, and eight-nucleate embryo sacs, and in cells (synergids, egg, and antipodal cells) of the organized embryo sac (Figure 7.6C to E) (Carroll and Mayhew, 1976b; Mayhew and Carroll, 1974). After fertilization, clusters of BSMV particles are seen

TABLE 7.4
Virus Distribution in Floral Parts and Ovule and Seed of Crop Plants

Virus	Host	Parts	Important References
Alfalfa mosaic (AMV)	*Medicago sativa*	Anther, pollen, ovary wall, ovule integument; seed coat, embryo	Wilcoxson et al., 1975; Pesie and Hiruki, 1986; Pesic et al., 1988; Bailiss and Offei, 1990
Bean common mosaic (BCMV)	*Phaseolus vulgaris*	Bud, ovary, ovule (not in embryo sac)	Crowley, 1957; Schippers, 1963
		Seed, embryo	Ekpo and Saettler, 1974
	Vigna mungo	Embryo	Agarwal et al., 1979
	V. unguiculata	Seed coat, embryo	Patil and Gupta, 1992; Gillaspie et al., 1993
	Arachis hypogaea	Seed coat, embryo (cotyledon and plumule)	Demski and Lovell, 1985; Xu et al., 1991
Barley stripe mosaic (BSMV)	*Hordeum vulgare*	Anther cells, pollen grains including sperms; ovule-integument, nucellus, female archesporium and gametophytes including egg, synergids, antipodals, zygote, endosperm, embryo-plumule, scutellum	Carroll, 1969, 1974, 1981; Carroll and Mayhew, 1976a,b; Mayhew and Carroll, 1974; Brlansky et al., 1986
Bean pod mottle (BPMV)	*Phaseolus vulgaris*	Seed coat, embryo	Zaumeyer and Thomas, 1948
Cowpea mosaic (CPMV)	*Vigna unguiculata*	Seed coat, embryo (cotyledons, embryonic axis)	Patil and Gupta, 1992
Lettuce mosaic (LMV)	*Lactuca sativa*	Ovary wall, integument, endosperm, embryo, pericarp	Hunter and Bowyer, 1991, 1993, 1994
Maize dwarf mosaic (MDMV)	*Zea mays*	Anther (not in pollen grains), silks, unfertilized ovaries, immature kernels (pericarp, endosperm, embryo; mature kernel) rarely in endosperm, pericarp, not in embryo	Mikel et al., 1982 Nemchnov et al., 1990
Pea seed-borne mosaic (PSbMV)	*Pisum sativum*	Flower parts, anther, ovary, immature seed-embryo, endosperm, seed coat; mature embryo, seed coat	Wang and Maule, 1992, 1994
Plum pox (PPV)	*Prunus* sp.	Seed coat, rarely embryo	Eynard et al., 1991

(continued)

TABLE 7.4 (CONTINUED)
Virus Distribution in Floral Parts and Ovule and Seed of Crop Plants

Virus	Host	Parts	Important References
Prune dwarf (PDV)	*Prunus avium*	Pollen grains, seed — seed coat, endosperm, embryo (cotyledons, hypocotyl) radicle axis	Kelley and Cameron, 1986
Soybean mosaic (SMV)	*Glycine max*	Flowers, green pods, immature seeds — seed coat, embryo; mature embryo	Galver, 1963; Bowers and Goodman, 1979
Southern bean mosaic (SBMV)	*Phaseolus vulgaris*	Flower, fruit all parts, immature seed — seed coat, embryo; mature seed — seed coat	Cheo, 1955; Crowley, 1957; McDonald and Hamilton, 1972
Squash mosaic (SqMV)	*Cucumis melo*	Seed coat, endosperm, embryo	Alvarez and Campbell, 1978
Tobacco mosaic (TMV)	*Lycopersicon esculentum*	Seed coat, endosperm	Taylor et al., 1961; Broadbent, 1965
	Capsicum annuum	Seed coat, rarely endosperm, embryo	Demski, 1981
	Malus sylvestris	Seed coat, embryo	Gilmer and Wilks, 1967
Tobacco ringspot (TRSV)	*Glycine max*	Pollen grain-intine, generative cell, ovule-integument, nucellus, wall and cells of embryo sac, embryo	Athow and Bancroft, 1959; Yang and Hamilton, 1974
Tobacco rattle (TRV)	*Lycopersicon esculentum*	Microspore mother cells, pollen grains	Gasper et al., 1984
Tobacco streak (TSV)	*Glycine max*	Seed coat, embryo	Fagbenle and Ford, 1970; Ghanekar and Schwenk, 1974

in cytoplasm of the zygote and in synergids. Virus particles have been recorded in the endosperm, embryo, and pericarp (Carroll, 1969, 1972). Although the chances for infection of the egg and zygote from pollen tube discharge is not completely ruled out, the presence of virus in the egg secures its transmission from the female gametophyte.

Contrary to the seed-transmitting strain, the plants mechanically infected with the NSP strain of BSMV usually did not reveal the presence of virus particles in young and mature anthers, ovules, embryo sacs, and embryos (Carroll and Mayhew, 1976a,b).

Seed-transmitted strains of AMV seem to resemble BSMV (MI-1) in their behavior. Wilcoxson, Johnson, and Frosheiser (1975) detected AMV particles in the epidermis, parenchyma, and vascular parenchyma of the ovary wall, anthers, pollen grains, and embryo (cotyledons). Using ELISA and ISEM, the occurrence of AMV

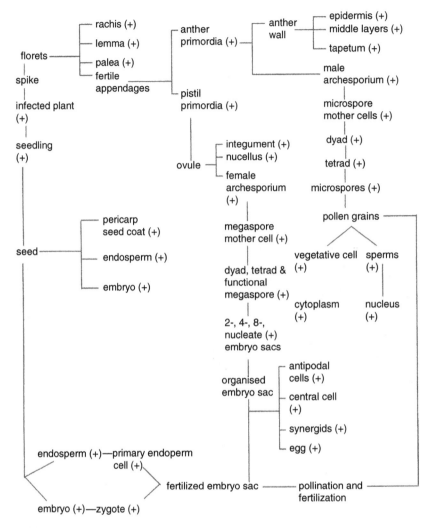

FIGURE 7.3 Schematic representation of infection cycle of BSMV seed-transmitted strain in barley. (+) = presence of virus. (-) = absence of virus. (Based on the publications of Carroll and co-workers.)

antigen has been observed in the ovule, microspores, pollen grains, and anther tapetum, but not in the embryo sac (Pesic, Hiruki, and Chen, 1988) and seed coat (Pesic and Hiruki, 1986). Baillis and Offei (1990) have concluded that AMV invades the female reproductive organs of most flowers systemically.

Similarly, southern bean mosaic virus (SBMV) and soybean mosaic virus (SMV) occur in all flower parts, green pods, immature seeds, and seed coat and embryo of mature seeds (Cheo, 1955; Bowers and Goodman, 1979). According to Yang and Hamilton (1974) tobacco ringspot virus (TRSV) particles are contained in the pollen, integuments, nucellus, embryo sac wall, and cells in soybean. Earlier Athow and Bancroft (1959) found TRSV in embryo, but not in the seed coat of soybean seed.

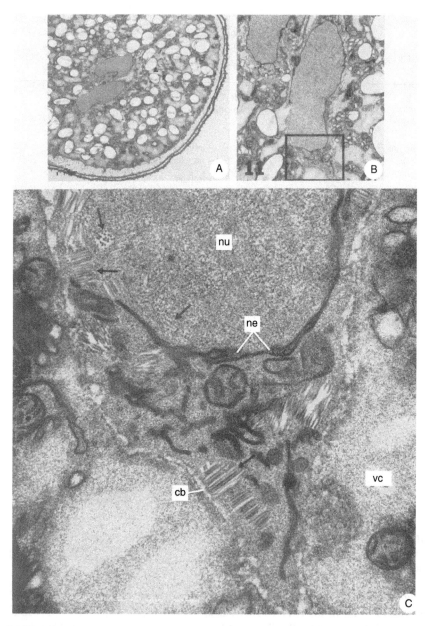

FIGURE 7.4 TEM micrographs of barley stripe mosaic virus (BSMV)-infected pollen grains of barley. A, Part of mature pollen grain containing two sperm cells having BSMV particles. B, Part from A magnified showing slightly enlarged view of the sperm cell. C, Portion outlined in B magnified to show BSMV infection in sperm cell. Note virions (arrows) in the nucleus and cytoplasm of the sperm cell. (Abbreviations: cb, cell boundary; nu, nucleus; ne, nuclear envelope, all of the sperm cell; vc, vegetative cell.) (From Carroll, T.W. 1974. *Virology* 60: 21–28. With permission.)

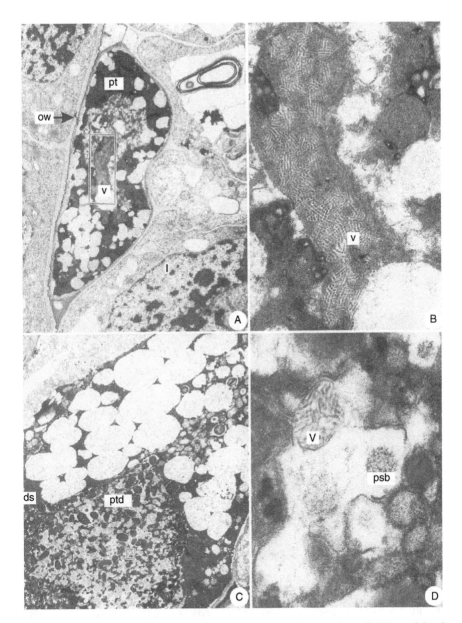

FIGURE 7.5 TEM of pollinated ovaries before fertilization showing BSMV particles in pollen tube and its discharge in embryo sac. A, Large aggregate of virus particles in a segment of pollen tube penetrating the ovary. B, Magnified view of large aggregate of virus particles from A. Particles are surrounded by dense cytoplasmic material. C, Pollen tube discharge within the degenerating synergid. D, Portion from C enlarged to show aggregate of virus particles in the discharge. (Abbreviations: ds, degenerating synergid; i, integument; ow, ovary wall; psb, polysaccharide body; pt, pollen tube; ptd, pollen tube discharge; v, virus particles.) (From Brlansky, R.H., Carroll, T.W., and Zaske, S.K. 1986. *Can. J. Bot.* 64: 853–858. With permission.)

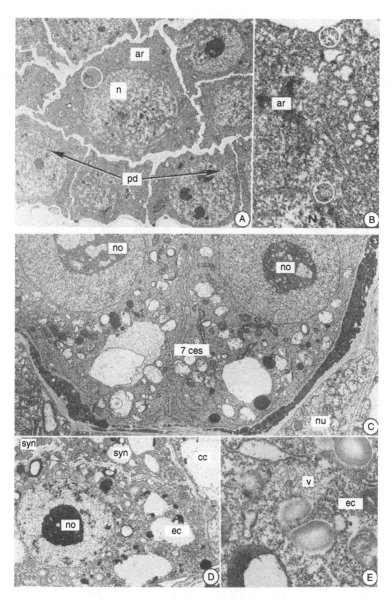

FIGURE 7.6 TEM of female archesporium and cells of female gametophyte in ovules of barley infected with BSMV, MI-1 strain. A, Female archesporial cell beneath the protoderm in ovule primordium. B, Portion outlined in A magnified showing virus particles in the cytoplasm (circles). C, Micropylar end of a seven-celled embryo sac with the surrounding nucellar cells. D, Portion of the embryo sac showing parts of egg, synergid, and central cell. E, Part of the egg cell showing encapsulated virus particles within the cytoplasm. (Abbreviations: ar, archesporium; cc, central cell; 7-ces, part of seven-celled embryo sac; ec, egg cell; n, nucleus; no, nucleolus; nu, nucellar cells; pd, protoderm layer of ovule primordium-epidermis; syn, synergid; v, virus.) (From Carroll, T.W. and Mayhew, D.E. 1976b. *Can. J. Bot.* 54: 2497–2512. With permission.)

7.4.2 BEAN COMMON MOSAIC VIRUS (BCMV)

Early infection of BCMV on bean plants causes necrosis and dropping of buds. The pistils at the ovule primordial stage lack virus particles, but these could be detected 4 or 5 days before flowering when the ovules contain the eight-nucleate embryo sac. Infectious virus reaches the ovules 2 or 3 days before flowering. Cross-pollination experiments using pollen and pistils of infected plants have shown that the embryo might get infection from infected eggs or pollen grains (Schippers, 1963). BCMV has been reported in immature and mature seeds from infected plants. It occurs in the cotyledons and plumule (Ekpo and Saettler, 1974).

BCMV has also been reported in embryos of urdbean (*Vigna mungo*) and cowpea (*Vigna unquiculata*) by Agarwal, Nene, and Beniwal (1979) and Patil and Gupta (1992), respectively.

7.4.3 LETTUCE MOSAIC VIRUS (LMV)

LMV particles are seen throughout the ovary and ovular tissues, except the embryo sac, in infected plants of lettuce cultivar Salinas. High levels of LMV infection occur in the integumentary tapetum (Hunter and Bowyer, 1994). In mature seeds (cypsil), LMV particles occur in the endosperm (Figure 7.7D), cotyledons (Figure 7.7E), hypocotyl-radicle axis, and pericarp (Figure 7.7A to C). The virus particles are either scattered in the cytoplasm (Figure 7.7C) or seen in association with protein bodies. Cylindrical and pinwheel-shaped cytoplasmic inclusions have been observed in the cells of the integument, ovary wall, endosperm (Figure 7.7D), and embryo (Figure 7.7E, F), but not in those of the pericarp (Hunter and Bowyer, 1993, 1994).

7.4.4 PEA SEED-BORNE MOSAIC VIRUS (PSBMV)

Figure 7.8 gives a schematic representation of the infection cycle of PSbMV in a high seed-transmitting cultivar of pea. The virus infection in unfertilized flowers occurs in sepals, petals, anther-epidermis, and carpel, but none in the pollen grains and ovule except the funiculus (Wang and Maule, 1992, 1994). In fertilized ovules, the virus antigen was initially detected in cells close to the vascular supply of the ovule, from where it spread throughout the integument. Subsequently, it reached the endospermic fluid and suspensor of the embryo. In the embryo proper, in spite of abundant virus in the embryo sac fluid, the infection was seen initially in cells that were in contact with the suspensor, indicating that it had traversed through the suspensor (Wang and Maule, 1994). The spread of infection of PSbMV from carpel to embryo clearly shows that the virus enters the ovule through the funiculus, probably via its vascular supply. From the ovular supply it spreads in the integument in permissive cultivar–virus interaction systems as in cultivar Vedette. Finally, the virus enters the embryo with the suspensor as its main route (Figure 7.9A). In a cultivar with a nonpermissive interaction, e.g., cultivar Progreta, virus enters the ovule through the funicular vascular supply, but is unable to invade the cells of the integument in the nonvascular region. Thus, the infection fails to reach the micropyle region crucial for the transmission of virus to the embryo via suspensor (Figure 7.9B). Although it is shown that the transmission of PSbMV occurs exclusively through

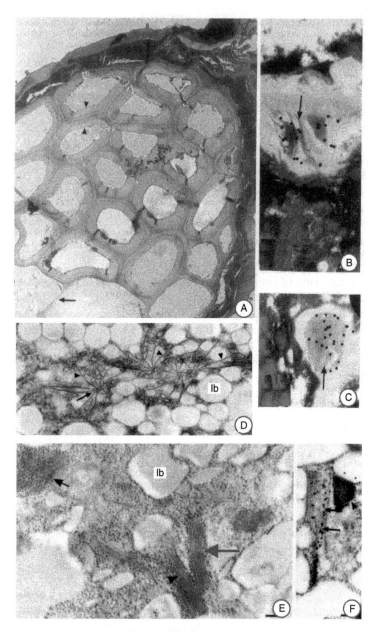

FIGURE 7.7 TEM of lettuce seed infected with lettuce mosaic virus (LMV). A, Pericarp
from rib area showing thick-walled sclerenchyma cells. B, C, Immunogold labeling of LMV
particles (arrows) in cells of outer pericarp. D, Infected endosperm cells showing *pinwheel*
inclusion (arrowheads) with associated LMV particles. E, F, Cells of cotyledons of ungerminated LMV-infected seed. Virus particles (arrow) interspersed with ribosomes and tightly
packed particles (arrowheads) in E. Immunogold labeling of aggregated (arrowhead) and
single (arrows) virus particles. (Abbreviations: lb, lipid body.) (From Hunter, D.G. and
Bowyer, J.W. 1993. *J. Phytopathol.* 137: 61–72. With permission.)

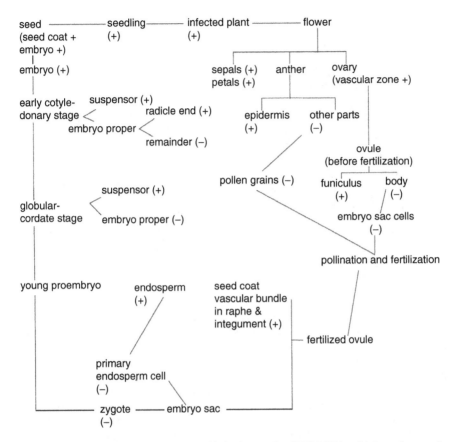

FIGURE 7.8 Schematic representation of infection cycle of PSbMV in a high seed transmitting cultivar in pea. (+) = presence of virus. (−) = absence of virus. (Based on the publications of Wang and Maule.)

direct embryo invasion in pea, it varies with virus–host cultivar combination. The embryonic suspensor is the main conduit for virus invasion of the embryo proper. The development of wall ingrowths in cells of the suspensor adjacent to the embryo sac boundary and the occurrence of plasmodesmata between the cells of the suspensor and those of the suspensor and endosperm provide support to this pathway (Johansson and Walles, 1993).

There are many other viruses reported in the embryo (Table 7.4). It may be pointed out that most viruses found in the embryo are also located in the endosperm and seed coat and pericarp as seen for BSMV in barley, LMV in lettuce, and PSbMV in pea.

7.5 CYTOPATHOLOGICAL EFFECTS

Viruses are known to induce histological as well as cytological changes including formation of *in vivo* cytoplasmic and nuclear inclusions in cells of infected plants.

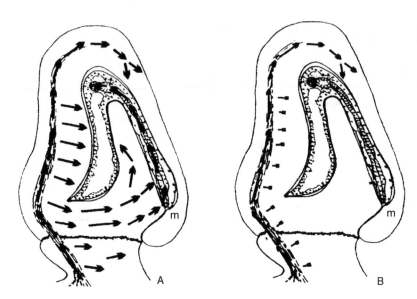

FIGURE 7.9 (Color figure follows p. 146.) Diagrammatic analysis of the distribution of pea seed-borne mosaic virus (PSbMV) in longitudinal section through immature pea seeds by immunocytochemistry using a monoclonal antibody to PSbMV coat protein in cultivar Vedette permissive for seed transmission and cultivar Progreta with nonpermissive interaction. A, Diagrammatic Ls of immature seed of cultivar Vedette showing widespread accumulation of the virus in testa tissues reaching near the micropyle providing point of contact between the testa tissues and the embryonic suspensor through which the virus enters embryo. B, Same for cultivar Progreta showing entry of virus through the vascular supply of the seed but the virus is unable to invade the adjacent testa tissues — a nonpermissive interaction. (Abbreviation: m, micropyle.) (From Maule, A.J. and Wang, D. 1996. *Trends Microbiol.* 4: 153–158. With permission.)

These have been widely observed in cells of vegetative parts by light microscopy and in recent years by electron microscopy, and the data are summarized by Mathews (1991) and Shukla, Ward, and Brunt (1994). The effects of virus infection occur on the nuclei, mitochondria, chloroplasts, and cell wall and are also observed as formations of cytoplasmic inclusions (cylindrical, crystalline, and pinwheel-shaped), and formations of nuclear inclusions (crystalline, amorphous, and tubular). The information on virus-induced inclusions in seed tissues is limited, and the effects on the nuclei and cell organelles have not been reported. LMV-induced cylindrical inclusions have been found throughout the ovular tissues, and are particularly abundant in the integumentary tapetum, residual integumentary cells, ovary wall (Hunter and Bowyer, 1994), in the embryo, endosperm, and pericarp (Hunter and Bowyer, 1993). Cytoplasmic pinwheel inclusions with the most characteristic feature of a central tubule, from which curved *arms* radiate (Figure 7.7D), are observed in the integumentary tapetum and embryonic tissues, but not in the cells of the pericarp. Lamellar structures with LMV particles in association are seen in integumentary tapetal cells, but Hunter and Bowyer (1994) consider them of unknown origin.

AMV forms rafts of short virus particles in the anthers, pollen, and cotyledons, and large crystalline bodies of long particles in the ovary wall, bud receptacle, and cotyledons. The crystalline bodies are crescent-shaped in cells of cotyledons, while in pollen grains they form star-like aggregations (Wilcoxson, Johnson, and Frosheiser, 1975). Pesic, Hiruki, and Chen (1988) observed large crystalline bodies in the cytoplasm of tapetal cells and pollen grains in anthers infected with AMV. Hoch and Provvidenti (1978) have reported areas of granular or fibrillar appearance associated with virus arrays of BCMV particles, virus-associated paracrystalline bodies, and arrays or undulating filaments in cells of infected seeds.

7.6 INACTIVATION OR LONGEVITY OF VIRUSES IN SEED DURING MATURATION AND STORAGE

Stability of viruses during seed maturation and dehydration has been shown to vary with the virus, host, and its cultivar. Inactivation of viruses in the seed coat during maturation has been observed in numerous cases, including AMV (Bailiss and Offei, 1990), SMV (Bowers and Goodman, 1979), SBMV (Cheo, 1955), BSMV (Gold et al., 1954; Carroll, 1972), BCMV (Ekpo and Saettler, 1974), and PSbMV (Stevenson and Hagedorn, 1973).

AMV is seed-transmitted in lucerne, and infectivity and ELISA tests have shown that infective virus incidence decreases rapidly with seed maturation. But the antigen incidence (ELISA test) declined more slowly than the infective virus (Baillis and Offei, 1990). The incidence of infective virus and antigen was higher in seeds of cultivar Maris Kabul than cultivar Europe. Incidence of seeds with infective AMV decreased during maturation from 59 to 6.9% in cultivar Maris Kabul and 75 to 1.7% in cultivar Europe.

SMV is seed-transmitted at frequencies from 0 to 75%, depending on soybean cultivar and the virus strain (Tu, 1989). Bowers and Goodman (1979) found infection in 94% and 58% of seeds of Midwest (highly seed transmitting cultivar) and Merit (poorly seed transmitting) at physiological seed maturity and 66% and 0.8% at harvest maturity, respectively. SMV was detected in the seed coat and embryos from immature seeds of both the cultivars, but in mature seeds it was detected in embryos of Midwest seeds, but not in those of Merit. SMV transmission showed only a slight decline after storage for 6 months at 14°C.

Cheo (1955) observed that during later stages of seed development, SBMV decreased in the pods and seed coat, but increased in the embryo in *Phaseolus vulgaris*. During dehydration, the virus in the embryo was inactivated very rapidly, but not in the seed coat. In contrast, BCMV was inactivated in the pod wall and seed coat during maturation and drying in bean. It remained unaffected in the embryo during maturation, drying, and also storage (Schippers, 1963). BCMV probably survives in seeds of navy bean as long as the seeds remain viable.

The maize dwarf mosaic virus (MDMV) occurs in large proportions in immature kernels (70%) from infected plants. MDMV was not detected in mature kernels and the frequency of detection in the pericarp fell to 2% (Mikel et al., 1982).

The longevity of viruses in seed during storage is also variable. Bennett (1969) has reported that the longevity of different viruses varies from 2 months to 6.5 years. TMV infection in tomato seeds declined rapidly in the first year, but was still present after 3 years of storage (Alexander, 1960). Frosheiser (1974) found little loss of AMV in infected alfalfa seed after 5 years at room temperature. Scott (1961) found 64% BSMV-infected seeds in seeds of barley stored for 6.5 years. An extreme example is the detection of BCMV in a bean seedling grown from seeds after 30 years of storage (Pierce and Hungerford, 1929).

7.7 CONCLUDING REMARKS

The number of seed-borne and seed-transmitted viruses and viroids is quite large, and the present list (Tables 7.1 to 7.3) includes more than 120 of them. The histopathological investigations of virus-infected seeds using ultra-thin sections and electron microscopy are not only meager, confined to perhaps a dozen virus–seed interfaces, but except for BSMV in barley, the information is insufficient in most cases and often not supported by adequate photomicrographs or diagrams. Although cryptoviruses are only seed- and pollen-transmitted, and seed transmission of viroids is linked with their affinity to meristematic cells, there is virtually no information on how these pathogens are introduced into the seed. Entry into seed is important because the propagation and distribution of cryptic viruses occur only through cell multiplication. Likewise, plant infection of more than a dozen viruses results in the production of symptomatic seeds (Phatak, 1974; Bos, 1977), but their histopathology has remained uninvestigated.

The ovules and seeds, borne on the placenta, have contacts with tissues including the vascular supply of the pistil. The ovule has both symplastic and apoplastic systems of transport of nutrients. Adequate histopathological data of infected reproductive shoot (flower) and fertile floral appendages throughout their development allow proper understanding of the infection cycle of seed-transmitting viruses through the host (Figures 7.3 and 7.8). The infection of seed-transmitting strains of BSMV spreads systemically in floral parts and exists in the egg, which develops into an embryo (Carroll and Mayhew, 1976b; Carroll, 1981). The delayed invasion of ovule and seed by PSbMV is also direct from the mother plant. PSbMV appears to be slow growing, does not reach the primordia of fertile appendages, and reaches the ovule primarily through vascular tissue. The embryonic invasion takes place through the suspensor (Wang and Maule, 1994; Maule and Wang, 1996). How the virus antigen traverses the embryo sac wall or the suspensor cell walls from the integument remains unexplained; Wang and Maule (1994) think that it is unlikely to occur apoplastically. The lack of histopathological investigation in such cases is sorely felt because it could certainly yield corroborative evidence. The histopathological observations on the infection cycle of seed-borne viruses in host plants, particularly the events leading to the infection of sporogenous cells, their survival, the infection of male and female gametophytic phases, and the passage of virus into the seed and embryo are important, not only to gain insight into host–parasite interactions, but also to determine strategies to check the spread of viruses.

REFERENCES

Agarwal, V.K., Nene, Y.L., and Beniwal, S.P.S. 1979. Location of bean common mosaic virus in urdbean seed. *Seed Sci. Technol.* 7: 455–458.

Agarwal, V.K. and Sinclair, J.B. 1997. *Principles of Seed Pathology*, 2nd ed. CRC Press, Boca Raton, FL.

Agrios, G.N. 1988. *Plant Pathology*, 3rd ed. Academic Press, San Diego.

Alexander, L.J. 1960. Inactivation of tobacco mosaic virus from tomato seed. *Phytopathology* 50: 627.

Alvarez, M. and Campbell, R.N. 1978. Transmission and distribution of squash mosaic virus in seeds of cantaloupe. *Phytopathology* 68: 257–263.

Athow, K.L. and Bancroft, J.B. 1959. Development and transmission of tobacco ringspot virus in soybean. *Phytopathology* 49: 697–701.

Baillis, K.W. and Offei, S.K. 1990. Alfalfa mosaic virus in lucerne seed during seed maturation and storage, and in seedlings. *Plant Pathol.* 39: 539–547.

Bennett, C.W. 1940. The relation of viruses to plant tissues. *Bot. Rev.* 6: 427–473.

Bennett, C.W. 1969. Seed transmission of plant viruses. *Adv. Virus Res.* 14: 221–259.

Boccardo, G., Lisa, V., Luisoni, L., and Milne, R.G. 1987. Cryptic plant viruses. *Adv. Virus Res.* 32: 171–174.

Bos, L. 1977. Seed-borne viruses. In *Plant Health and Quarantine in International Transfer of Genetic Resources*. Hewitt, W.B. and Chiarappa, L., Eds. CRC Press, Boca Raton, FL, pp. 39–65.

Bowers, G.R. and Goodman, R.M. 1979. Soybean mosaic virus: infection of soybean seed parts and seed transmission. *Phytopathology* 69: 569–572.

Brants, D.H. 1964. The susceptibility of tobacco and bean leaves to tobacco mosaic virus infection in relation to the condition of ectodesmata. *Virology* 23: 588–594.

Broadbent, L. 1965. The epidemiology of tomato mosaic. XI. Seed-transmission of TMV. *Ann. Appl. Biol.* 56: 177–205.

Broadbent, L. 1976. Epidemiology and control of tomato mosaic virus. *Ann. Rev. Phytopathol.* 14: 75–95.

Brlansky, R.H., Carroll, T.W., and Zaske, S.K. 1986. Some ultrastructural aspects of the pollen transmission of barley stripe mosaic virus in barley. *Can. J. Bot.* 64: 853–858.

Caldwell, J. 1962. Seed-transmission of viruses. *Nature* 193: 457–459.

Carroll, T.W. 1969. Electron microscopic evidence for the presence of barley stripe mosaic virus of barley embryos. *Virology* 37: 649–657.

Carroll, T.W. 1972. Seed transmissibility of two strains of barley stripe mosaic virus. *Virology* 48: 323–336.

Carroll, T.W. 1974. Barley stripe mosaic virus in sperm and vegetative cells of barley pollen. *Virology* 60: 21–28.

Carroll, T.W. 1981. Seedborne viruses: virus-host interactions. In *Plant Diseases and Vectors: Ecology and Epidemiology.* Maramorosch, K. and Harris, K.J., Eds. Academic Press, London, pp. 293–317.

Carroll, T.W. and Mayhew, D.E. 1976a. Anther and pollen infection in relation to the pollen and seed transmissibility of two strains of barley stripe mosaic virus. *Can. J. Bot.* 54: 1604–1621.

Carroll, T.W. and Mayhew, D.E. 1976b. Occurrence of virions in developing ovules and embryo sacs of barley in relation to the seed transmissibility of barley stripe mosaic virus. *Can. J. Bot.* 54: 2497–2512.

Cheo, P.C. 1955. Effect of seed maturation on inhibition of southern bean mosaic virus in bean. *Phytopathology* 45: 17–21.

Crowley, N.C. 1957. Studies on the seed transmission of plant virus diseases. *Austr. J. Biol. Sci.* 10: 449–464.

Dawson, W.O., Bubrick, P., and Granthan, G.L. 1988. Modification of tobacco mosaic virus coat protein gene affecting replication, movement, and symptomatology. *Phytopathology* 78: 783–789.

Demski, J.W. 1981. Tobacco mosaic virus is seedborne in pimiento peppers. *Plant Dis.* 65: 723–724.

Demski, J.W. and Lovell, G.R. 1985. Peanut stripe virus and its distribution in peanut seed. *Plant Dis.* 69: 734.

Ekpo, E.J.A. and Saettler, A.W. 1974. Distribution of bean common mosaic virus in developing bean seed. *Phytopathology* 64: 269–270.

Esau, K. 1953. *Plant Anatomy.* John Wiley & Sons, New York.

Esau, K. 1967. Anatomy of plant virus infections. *Ann. Rev. Phytopathol.* 5: 45–86.

Esau, K., Cronshaw, J., and Hoefert, L.L. 1967. Relation of beet yellows virus to the phloem and to movement in the sieve tube. *J. Cell Biol.* 32: 71–87.

Eynard, A., Roggero, P., Lenzi, R., Conti, M., and Milne, R.G. 1991. Test for pollen and seed transmission of plum pox virus (sharka) in two apricot cultivars. *Adv. Hort. Sci.* 5: 104–106.

Fagbenle, H.H. and Ford, R.E. 1970. Tobacco streak virus isolated from soybean, *Glycine max. Phytopathology* 60: 814–820.

Fahn, A. 1974. *Plant Anatomy.* Pergamon Press, Oxford, U.K.

Fauquet, C.M. and Martelli, G.P. 1995. Updated ICTV list of names and abbreviations of viruses, viroids and satellites infecting plants. *Arch. Virol.* 140: 393–413.

Fauquet, C.M. and Mayo, M.A. 1999. Abbreviations for plant virus names — 1999. *Arch. Virol.* 144: 1249–1273.

Frosheiser, F.I. 1974. Alfalfa mosaic virus transmission to seed through alfalfa gametes and longevity in alfalfa seed. *Phytopathology* 64: 102–105.

Galver, G.E. 1963. Host-range, purification, and electron microscopy of soybean mosaic virus. *Phytopathology* 53: 388–393.

Gaspar, J.O., Vega, J., Camargo, I.J.B., and Costa, A.S. 1984. An ultrastructural study of particle distribution during microsporogenesis in tomato plants infected with the Brazilian tobacco rattle virus. *Can. J. Bot.* 62: 372–378.

Ghanekar, A.M. and Schwenk, F.W. 1974. Seed transmission and distribution of tobacco streak virus in six cultivars of soybeans. *Phytopathology* 64: 112–114.

Gillaspie, A.G., Jr., Hopkins, M.S., and Pinrow, D.L. 1993. Relationship of cowpea seedpart infection and seed transmission of blackeye cowpea mosaic potyvirus in cowpea. *Plant Dis.* 77: 875–877.

Gilmer, R.M. and Wilks, J.M. 1967. Seed transmission of tobacco mosaic virus in apple and pear. *Phytopathology* 57: 214–217.

Gold, A.H., Suneson, C.A., Houston, B.R., and Oswald, J.W. 1954. Electron microscopy and seed and pollen transmission of rod-shaped particles associated with the false stripe virus disease of barley. *Phytopathology* 44: 115–117.

Halk, E.L. and McGuire, J.M. 1973. Translocation of tobacco ringspot virus in soybean. *Phytopathology* 63: 1291–1300.

Helms, K. and Wardlaw, L.F. 1976. Movement of viruses in plants: long distance movement of tobacco mosaic virus in *Nicotiana glutinosa*. In *Transport and Transfer Processes in Plants*. Wardlaw, L.F. and Passioura, J.B., Eds. Academic Press, New York.

Hoch, H.C. and Provvidenti, R. 1978. Ultrastructural localization of bean common mosaic virus in dormant and germinating seeds of *Phaseolus vulgaris*. *Phytopathology* 68: 327–330.

Hunter, D.G. and Bowyer, J.W. 1991. Location of lettuce mosaic virus in mature lettuce seed tissues by immunogold cytochemistry. *Aust. Plant Path.* 20: 3–5.

Hunter, D.G. and Bowyer, J.W. 1993. Cytopathology of lettuce mosaic virus–infected lettuce seeds and seedlings. *J. Phytopathol.* 137: 61–72.

Hunter, D.G. and Bowyer, J.W. 1994. Cytopathology of mature ovaries from lettuce plants infected by lettuce mosaic potyvirus. *J. Phytopathol.* 140: 11–18.

Johansen, E., Edwards, M.C., and Hampton, R.O. 1994. Seed transmission of viruses: Current perspectives. *Ann. Rev. Phytopathol.* 32: 363–386.

Johansson, M. and Walles, B. 1993. Functional anatomy of the ovule in broad bean (*Vicia faba* L.). I. Histogenesis prior to and after pollination. *Int. J. Plant Sci.* 154: 80–89.

Kelley, R.D. and Cameron, H.R. 1986. Location of prune dwarf and prunus necrotic ring spot viruses associated with sweet cherry pollen and seed. *Virology* 76: 317.

Kitajima, E.W. and Lauritus, J.A. 1969. Plant virions in plasmodesmata. *Virology* 37: 681–684.

Lucas, W.J. and Gilbertson, R.L. 1994. Plasmodesmata in relation to viral movement within leaf tissues. *Ann. Rev. Phytopathol.* 32: 387–411.

Mandahar, C.L. 1981. Virus transmission through seed and pollen. In *Plant Diseases and Vectors. Ecology and Epidemiology*. Maramorosch, K. and Harris, K.F., Eds. Academic Press, New York, pp. 241–292.

Mandahar, C.L. and Gill, P.S. 1984. The epidemiological role of pollen transmission of viruses. *J. Plant Dis. Prot.* 91: 246–249.

Mathews, R.E.F. 1991. *Plant Virology*, 3rd ed. Academic Press, New York.

Maule, A.J. 1991. Virus movement in infected plants. *Crit. Rev. Plant Sci.* 9: 457–473.

Maule, A.J. and Wang, D. 1996. Seed transmission of plant viruses: a lesson in biological complexity. *Trends Microbiol.* 4: 153–158.

Mayhew, D.E. and Carroll, T.W. 1974. Barley stripe mosaic virus in the egg cell and egg sac of infected barley. *Virology* 58: 561–567.

McDonald, J.G. and Hamilton, R.L. 1972. Distribution of southern bean mosaic virus in seed of *Phaseolus vulgaris*. *Phytopathology* 62: 387–389.

Meshi, T., Watanabe, Y., Sato, T., Sugimoto, A., Maeda, T., and Okada, Y. 1987. Function of the 30 Kd protein of tobacco mosaic virus involvement in cell to cell movement and dispensability for replication. *EMBO J.* 6: 2557–2563.

Mikel, M.A., D'Arcy, C.J., Rhodes, A.M., and Ford, R.E. 1982. Seed transmission of MDMV in sweet corn. *Phytopathology* 72: 1138 (Abstr.).

Mink, G.I. 1993. Pollen and seed-transmitted viruses and viroids. *Ann. Rev. Phytopathol.* 31: 375–402.

Neergaard, P. 1979. *Seed Pathology*. Vols. 1 and 2. Macmillan Press, London.

Nemchnov, L.G., Tertyak, D.D., and Satya Prasad, M. 1990. Localization of dwarf mosaic virus in reproductive organs of infected maize plants during ontogenesis. *Set'skokhoz-yastvennaya Biologiya* 3: 192–195.

Offler, C.E. and Patrick, J.W. 1984. Cellular structures, plasma membrane surface areas and plasmodesmatal frequencies of seed coats of *Phaseolus vulgaris* L. in relation to photosynthate transfer. *Aust. J. Bot.* 11: 79–99.

Offler, C.E., Nerlich, S.M., and Patrick, J.W. 1989. Pathway of photosynthate transfer in the developing seed of *Vicia faba* L. transfer in relation to seed anatomy. *J. Exper. Bot.* 40: 769–780.

Patil, M.D. and Gupta, B.M. 1992. The location of two mosaic viruses in cowpea seeds. *J. Turk. Phytopathol.* 21: 21–23.

Pesic, Z. and Hiruki, C. 1986. Differences in the incidence of alfalfa mosaic virus in seed coat and embryo of alfalfa seed. *Can. J. Plant Pathol.* 8: 39–42.

Pesic, Z., Hiruki, C., and Chen, M.H. 1988. Detection of viral antigen by immunogold cytochemistry in ovules, pollen and anthers of alfalfa infected with alfalfa mosaic virus. *Phytopathology* 78: 1027–1032.

Pierce, W.H. and Hungerford, C.W. 1929. Symptomatology, transmission, infection and control of bean mosaic in Idaho. *Idaho Exp. Sta. Res. Bull.* 7: 1–37.

Phatak, H.C. 1974. Seed-borne plant viruses identification and diagnosis in seed health testing. *Seed Sci. Technol.* 2: 3–155.

Saito, T., Yamanaka, K., and Okada, Y. 1990. Long-distance movement and viral assembly of tobacco mosaic virus mutants. *Virology* 176: 329.

Samuel, G. 1934. Movement of tobacco mosaic virus within the plant. *Ann. Appl. Biol.* 21: 90–111.

Schippers, B. 1963. Transmission of Bean common mosaic virus by seeds of *Phaseolus vulgaris* L. cultivar Beka. *Acta Bot. Neerl.* 12: 433–497.

Schneider, I.B. and Worley, J.F. 1959. Rapid entry of infectious particles of southern bean mosaic virus into living cells following transport of the particles in the water stream. *Virology* 8: 243.

Scott, H.A. 1961. Serological detection of barley stripe mosaic virus in single seeds and dehydrated leaf tissues. *Phytopathology* 51: 200–201.

Shukla, D.D., Ward, C.W., and Brunt, A.A. 1994. *The Potyviridae.* CAB International, Wallingford, U.K.

Siegel, A., Zaitlin, M., and Sehgal, O.M. 1962. The isolation of defective tobacco mosaic virus strains. *Proc. Natl. Acad. Sci. U.S.A.* 48: 1845.

Singh, D. 1998. Seed development and nutrition. In *Plant Reproduction, Genetics and Biology.* Gohil, R.N., Ed. Scientific Publishers (India), Jodhpur, pp. 137–154.

Stace-Smith, R. and Hamilton, R.L. 1988. Inoculum thresholds of seedborne pathogens, Viruses. *Phytopathology* 78: 875–880.

Stevenson, W.R. and Hagedorn, D.J. 1973. Further studies on seed transmission of pea seedborne mosaic virus in *Pisum sativum. Plant Dis. Rep.* 57: 248–252.

Taylor, R.H., Grogan, R.G., and Kimble, K.A. 1961. Transmission of tobacco mosaic virus in tomato seed. *Phytopathology* 51: 837–842.

Thomas, P.E. and Fulton, R.W. 1968. Correlation of ectodesmata number with non-specific resistance to initial virus infection. *Virology* 34: 459.

Thorne, J.H. 1980. Kinetics of ^{14}C photosynthate uptake by developing soybean fruit. *Plant Physiol.* 45: 975–979.

Thorne, J.H. 1981. Morphology and ultrastruture of maternal seed tissues of soybean in relation to import of photosynthate. *Plant Physiol.* 47: 1016–1025.

Tu, J.C. 1989. Effect of different strains of soybean mosaic virus on growth, maturity, yield, seed mottling and seed transmission in several soybean cultivars. *J. Phytopathol.* 126: 231–236.

Wang, D. and Maule, A.J. 1992. Early embryo invasion as a determinant in pea of the seed transmission of pea seed-borne mosaic virus. *J. Gen. Virol.* 73: 1615–1620.

Wang, D. and Maule, A.J. 1993. Seed transmission of pea seed-borne mosaic virus in pea — a process full of surprises. 9th International Congress of Virology, Glasgow, Scotland, Abstr. 64–68.

Wang, D. and Maule, A.J. 1994. A model for seed transmission of a plant virus: genetic and structural analyses of pea embryo invasion by pea seed-borne mosaic virus. *Plant Cell* 6: 777–787.

Wilcoxson, R.D., Johnson, L.E.B., and Frosheiser, F.L. 1975. Variation in the aggregation forms of alfalfa mosaic virus strains in different alfalfa organs. *Phytopathology* 65: 1249–1254.

Wolf, S., Deom, C.M., Beachy, R.M., and Lucas, W.J. 1989. Movement protein of tobacco mosaic virus modifies plasmodesmatal size exclusion limit. *Science* 246: 377–379.

Xu, Z., Chen, K., Zhang, Z., and Chen, J. 1991. Seed transmission of peanut stripe virus in peanut. *Plant Dis.* 75: 723–726.

Yang, A.F. and Hamilton, R.I. 1974. The mechanism of seed transmission of tobacco ringspot virus in soybean. *Virology* 62: 26–37.

Zaumeyer, W.J. and Thomas, H.R. 1948. Pod mottle, a virus disease of beans. *J. Agric. Res.* 77: 80–96.

8 Seed Infection by Nematodes

Nematodes, also called eel-worms, are wormlike in appearance, but quite distinct taxonomically from true worms. They live freely in fresh or salt water or in soil. They are saprozoic and also parasitize animals (including man) and plants. Many species of nematodes parasitize higher plants and cause disease symptoms. Plant parasitic nematodes are small (invisible to the naked eye), long, tubular, round in the cross section, unsegmented, and smooth invertebrates. They inhabit different parts of plants, namely, roots, stems, flower buds, and seeds. Depending on the host–parasite relationship, plant parasitic nematodes may be sedentary or migratory, and ectoparasitic and/or endoparasitic in their feeding habits.

Seed-borne nematodes may be internal or may occur as seed infestation. Several saprozoic soil nematodes occur in beds of soil mixed with seed. Seed infestation by nematodes does not produce any specific symptoms on seed, but in those that occur as endoparasites, the floral structure may be modified into a seed gall (*Anguina* spp.), may produce symptoms such as discoloration of testa in groundnut (*Ditylenchus destructor*), or may be symptomless as in paddy kernels infected by *Aphelenchoides besseyi*. Neergaard (1979) listed five genera — *Anguina, Aphelenchoides, Ditylenchus, Heterodora,* and *Rhadinaphelenchus* — that are seed-transmitted. *Heterodora* and *Rhadinaphelenchus* occur as endoparasites in roots, stems, and leaves. The latter also affects the growing apex, inflorescence, and husks of dropped nuts (Fenuwick, 1957). Table 8.1 provides a list of nematodes recorded as seed-borne in crop plants.

Detailed histopathological information is available on seed infection caused by different species of *Anguina*. Information on the infection of other nematodes is scanty. The available histopathological observations on seeds infected by endoparasitic nematodes are given under separate heads. A brief account at the end of this chapter discusses the interaction of seed nematodes and bacteria for which histopathological observations are available.

8.1 PENETRATION BY NEMATODES

Nematodes have a well-developed sensory and behavioral system that enables them to locate and attack specific parts of the plants. All plant parasitic nematodes possess a hollow buccal stylet that extrudes from the buccal cavity to puncture the cell wall. The nematodes vigorously attack the cell wall or the plant surface by repeated thrusts of the stylet. They puncture the cell wall, inject saliva into the cell, suck a part of

TABLE 8.1
Seed Infection of Nematodes

Nematodes	Host	Part(s) of Seed Infected	Important References
Anguina tritici	*Triticum aestivum, Avena sativa, Secale cereale,* and other grasses	Seed galls	Byars, 1920; Gupta and Swarup, 1968; Southey, 1972; Agarwal, 1984; Midha and Swarup, 1974
A. agrostis	*Agrostis* spp. and other grasses	Seed galls	Courtney and Howell, 1952; Stynes and Bird, 1982; Southey, 1973
A. agropyronifloris	*Agropyron smithii*	Seed galls	Norton and Sass, 1966
Ditylenchus destructor	*Arachis hypogaea*	Hull, seed coat, and surface of cotyledons	De Waele et al.,1989; Jones and De Waele, 1990; Hooper, 1973
D. africanus	*Arachis hypogaea*	Hull, seed coat, and embryo	Venter et al., 1995
D. dipsaci	*Vicia faba*	Seed coat	Neergaard, 1979
	Allium sativum	Hilum, below seed coat	Goodey, 1945; Southey, 1965
Aphelenchoides besseyi	*Oryza sativa*	Beneath hull	Thorne, 1961; Huang and Huang, 1972; Franklin and Siddiqui, 1972
A. arachidis	*Arachis hypogaea*	Seed coat	Bos, 1977; McDonald et al., 1979
A. ritzembosi	*Callistephus* sp.	Below seed coat	Siddiqui, 1974; Caubel, 1983
A. blastophothorus	*Callistephus* sp.	Below seed coat	Caubel, 1983
Pratylenchus brachyurus	*Arachis hypogaea*	Hull	Good et al., 1958; Corbett, 1976

the cell contents, and move on within a few seconds. Nematodes may also enter tissues through natural openings, such as stomata, lenticels, and cracks in the surface. Dropkin (1969, 1977) has given an account of the process of infection by plant parasitic nematodes and the various cellular responses. Species of *Anguina* stimulate gall formation in flower parts of grasses. During gall formation hypertrophy and hyperplasia of the parenchyma cells of the pericarp take place, and the central cavity harbors the nematodes. No syncytia (giant cells) develop in tissues attacked by nematodes.

8.2 HISTOPATHOLOGY

Of the genera that cause internal infection of seeds, *Ditylenchus, Aphelenchoides,* and *Pratylenchus* are soil- as well as seed-borne. They are facultative endoparasites and produce most stages in soil. Usually the preadult stage (fourth stage) is infective. Other stages, except egg and first-stage larvae, are also known to cause infection. *Anguina* shows more marked adaptation to parasitism. They are obligate parasites and normally complete their development only after invading the inflorescence, but, rarely, galls containing adults may be formed on leaves. Second-stage larvae that remain in seed galls are released in soil. These become associated with the host seedlings on which they feed ectoparasitically until they invade the floral parts.

8.2.1 *ANGUINA TRITICI* (STEINBUCK) CHITWOOD (EAR COCKLE DISEASE)

The ear cockle disease of wheat caused by *Anguina tritici* was the first plant nematode disease that was discovered. It was observed by Needham, a Dutch clergyman, in 1743. The nematode is known to infect bread wheat in most wheat-growing countries of the world (Southey, 1972). It also infects emmer wheat (*Triticum dicoccum*), spelt wheat (*Triticum spelta*), rye (*Secale cereale*), oat (*Avena sativa*), and certain grasses. The nematode produces seed galls, which are the major source of inoculum.

8.2.1.1 Disease Cycle (Figure 8.1)

The galls contain second-stage infective quiscent larvae. On sowing, the galls, along with healthy seed, reach the soil. The galls absorb moisture, and the larvae become active and are released. They move in a thin film of water and reach different parts of the seedling, including roots and parts of seed, aerial parts (shoot), and spaces between coleoptile and leaf and between those of the leaf bases. The nematodes that become lodged in seed tissues are of little consequence in disease development, but those that reach aerial parts are then carried up with the growing seedling. The larvae feed ectoparasitically, affect growth, and cause various symptoms in plant parts. After the plant enters the reproductive phase, the larvae in large numbers lie around the differentiating florets in the spike enclosed by the boot leaf. Subsequently, they enter the primordia of floral structures at the early stages of development. The infected structures (stamen and/or carpel) develop into galls, which are green initially and turn brown-black at maturity.

After becoming endoparasitic, the second-stage larvae molt and differentiate into adult males and females. The nematode is amphimictic. The females lay eggs inside the galls. The eggs undergo embryogenesis and produce second-stage larvae in the green galls. These larvae remain in quiescent form in dried galls. The total life period is between 106 to 113 days (Swarup, Dasgupta, and Koshy, 1989). Only one generation is completed in a crop season.

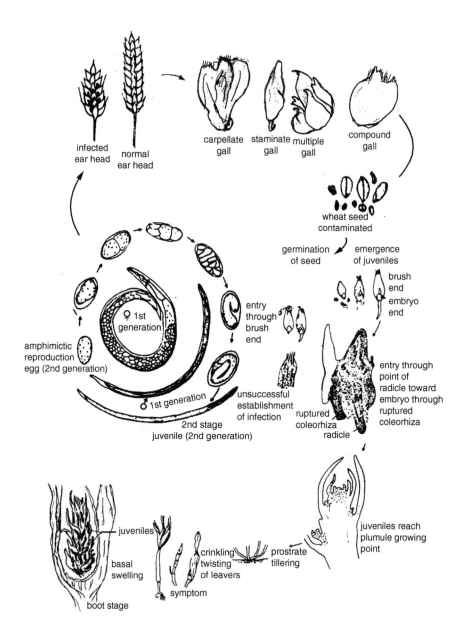

FIGURE 8.1 Life cycle of *Anguina tritici* in wheat. (Adapted and redrawn after Swarup, G., Sethi, C.L., and Gokte, N. 1993. In *Seed-Borne Diseases and Seed Health Testing of Wheat.* Mathur, S.B. and Cunfer, B.M., Eds. Danish Government Institute of Seed Pathology for Developing Countries, Copenhagen. Denmark.)

8.2.1.2 Origin of Galls

Ater macroscopic studies, Marcinowski (1910) concluded that galls could arise from undifferentiated as well as differentiated tissues of spikelets. Byars (1920) agreed largely with Marcinowski that occasionally one or two stamens of a flower may be normal while others are transformed into galls. Midha and Swarup (1974) observed that it is the staminate tissue that is primarily involved in gall formation. Carpellate galls are also formed. On the basis of macroscopic and microscopic observations, Agarwal (1984) traced the origin of seed galls to the anther, ovary, or anther–ovary together (Figure 8.2A, B to F). In all, seven types of galls are identified, divided into three broad categories.

Category 1: Staminate galls
    (i) single anther gall (Figure 8.2B)
    (ii) double anther gall
    (iii) triple anther gall

    Galls, (i) to (iii) are sessile, but since sporogenous tissue is seen in such developing galls, they are called anther galls rather than staminate galls. The basal part of primordium forming filament fails to develop after the infection has taken place. However, when the infection is delayed or gall formation is slow, partial galls with distinct filament develop.

Category 2: Carpellate galls
    (iv) ovary gall (Figure 8.2C)

    Usually these galls do not show any evidence of style–stigma differentiation. However, the normal half of the carpel of the partial ovary galls show well-developed feathery stigma.

Category 3: Staminate–carpellate galls
    (v) single anther–ovary gall (Figure 8.2D)
    (vi) double anther–ovary gall.
    (vii) triple anther–ovary gall (Figure 8.2F)

The development of various types of galls may take place in the same spikelet. Once a particular structure becomes infected in a spikelet, other floral parts also continue to develop and can be observed as whitish, pale, and sterile structures (Figure 8.2B to F).

Agarwal (1984) made an extensive study on mature infected field crop and experimental potted plants in order to collect evidence on the types of galls produced in wheat. Out of the 1257 green galls studied, triple anther–ovary galls were predominant (74.62%), and the single anther galls were the second dominating type (11.06%).

Besides the anther and ovary galls, galls are rarely formed on leaves, glumes, palea, lemma, and its awn and lodicules. In exceptional cases, as many as four or five galls are seen in a spikelet.

8.2.1.3 Histology of Developing and Mature Galls

The histology of developing galls permits identification of galls developing from anther or ovary; therefore, a separate account is given for them.

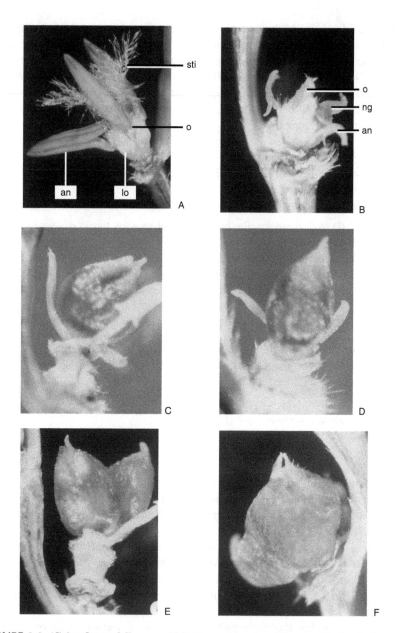

FIGURE 8.2 (Color figure follows p. 146.) Normal floret and florets with developing galls. Glumes, lemma, and palaea removed in all cases. A, Normal floret showing three stamens, ovary with feathery stigma, and lodicules. B, Floret showing developing single anther gall with two stamens and the ovary. C, Floret showing ovary gall and three stamens. D, Compound green gall formed by the ovary and one stamen. E, Two green galls in a floret, a double anther gall (left) and ovary gall (right). F, A compound triple anther and ovary gall. (Abbreviations: an, anther; lo, lodicules; ng, nematode gall; o, ovary; sti, stigma.) (From Agarwal, K. 1984. Ph.D. thesis, University of Rajasthan, Jaipur, India.)

Anther Gall: The young anther galls are oval to round with a prominent cavity occupied by a large number of second stage larvae and the degenerating sporogenous cells (Figure 8.3A to C). The wall comprises four to seven layers, differentiated into two zones — an outer zone of two to three layers and an inner zone of two to four layers of richly protoplasmic cells. The sporogenous tissue is completely absorbed by the time the final molting takes place and males and females are formed (Figure 8.3D). Subsequently, the females lay eggs, which undergo embryogenesis and molt to form second-stage larvae. Simultaneously, the wall cells undergo hyperplasia and become 12- to 14-layered. Inner layers remain parenchymatous, and they become absorbed during maturity. The outer layers become thick-walled and lignified. The mature galls are circular or slightly angular in transection, and the cavity has numerous second-stage larvae (Figure 8.3E).

Ovary Gall: The young ovary galls are more or less circular in outline in transections with a small cavity occupied by a few nematodes. The cavity enlarges and the larvae begin to molt, forming adult male and female nematodes. The wall cells undergo hyperplasia. They are thin-walled and parenchymatous. The females lay eggs (Figure 8.3F, G), which develop and molt to form second-stage larvae. The inner layers remain thin-walled and are gradually absorbed, while the outer ones become thick-walled and lignified (Figure 8.3H).

At maturity the ovary galls closely resemble the anther galls and are brown-black. The mature seed galls do show some differences in their morphology and color. In dry seed inspection, three categories of galls — bold, shriveled, and compound — are recognized. Longitudinal sections of compound galls often show a multicavity condition. No distinction is observed in the portion of the wall contributed by different organs.

8.2.2   *ANGUINA AGROSTIS* (STEINBUCH) FILIPJEV (BENT GRASS GALL NEMATODE)

This nematode was recorded for the first time in 1797 from galls in spikes of colonial bent grass (*Agrostis tenuis*) and has been subsequently recorded in many grasses (Thorne, 1961; Southey, 1973). It is a serious pest of *Lolium rigidum* (Stynes and Bird, 1982) and *Leymus chinensis* (Ma and Gu, 1987). The mode of infection and development of galls in the inflorescence of grasses by *A. agrostis* is similar to that of *A. tritici* (Courtney and Howell, 1952; Stynes and Bird, 1982). Spindle-shaped purple galls are produced in the inflorescence of the infected plants in *A. tenuis* and other grasses.

8.2.2.1   Histology and Development of Galls

The infective second-stage larvae of *A. agrostis* colonizes the plant in the vegetative phase, migrates toward the main and lateral shoot apices, and remains in their vicinity. A large number of nematodes collect around the floret primordia (Figure 8.4A). Stynes and Bird (1982) believed that the nematodes feeding on the primordium of the gynoecium induce rapid cell division and enlargement. This growth and simultaneous local destruction of cells produce a depression, and then finally a gall with

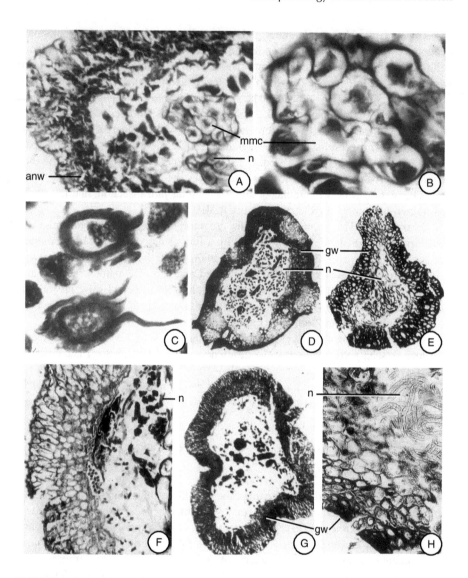

FIGURE 8.3 Histology of developing and mature galls caused by *Anguina tritici* in wheat. A to E, Anther gall. A, Ts part of young green gall showing nematode larvae and degenerating sporogenous tissue. B, C, Parts from A magnified to show degenerating sporogenous cells and second stage larvae. D, Ts green gall showing adult nematodes and eggs. E, Ts mature gall, cavity full of second-stage larvae and the wall of four to six lignified layers. F to H, Ovary gall. F, Part of gall showing eggs and adults. G, Ts gall at egg-laying stage. H, Ts part of mature gall showing second-tage larvae and thick and lignified cells of wall. (Abbreviations: anw, anther wall; gw, gall wall; mmc, microspore mother cells; n, nematode.) (Agarwal, K. 1984. Ph.D. thesis, University of Rajasthan, Jaipur, India.)

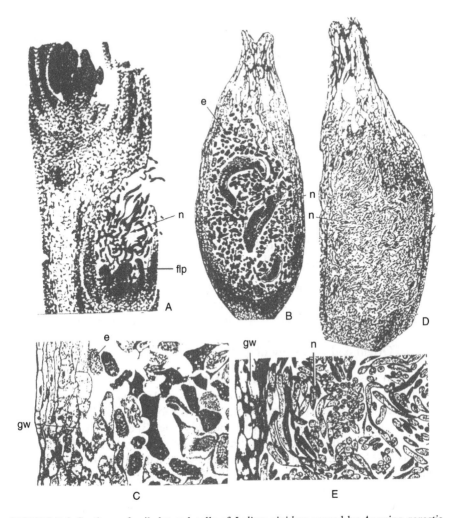

FIGURE 8.4 Sections of spikelet and galls of *Lolium rigidum* caused by *Anguina agrostis*. A, Ls part of spikelet showing large number of nematodes (second-stage larvae) around a floret primordium. B, Ls of gall showing adult nematodes and large number of eggs. C, A part from B magnified. D, Ls almost mature gall showing gall cavity full of larvae and thin wall. E, Part from D magnified showing nematodes (second-stage larvae) and thick-walled cells of gall wall. (Abbreviations: e, eggs; flp, floret primordium; gw, gall wall; n, nematode.) (A to E, From Stynes, B.A. and Bird, A.F. 1982. *Phytopathology* 72: 336–346. With permission.)

a cavity and a distinct opening. Such galls that develop in place of *ovules* and less commonly in place of stamens contain usually three or four, rarely more, nematodes. Both Stynes and Bird (1982) and Courtney and Howell (1952) concluded that the galls develop in the place of ovules and called them *ovular galls*. The authors would like to point out here that the developing primordium of the gynoecium in its ontogeny is open at the apex, and the cavity is enclosed by the ovary wall. The galls formed are the *ovary galls* since their wall is made up by the pericarp (ovary wall) rather than any part of the ovule. The nematodes may enter through the opening at the top or they may feed and enter after puncturing cells using their buccal stylet.

After the gall initiation has taken place, the nematode larvae molt, forming males and females (Figure 8.4B). Similar to *A. tritici*, this nematode is also amphimictic. The females lay eggs (Figure 8.4B, C) and finally second-stage larvae are hatched. Simultaneously, the galls enlarge, become 3 to 5 mm long and purplish due to the formation of anthocyanin pigment. Adults die in more advanced galls. The larvae are closely pressed against each other in mature dried galls. The cells of the gall wall lose cytoplasmic contents, collapse, and develop weak thickenings. The cells lining the cavity have callose deposition (Figure 8.4D, E).

In some florets, two galls occasionally develop. The additional gall may arise from stamen, glumes, or rachis. The structure of mature galls incited by *A. agrostis* on other hosts is also similar to that described here (Stynes and Bird, 1982).

8.2.3 ANGUINA AGROPYRONIFLORIS NORTON (WESTERN WHEATGRASS NEMATODE)

Anguina agropyronifloris produces galls in the inflorescence of western wheat grass, *Agropyron smithii* (Lakon 1953; Norton and Sass, 1966) and in golden oatgrass, *Trisetum flavescens* (Lakon, 1953). The galls are elongated with a brittle wall.

8.2.3.1 Origin and Histology of Galls

Norton and Sass (1966) studied plant infection and gall formation in artificially inoculated seeds by placing 300 to 1000 juveniles of *Anguina agropyronifloris* per seed in the greenhouse. The nematodes invade the roots and scutellar node as both ecto- and endoparasites. They also reach the axils of leaf sheaths. The endoparasitic infection of root and scutellar node does not cause floret infection. Ectoparasitic nematodes in leaf sheaths migrate to the apex and finally the inflorescence. They penetrate the undeveloped ovary as the earhead emerges from the boot leaf. Only fertile florets are invaded. Once the ovary is infected, other parts cease to develop. The larvae molt to produce adult males and females. The females lay eggs, which molt and form the second-stage larvae. The cells of the ovary wall elongate along the long axis. The wall is five- to six-layered, greatly compressed, weakly thick-walled, and brittle at maturity. The cavity is occupied by second-stage juvenile larvae (Norton and Sass, 1966).

8.2.4 DITYLENCHUS DESTRUCTOR THORNE (POTATO NEMATODE IN GROUNDNUT)

Ditylenchus destructor is an important pest of groundnut (*Arachis hypogaea*) in the Republic of South Africa (De Waele et al., 1989; Jones and De Waele, 1988). The groundnut hulls have brown necrotic tissue at the point of connection with peg and black discoloration along the longitudinal veins. The infected seeds are shrunken, and the testa and embryo show yellow to brown or black discoloration.

Jones and De Waele (1990) determined the time and mode of entry, and the spread of *D. destructor* in pod and seed of groundnut under field and greenhouse conditions. The developing pods are invaded after the fruiting pegs have penetrated the soil. The nematodes enter the pod near the point of connection with the peg (Figure 8.5D). They invade the peg and feed on the parenchyma cells, which collapse later. The nematodes enter the exocarp, feed on the parenchyma cells, and migrate to the base of the mesocarp (Figure 8.5A, E). They remain confined to the sclerenchymatous mesocarp for quite some time, and its cells (in contact with nematodes) become discolored. As the mesocarp breaks down, the nematodes reach the endocarp. They enter the seed through the micropyle, invade the seed coat (Figure 8.5F) and embryo (Figure 8.5C), and lodge on the surface of the cotyledons. At maturity the testa carry eggs and nematodes in the testa (Figure 8.5B, H). In the testa, nematodes may also occur in vascular bundles (Figure 8.5G).

Venter, McDonald, and van der Merwe (1995) have reported that *Ditylenchus africanus* Wendt et al. (peanut pod nematode) is both seed- and soil-borne. The nematodes survive in the testa and embryo of the seed and in the hull. *Ditylenchus angustus* has been found in dried seeds of rice, located mainly in the germ portion (Prasad and Varaprasad, 2002).

8.2.5 *Aphelenchoides besseyi* Christie (White Tip Nematode of Rice)

The disease was first described in the United States and Japan, but it is now known in most of the rice-growing countries of the world (Franklin and Siddiqui, 1972; Franklin, 1982). The nematode is an ectoparasite that is carried beneath the hull in the rice kernel. In India the disease has also been reported on *Setaria italica*, the fox tail millet (Lal and Mathur, 1988) and *Panicum melacium* (Gokte et al., 1990). Its other important host is strawberry in the United States and Australia and several other flowering plants in Hawaii, the Philippines, and Japan (Franklin, 1982).

Detailed histological observations are lacking. The infested seeds, upon reaching soil, absorb moisture and the preadult nematodes emerge and move in the thin film of water to reach different plant parts. They feed ectoparasitically on vegetative tissues, migrate to the growing panicle, puncture the inflorescence, reach the florets, and develop beneath the glumes. The nematode is amphimictic, and its density increases before anthesis. As the grain dries, second-stage larvae undergo anhydrobiosis and persist in the paddy kernel (Huang and Huang, 1972; Nandkumar et al., 1975). Nandkumar et al. (1975) have observed that in dry seed the nematodes remain coiled up inside the palea and on the surface of the lodicules, which become papery as the kernel matures.

Aphelenchoides besseyi was detected beneath the glumes in anhydrobiotic state in *P. miliaceum*. The maximum number of nematodes obtained from a seed was 16, and the average number was 1.8 (Gokte et al., 1990).

8.2.6 *Aphelenchoides arachidis* Bos (Testa Nematode of Groundnut)

The *testa* nematode of groundnut (*A. arachidis*) is a facultative endoparasite of roots, hypocotyl, pods, and testa (Bos, 1977). The nematode penetrates young roots and

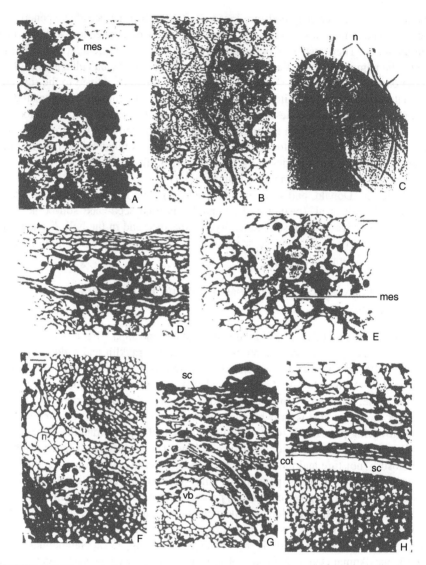

FIGURE 8.5 Histopathology of peanut pod infected with *D. destructor*. A to C, Naturally infected. A, Ts through the mesocarp at the base of the young pod showing an opening created by nematodes. B, Eggs and nematodes in testa of mature seed. C, Ls part of cotyledon showing nematodes and eggs. D to H, Artificially infected. D, Ts peg showing nematodes in parenchyma tissue. E, Ts mesocarp at the base of a young pod showing discolored sclerenchyma tissue and openings that developed following the presence of nematodes. F, G, Ts immature peanut seed showing nematodes in parenchyma and vascular bundle in seed coat. H, Ts mature seed showing nematodes in the parenchyma tissue of seed coat. (Abbreviations: cot, cotyledon; mes, mesocarp; n, nematode; sc, seed coat; vb, vascular bundle.) (From Jones, B.L. and De Waele, D. 1990. *J. Nematol.* 22: 268–272. With permission.)

hypocotyls, and it reproduces itself in very young plants. Affected seeds have translucent testa with dark brown vascular strands, as seen in fresh seeds. In dry seeds the testa is dark brown, wrinkled, and thicker than the normal testa. The epidermal layer of the testa is reduced, and the internal tissue is disorganized (McDonald, Bos, and Gumel, 1979). More than 25,000 nematodes are often found per testa. Bridge et al. (1977) reported presence of nematodes in the spaces between the shell, testa, and cotyledons, but not in the embryo (Bos, 1977; Bridge et al., 1977).

8.2.7 *Pratylenchus brachyurus* (Godfrey) Filipjev

Pratylenchus brachyurus also occurs in the pegs and hulls of groundnut (Good et al., 1958). Only macroscopic observations have been made. In pod shells nematodes are located in the tissue between the vascular network of the pericarp. The nematode has never been observed to enter the seed.

8.3 ASSOCIATION OF NEMATODE AND BACTERIA

Nematodes as vectors of fungi, bacteria, and viruses are reported in many publications. Several reviews have also appeared on the interaction of nematodes and other pathogens (see Southey, 1982). *Anguina* spp. are known to associate with bacteria belonging to the genus *Rathayibacter (Corynebacterium)*. In wheat, *Anguina tritici* acts as a vector carrying the bacterium on the external body surface (Gupta and Swarup, 1972; Suryanarayana and Mukhopadhaya, 1971), but in the case of *A. agrostis*, it is still not clear how *R. rathayi* appears in the galls. Under favorable conditions in wheat, *R. tritici (C. tritici)* multiplies very quickly in the young gall, forming a thick viscous fluid in which the nematode larvae are unable to survive. The emerging ears are sterile and covered with the yellow shiny mass of the bacterium. The disease is called *yellow slime rot* or *tundu disease*. Galls are not formed in affected ears under such conditions. However, under less favorable conditions for the bacterium, partial ear cockle and yellow ear rot symptoms are produced.

Anguina agrostis and its associate, *R. rathayi (C. rathayi)*, form distinct galls in *Lolium rigidum*. Two types of galls produced are readily recognized. A detailed histopathological study on these galls was carried out by Bird, Stynes, and Thomson (1980).

Anguina agrostis and *R. rathayi (C. rathayi)* incite nematode and bacterial galls, respectively, in the inflorescence of *L. rigidum*. Two types of seed galls are formed. They are almost of the same size (Figure 8.6A, B), but those containing nematodes are dark brown, and others colonized by bacteria are bright yellow. The walls of galls containing nematodes are about twice as thick (Figure 8.6C) as the walls of the bacterial galls (Figure 8.6E). The gall cavity in nematode galls is occupied by numerous coiled, closely packed infective second-stage larvae (Figure 8.6D). The interior of the bacterial gall is full of *R. rathayi*, arranged in a regular array, both along the wall and within the cavity of the gall (Figure 8.6E). In addition to bacteria, freeze-fracture preparations viewed under TEM show numerous particles 25 to 30 nm in diameter. These particles occur in association with the bacteria and the cell walls of galls (Figures 8.6F, G). Their presence was also recorded in freeze-etch preparations of cultured *R. rathayi*. Bird, Stynes, and Thomson (1980) suspect them to be corynephages. It is well known that the galls induced by *A. agrostis* in *L. rigidum* become toxic to animals when further colonized by *R. rathayi* (Stynes et al., 1979).

8.4 SURVIVAL IN SEED

In general, nematodes occur as endophytes in seed galls, and they are quiescent, having a very low metabolic activity. These features enable them to overcome unfavorable environmental conditions and to extend their survival during storage. However, information on their survival and the mode of anhydrobiosis in seed is not available in all cases.

The second-stage juveniles of *A. tritici* in seed galls of wheat are able to remain dormant for many years. Byars (1920) reported successful reactivation of larvae in 1920 from galls imported from Turkistan in 1910. He cited a case of resumed vital activity after a dormancy of 27 years. Fielding (1951) found 100% revival of larvae from galls stored for 28 years under dry conditions while Reeder (1954) observed that galls can survive up to 8 or 9 or even 14 years under moist conditions. The dormant larvae in galls are also resistant to temperature changes, but they are unable to withstand very high temperatures. In galls kept at 60°C for 10 min, all the larvae were killed (Heald, 1933). Mukhopadhyaya, Chand, and Suryanarayana (1970) found that only 30% of larvae survived in the galls placed at 42°C for 7 days. They also studied the survival of nematodes inside the galls at various soil depths and conditions in India. Many larvae survived up to 3 months at 20 cm depth and few survived at any depth after 4 months.

The dry nematode galls in *L. rigidum* contain coiled and closely apposed infective anhydrobiotic larvae of *A. agrostis*. Bird, Stynes, and Thomson (1980) found that these larvae in dry galls can withstand the dryness and heat of the Australian summer.

The white tip nematode of rice (*Aphelenchoides besseyi*) could survive in dehydrated rice grains from 23 months to 8 years (Todd and Atkins, 1959). Yoshu and Yamoto (1950) found 75% survival of *A. besseyi* in grains kept for 3 years.

Basson, De Waele, and Meyer (1993) studied the survival of *Ditylenchus destructor* in soil, hull, and seeds of groundnut cultivar Sellie. *Ditylenchus destructor* can survive in the field in the absence of host plants and in hulls left in the field after harvest for at least 7 to 8 months, but very few nematodes survived in whole seeds stored in paper bags at 10°C. The surviving nematodes, however, were able to build up large populations. A gradual decrease in the survival of nematodes with increasing time, especially during the first 3 months, was recorded in fragmented hulls and seeds stored at 22°C.

Ditylenchus dipsaci survives adverse conditions by forming *nematode wool*. The nematode loses up to 99% of its body water and undergoes anhydrobiosis. The nematode larvae have been found to survive in dried teasel for 23 years (Fielding, 1951).

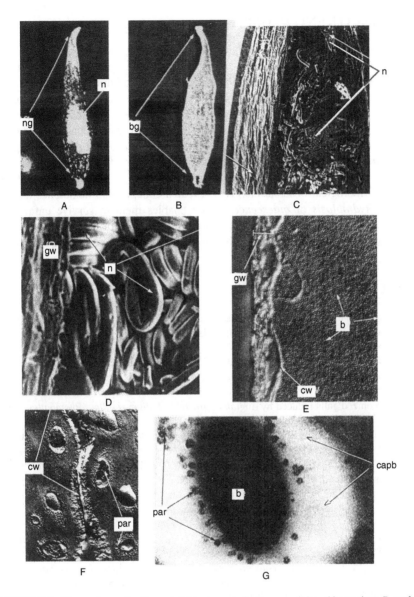

FIGURE 8.6 Comparison of galls containing nematode (*A. agrostis*) and bacterium *R. rathayi* (*C. rathayi*) in *L. rigidum*. A, Seed gall with wall cut longitudinally to show nematodes. B, Gall colonized by bacteria. C, D, Scanning electron micrographs showing parts of nematode gall, and enlarged part showing the regularly packed infective larvae, respectively. E, Nomarski differential interference contrast photograph of a section of a gall colonized by bacteria. F, TEM of a freeze-etched hydrated gall colonized by bacteria. G, A magnified bacterium showing bacterial capsule and adherent particles. (Abbreviations: b, bacteria; bg, gall containing bacteria; capb, capsule of bacterium; cw, cell wall; gw, gall wall; n, nematodes; ng, nematode gall; par, particle.) (From Bird, A.F., Stynes, B.A., and Thomason, W.W. 1980. *Phytopathology* 70: 1104–1109. With permission.)

8.5 CONCLUDING REMARKS

Critical morphological and histological studies of nematode infections of seeds, wherever completed, have conclusively shown the path of penetration, the parts of the floret forming galls, location in the seed, and the conditions under which nematodes occur in seed. There are several examples in which close association of nematodes and seed tissues are known, e.g., *D. dipsaci* in lucerne, clover, onion, faba bean, and broad bean; *D. africanus* and *P. brachyurus* in groundnut; *A. besseyi* in rice; *A. africanus* in groundnut; *A. ritzembosi* and *A. blastophorus* in *Callistephus;* *Heterodora* in bean, pea, and sugarbeet; and *Rhadinaphaenchus* in coconut. Detailed information on the nematode association in the majority of these cases is lacking.

Histopathological observations have revealed changes that nematodes undergo during *anhydrobiosis* or *quiescence,* which enables them to survive unfavorable conditions. Information on location and period of survival of nematodes in seed is useful for seed certification and quarantine, and for developing control strategies.

REFERENCES

Agarwal, K. 1984. Studies on Seed-Borne Mycoflora and Some Important Seed-Borne Diseases in Wheat Cultivars Grown in Rajasthan. Ph.D. thesis, University of Rajasthan, Jaipur, India.

Basson, S., De Waele, D., and Meyer, A.J. 1993. Survival of *Ditylenchus destructor* in soil, hulls and seeds of groundnut. *Fundam. Appl. Nematol.* 16: 79–85.

Bird, A.F., Stynes, B.A., and Thomson, W.W. 1980. A comparison of nematode and bacteria colonized galls induced by *Anguina agrostis* in *Lolium rigidum*. *Phytopathology* 70: 1104–1109.

Bos, W.S. 1977. *Aphelenchoides arachidis* n.sp. (Nematoda: Aphelenchoidea), an endoparasite of the testa of groundnuts in Nigeria. *Z. Pflanzenkr. Pflanzenschutz* 84: 95–99.

Bridge, J., Bos, W.S., Page, L.J., and McDonald, D. 1977. The biology and possible importance of *Aphalenchoides arachidis*: a seed-borne endoparasitic nematode of groundnuts from northern Nigeria. *Nematologica* 23: 255–259.

Byars, L.P. 1920. The nematode disease of wheat caused by *Tylenchus tritici*. *Bull. U.S. Dept. Agric.* 1042: 40.

Caubel, G. 1983. Epidemiology and control of seed-borne nematodes. *Seed Sci. Technol.* 11: 989–996.

Corbett, D.C.M. 1976. *Pratylenchus brachyurus*. C.I.H. descriptions of plant-parasitic nematodes. Set 6, No. 89. CAB International, Wallingford, U.K.

Courtney, W.D. and Howell, H.B. 1952. Investigation of the bent grass nematode, *Anguina agrostis* (Steinbuch, 1799) Filipjev, 1936. *Plant Dis. Rep.* 36: 75–83.

De Waele, D., Jones, B.L., Bolton, C., and Van den Berg, E. 1989. *Ditylenchus destructor* in hulls and seeds of peanut. *J. Nematol.* 21: 10–15.

Dropkin, V.H. 1969. Cellular reponses of plants to nematode infections. *Ann. Rev. Phytopathol.* 7: 101–122.

Dropkin, V.H. 1977. Nematode parasites of plants, their ecology and the process of infection. In *Encyclopaedia of Plant Physiology* (New Series). Heitefuss, R. and Williams, P.H., Eds. Springer-Verlag, Berlin. Vol. 4, pp. 222–242.

Fenuwick, D.W. 1957. Red ring disease of coconuts in Trinidad and Tobago. Colonial Office Report, London, p. 55.

Fielding, M.J. 1951. Observations on the length of dormancy in certain plant infecting nematodes. *Proc. Helminth. Soc. Wash.* 18: 110–112.

Franklin, M.T. and Siddiqui, M.R. 1972. *Aphelenchoides besseyi.* C.I.H. descriptions of plant — parasitic nematodes. Set 1, No. 4. CAB International, Wallingford, U.K.

Franklin, M.T. 1982. *Aphelenchoides* and Related Genera. In *Plant Nematology*, Southey, J.F., Ed. Her Majestry's Stationery Office, London, pp. 172–187.

Gokte, N., Mathur, V.K., Rajan, and Lal, A. 1990. *Panicum miliaceum* — A new host record for *Aphelenchoides besseyi. Indian J. Nematol.* 20: 111–112.

Good, J.M., Boyle, I.W., and Hammons, R.O. 1958. Studies of *Pratylenchus brachyurus* on peanuts. *Phytopathology* 48: 530–535.

Goodey, T. 1945. *Anguillulina dipsaci* on onion seed and its control by fumigation with methyl bromide. *J. Helminth.* 21: 45–59.

Gupta, P. and Swarup, G. 1968. On the earcockle and yellow ear rot diseases of wheat. I. Symptoms and histopathology. *Indian Phytopathol.* 21: 318–323.

Gupta, P. and Swarup, G. 1972. Earcockle and yellow ear rot disease of wheat. II. Nematode bacterial association. *Nematologica* 18: 320–324.

Heald, F.D. 1933. *Manual of Plant Diseases.* McGraw-Hill, New York.

Hooper, D.J. 1973. *Ditylenchus destructor.* C.I.H. descriptions of plant-parasitic nematodes. Set 2, No. 21. CAB International, Wallingford, U.K.

Huang, C.S. and Huang, S.P. 1972. White tip nematode in florets and developing grains of rice. *Bot. Bull. Acad. Sinica* (Taiwan) 193: 1–10.

Jones, B.L. and De Waele, D. 1988. First report of *Ditylenchus destructor* in pods and seeds of peanut. *Plant Dis.* 72: 453.

Jones, B.L. and De Waele, D. 1990. Histopathology of *Ditylenchus destructor* on peanut. *J. Nematol.* 22: 268–272.

Lakon, G. 1953. Die Alchen-Fruchtgallen der Gramineen. *Saatgutwirtschaft* 5: 257–258.

Lal, V. and Mathur, V.K. 1988. Record of *Aphelenchoides besseyi* on *Setaria italica. Indian J. Nematol.* 18: 131.

Ma, H.L. and Gu, X.Z. 1987. Discovery and preliminary study in *Anguina* in *Leymus chinense. Grassl. China* 4: 29–33.

Marcinowski, K. 1910. Parasitisch und semiparasitisch an Pflanzen lebende Nematoden. *Arb. Kais. Biol. Anst. Land. Forstwirtsch.* 7: 1–192.

McDonald, D., Bos, W.S., and Gumel, M.H. 1979. Effects of infestations of peanut (groundnut) seed by the testa nematode, *Aphelenchoides arachidis* on seed infection by fungi and on seedling emergence. *Plant Dis. Rep.* 63: 464–467.

Midha, S.K. and Swarup, G. 1974. Studies on the wheat gall nematode caused by *Anguina tritici. Indian J. Nematol.* 4: 53–63.

Mukhopadhyaya, M.C., Chand, J.N., and Suryanarayana, D. 1970. Studies on the longevity of earcockles. *Punjab Agric. Univ. J. Res.* 7: 625–627.

Nandkumar, C., Prasad, J.S., Rao, Y.S., and Rao, J. 1975. Investigations on the white-tip nematode *Aphelenchoides besseyi* Christie, 1942 of rice (*Oryza sativa* L.). *Indian J. Nematol.* 5: 62–69.

Needham, T. 1743. A letter concerning certain chalky tubulous concretions called malm, with some microscopical observations on the farina of red lily and of worms discovered in smutty corn. *Philos. Trans. Roy. Soc. London* 2: 173, 174, 634–641.

Neergaard, P. 1979. *Seed Pathology.* Vols. 1 and 2. Macmillan Press, London.

Norton, D.C. and Sass, J.E. 1966. Pathological changes in *Agropyron smithii* induced by *Anguina agropyronifloris. Phytopathology* 56: 769–771.

Prasad, J.S. and Varaprasad, K.S. 2002, Ufra nematode, *Ditylenchus angustus* is seed-borne. *Crop Prot.* 21: 75–76.

Reeder, J.N. 1954. A note on the longevity of wheat nematode, *Anguina tritici* Steinbuch. *Plant Dis. Rep.* 38: 268–269.

Siddiqui, M.R. 1974. *Aphelenchoides ritzemabosi* C.I.H. descriptions of plant-parasitic nematodes. Set 3, No. 32. CAB International, Wallingford, U.K.

Southey, J.F. 1965. The incidence and location of stem eelworms on onion seed. *Plant Pathol.* 14: 55–59.

Southey, J.F. 1972. *Anguina tritici*. C.I.H. descriptions of plant-parasitic nematodes. Set 1, No. 13. CAB International, Wallingford, U.K.

Southey, J.F. 1973. *Anguina agrostis*. C.I.H. descriptions of plant-parasitic nematodes. Set 2, No. 20. CAB International, Wallingford, U.K.

Southey, J.F., Ed. 1982. *Plant Nematology*. Her Majesty's Stationery Office, London, U.K.

Stynes, B.A. and Bird, A.F. 1982. Development of galls induced in *Lolium rigidum* by *Anguina agrostis*. *Phytopathology* 72: 336–346.

Stynes, B.A., Petterson, D.S., Lloyd, J., Payne, A.L., and Lanigan, W. 1979. The production of toxin in annual ryegrass, *Lolium rigidum*, infected with a nematode, *Anguina* sp. and *Corynebacterium rathayi*. *Aust. J. Agric. Res.* 30: 201–209.

Suryanarayana, D. and Mukhopadhyaya, M.C. 1971. Earcockle and tundu diseases of wheat. *Indian J. Agric. Sci.* 41: 407–413.

Swarup, G., Dasgupta, D.R., and Koshy, P.K. 1989. *Plant Diseases*. Anmol Publications, New Delhi, India.

Swarup, G., Sethi, C.L., and Gokte, N. 1993. Ear cockle (nematode gall) and yellow slime. In *Seed-Borne Diseases and Seed Health Testing of Wheat*. Mathur, S.B. and Cunfer, B.M., Eds. Danish Government Institute of Seed Pathology for Developing Countries, Copenhagen, Denmark.

Thorne, G. 1961. *Principles of Nematology*. McGraw-Hill, New York.

Todd, E.H. and Atkins, J.G. 1959. White tip disease of rice. II. Seed treatment studies. *Phytopathology* 49: 184–188.

Venter, C., McDonald, A.H., and van der Merwe, P.J.A. 1995. Integrated control of the peanut pod nematode on groundnuts. 12th Symposium Nematological Society of Southern Africa, March 1995. (Text of poster presented, personal communication.)

Yoshu, H. and Yamamota, S. 1950. A rice nematode disease "Senchu shingarebyo." I. Symptoms and pathogenic nematode. *J. Fac. Agric. Kyushu Univ.* 9: 210–222.

9 Physiogenic or Nonpathogenic Seed Disorders

Seed disorders in which no recognizable pathogenic incitant is associated are known as physiogenic or nonpathogenic diseases. Fungi, bacteria, viruses, mycoplasma, nematodes, and spermatophytic parasites are common pathogenic causes of plant diseases. Nonpathogenic abnormalities are caused by disorders in the physiology of plants due to unfavorable environments including soil conditions.

Physiogenic diseases may be induced by (1) unfavorable soil conditions that may be due to the deficiency of essential elements, macroelements (nitrogen, phosphorus, calcium, magnesium, potassium, and sulfur), or microelements (iron, manganese, boron, zinc, copper, and molybdenum), adverse water relations (draught, water logging, and impeded aeration), and adverse physiochemical conditions (alkalinity, acidity, and salinity); (2) climatic stresses, including temperature, light, and humidity; and (3) atmospheric environmental pollutants (gases such as ethylene, ammonia, sulfur dioxide, nitrogen dioxide, hydrogen fluorides, and particulates).

Wallace (1951) has given a good account of diseases caused by mineral deficiencies in plants. Scaife and Turner (1984) and Bennett (1993) have discussed nutrition deficiencies and toxicity in crop plants. Levitt (1972) has provided information on the responses of plants to environmental stresses. In recent years, considerable information has been generated on air pollution and injuries to plants. Some of the important publications are those of McMurtney (1953), Darley and Middleton (1966), Jacobson and Hill (1970), Lacasse and Treshow (1976), Pell (1979), Lawrence and Weinstein (1981), and Evans (1984). Agrios (1988) has included a chapter on environmental factors that cause plant diseases in his book, while Neergaard (1979) has given a good account of nonpathogenic/physiogenic seed disorders.

Aerial parts of plants, particularly growth and features of stem and leaves, are commonly affected by nonpathogenic factors. Some of them also affect the reproductive phase, causing failure of seed formation. Only a few cause notable disorders in seeds (Table 9.1). Seed disorders for which adequate histological information is available are described under separate headings.

TABLE 9.1
Major Seed Disorders of Nonpathogenic Origin

Cause	Plant species	Symptoms	Important References
Nutrient Deficiency			
Macronutrients			
Nitrogen	*Triticum aestivum*	Localized or entire grain with light yellowish spots — yellow berry of wheat	Neergaard, 1979
Potassium	*Pisum sativum, Cucumis sativus*	Failure of seed to mature, small seeds; abnormally tapering seeds in cucumber	Eckstein et al., 1937; Stapel and Bovien, 1943; Hoffman, 1933
Micronutrients			
Manganese	*Pisum sativum*, beans — broad bean, harricot bean, runner bean	Marsh spot or cotyledons with brown necrosis on adaxial faces	Mansholt, 1894; Perry and Howell, 1965; Singh, 1974
Boron	*Arachis hypogaea*	Hollow heart or cavitation of cotyledons on adaxial surface, region turns brown when seeds roasted	Harris and Gilman, 1957; Bennett, 1993
Zinc	*Vicia faba*	Seeds small with depression on either sides of hilum, variation in height and shape of papillae and deposition of wax on seed surface	Gupta et al., 1994
Humidity Effects			
Low humidity or physiological draught	*Pisum sativum, Cicer arietinum*	Cotyledons with transverse cracks	Neergaard, 1979; Jain, 1984
	Pisum sativum, Cicer arietinum, Vicia faba, Phaseolus vulgaris, Tropaeolum majus	Hollow heart or cavitation on adaxial face of cotyledons	Myers, 1947, 1948; Perry and Howell, 1965; Allen, 1961; Perry and Harrison, 1973; Mattusch, 1973; Singh, 1974; Jain and Singh, 1985
	Lactuca sativa	Physiological necrosis or spotted cotyledons	Dempsey and Harrington, 1951; Finley, 1959; Bass, 1970; Smith, 1989
High humidity	*Sinapsis alba, Raphanus sativus*	Gray discoloration of seeds	Neergaard, 1945; Jørgensen, 1967

9.1 MINERAL NUTRIENT DEFICIENCY

The deficiency in soil of a number of major or micronutrients leads to the development of visible symptoms in seeds (Table 9.1). The symptoms may be exomorphic, caused externally, e.g., yellow berry of wheat, tapering seeds in cucumber, and distorted seeds of *Vicia faba* due to the deficiency of nitrogen, potassium, and zinc, respectively, or endomorphic (internal), mostly seen on the adaxial surface of cotyledons, namely marsh spot of peas and beans, and hollow heart in groundnut caused by the deficiency of manganese and boron, respectively. The use of the term *hollow heart* for the disorder in groundnut is a misnomer since this term is widely used for an anomaly caused in peas due to low atmospheric humidity at the time of the maturity of the crop. Detailed histopathological information is available only on pea seeds showing marsh spot.

9.1.1 MARSH SPOT IN PEAS

Marsh spot is characterized by discolored brown areas in the center of the adaxial face of the cotyledons (Figure 9.1A, affected cotyledons [right] and normal [left]). The disorder was first described by Mansholt in 1894 and later observed in the Netherlands (De Bruijn, 1933), Britain (Lacey, 1934), and Finland (Jamalainen, 1936). It is common in peas grown in Romney Marsh, England, and in newly reclaimed sea silt soils of the Netherlands (Noble, 1960). Singh (1974) detected marsh spot in seeds of smooth seeded cultivars, namely Diktron, Marrow (the Netherlands) and Feltham First (Scotland). Manganese deficiency in soil is shown to be the cause of this disorder (De Bruijn, 1933; Löhnis, 1936; Pethybridge, 1936; Heintze, 1938, 1956; Cuddy, 1959), and manganese sulphate as a measure of control was established by these authors.

Wallace (1951) has figured brown lesion on adaxial surface of cotyledons, similar to marsh spot, in broad bean, harricot bean and runner bean. He also attributes this disorder to manganese deficiency of soil.

9.1.1.1 Histology

Anatomically the asymptomatic cotyledons consist of parenchyma cells, which, apart from the epidermis, are all polygonal (Figure 9.1B). None of the mesophyll layers are of the palisade type. Air spaces occur throughout the mesophyll (Figure 9.1B, C). The size and shape of the nuclei show great variations depending upon the size of the cell and the amount of the starch grains. The nuclei are hypertrophied, having diffused and granular intranuclear contents. An organized nucleus is rarely seen. The cells contain abundant simple starch grains (Figure 9.1B, C).

The affected cotyledons on the adaxial side reveal that the subhypodermal layers, three to four cells deep, show more serious symptoms of derangement. The cells and intercellular space are enlarged and the cell contents stained deeply with safranin. The cell walls become broad with granular material beneath. The cell contents, including starch grains, show corrosion and secretion of glistening dark brown oily droplets, which accumulate in intercellular air space (Figure 9.1D, E).

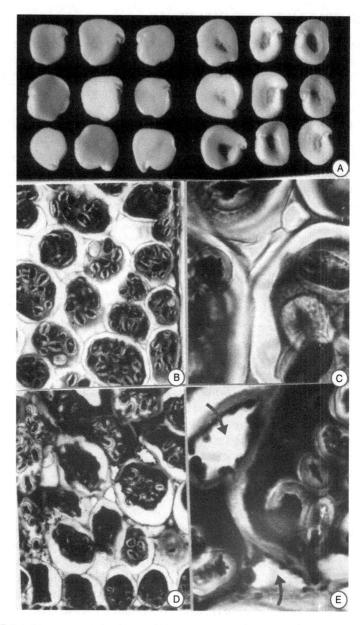

FIGURE 9.1 Photographs of normal and marsh spot-affected cotyledons of pea. A, Cotyledons showing adaxial surface from water-soaked split seeds of cultivar Koroza (Holland) — normal (left) and marsh spot affected (right). B to E, Histology of normal and marsh spot-affected cotyledons of cultivar Feltham First (Scotland). B, Ts part of normal cotyledon from adaxial side, showing clear air spaces and cells packed with reserve food contents. C, A portion from B magnified. D, E, Ts parts of marsh spot-affected cotyledon showing accumulation of pigmented material in air spaces (arrows) and depletion of cell contents. (From Singh, D. 1974. *Seed Sci. Technol.* 2: 443–456. With permission.)

A comparative account of ultrastructure of asymptomatic and symptomatic cotyledons is given by Singh and Mathur (1992). The cell walls are uniformly thick and more or less smooth in asymptomatic cotyledons. The fibrillar material is clear, and fine plasmodesmatal connections occur across the cell wall. The plasma membrane lies close to the primary wall, and the cytoplasmic net is well organized, showing cell organelles such as endoplasmic reticulum, mitochondria, vacuoles, ribosomes, protein bodies, and starch grains. The aggregation of endoplasmic reticulum, mostly rough endoplasmic reticulum, and mitochondria is much more in the peripheral cytoplasm. The protein bodies vary in size and are deeply stained (Figure 9.2A).

In symptomatic cotyledons the cell walls become broad, showing loosening of fibrillar material, and they are prominently sinuate with well-developed plasmodesmatal channels connecting adjoining cells and the intercellular spaces (Figure 9.2B). The plasma membrane is deeply invaginated. The ribosomes and mitochondria are disfigured. The protein bodies are membrane bound, but the deposition of reserve material is poor and vacuolation in these bodies produces bizarre shapes (Figure 9.2B). Initially, the cell contents do not reveal the occurrence of oil droplets, but their appearance in the intercellular spaces is indicative of secretion. The starch grains are greatly corroded.

Increased derangement leads to severe subcellular effects. The cell walls show further loosening of fibrils, broad plasmodesmatal channels, and accumulation of a large amount of granular material beneath the cell wall (Figure 9.2C). Oily material occurs interspersed among these granules. The plasma membrane is dissociated from the cell wall and shows prominent breaks. The cytoplasmic net shows disintegration, small fragments of endoplasmic reticulum randomly dispersed, weakening or breakdown of the unit membrane around the cell organelles, and protein bodies. The remains of protein bodies show great variation in shape (Figure 9.2B). Air spaces are greatly enlarged and show accumulation of a large amount of pigmented oily secretion (Figure 9.2D).

9.2 HUMIDITY EFFECTS

Climatic stresses, particularly humidity, low or high, cause visible symptoms in seeds of crop plants (Table 9.1). The symptoms may be exomorphic, e.g., gray seeds of radish and white mustard, or endomorphic, discoloration (lettuce), transverse cracks (pea and chickpea) and cavitation on adaxial face (pea and chickpea) of cotyledons. Histology of hollow heart in pea and gray discolored seeds in white mustard is known (Perry and Howell, 1965; Singh, 1974; Jørgensen, 1967).

9.2.1 Low Humidity Effects

9.2.1.1 Hollow Heart

Hollow heart or cavitation (Figure 9.3A) on the adaxial face of cotyledons of pea (*Pisum sativum*) was first observed by Myers (1947, 1948) in seeds from New Zealand, Australia, and the United States. Perry and Howell (1965) examined more

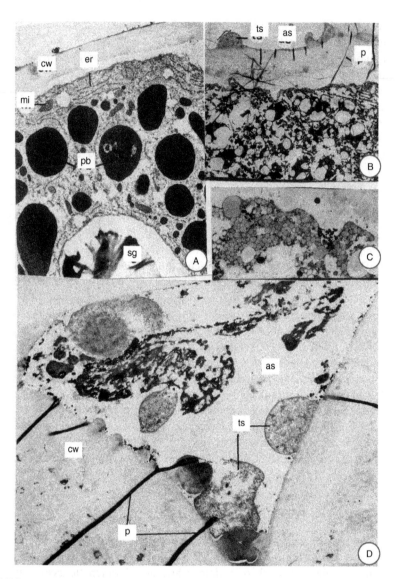

FIGURE 9.2 Ultrastructure of normal and marsh spot-affected cotyledons of pea. A, TEM photomicrograph showing part of the normal cotyledon cell. Note the uniformly thick cell wall, plasma membrane, and the cytoplasmic net with cell organelles. B, TEM of marsh spot-affected cotyledon cell showing changes in the cell wall, broadened plasmodemata, sinuate plasma membrane, vacuolated protein bodies, and pigmented contents in air spaces. C, D, TEM of severely affected cotyledons. C, Portion of cell beneath the cell wall showing aggregation of granular material. D, A portion showing broad cell wall with well-developed plasmodematal channels and air space with pigmented tanniferous material. (Abbreviations: as, air space; cw, cell wall; er, endoplasmic reticulum; mi, mitochondria; p, plasmodemata; pb, protein bodies; sg, starch grain; ts, tanniferous secretion.) (From Singh, D. and Mathur, R. 1992. *Phytomorphology* 42: 145–150.)

than 150 samples of pea seed from England, Hungary, the Netherlands, New Zealand, and the United States. and reported that hollow heart is a common disorder of wrinkle-seeded peas. Singh (1974), who tested seeds from Argentina, Czechoslovakia, Denmark, India, Lesotho, the Netherlands, New Zealand, Romania, Scotland, Turkey, and the United States, also concluded that the wrinkle-seeded cultivars have a moderate to heavy incidence of hollow heart whereas the smooth-seeded cultivars show either no or a very low incidence. Perry and Howell (1965) also observed that cultivar Alaska, a round-seeded pea, consistently had the least hollow heart.

Jain and Singh (1985) examined 304 chickpea seed samples belonging to 297 cultivars (254 cultivars desigram and 43 cultivars kabuligram) from India for hollow heart. A total of 156 cultivars (61%) of desigram and 37 cultivars (86%) of kabuligram showed hollow heart incidence of 0.5 to 20% and 0.5 to 62%, respectively. The cavities on the adaxial surface of cotyledons were usually identical and rarely dissimilar or on one cotyledon only.

Hollow heart in pea may be due to too quick drying of immature seed. Perry and Harrison (1973) experimentally showed that predisposition to hollow heart may be caused by high ambient temperature during maturation of seeds on plants, and by drying them when immature. Hollow heart increased with increasing temperature from 35 to 45°C and decreased as seed maturity advanced.

Hollow heart has also been reported in *Vicia faba* and *Phaseolus vulgaris* (Perry and Howell, 1965) and *Tropaeolum majus* (Heit and Crosier, 1961).

9.2.1.1.1 Hollow Heart and Shape and Size of Seeds

Gane and Biddle (1973) observed the relationship between shape of seed and incidence of hollow heart in seed samples of cultivars Lincoln and Kelvedon Wonder. In the former, triangular seeds showed a higher incidence of hollow heart, whereas this occurred in only two of the five samples of Kelvedon Wonder. Singh (1974) examined samples of four cultivars having seeds with three distinct shapes, namely, (1) squarish seeds with depression on two faces; (2) conical seeds with basal depression; and (3) irregular seeds. Squarish seeds formed the bulk in the samples. No positive correlation in the incidence of hollow heart and seed shape squarish, conical and irregular was found.

Seed samples of cultivar 4.12.01 7d Turkey had 44% large and heavy and 56% small and light seeds. The hollow heart incidence in small seeds was slightly higher (52%) than in the large seeds (42%) (Singh, 1974).

9.2.1.1.2 Histology

The histology of affected cotyledons has been examined in pea (Perry and Howell, 1965; Singh, 1974) and chickpea (Jain, 1984). Anatomically, asymptomatic cotyledons in wrinkle-seeded pea varieties consist of parenchyma cells as described under *marsh spot* (Figure 9.3B). But the starch grains are compound (Figure 9.3B, C) and the protoplasmic contents are evenly distributed in the cell.

In symptomatic cotyledons, the anatomical symptoms are prominent in the cells of mesophyll on the adaxial face, i.e., in the region of cavitation. The cells are loose with large air spaces. The protoplasmic contents are poor and aggregated in the center as if plasmolyzed (Figure 9.3D, E). However, the cells on the abaxial face

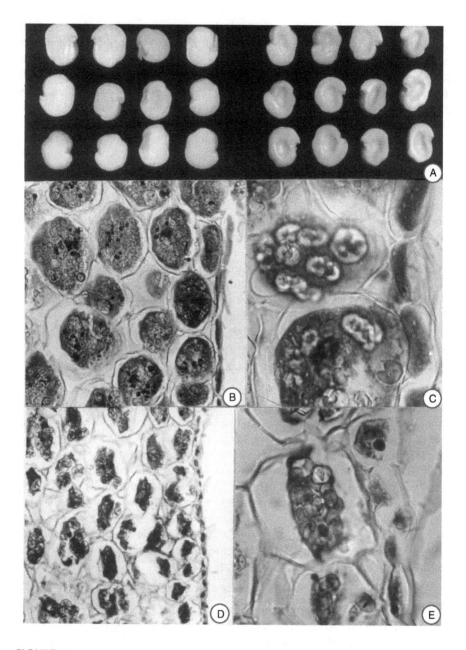

FIGURE 9.3 Morphology and histology of normal and hollow-hearted cotyledons of pea cultivar Dark Skin Perfection (Scotland). A, Cotyledon showing adaxial surface from water-soaked split seeds of hollow heart (right) and normal (left). B, C, Ts part of normal cotyledon from adaxial side. D, E, Ts from concavity in hollow-hearted cotyledons showing poor cell contents. (From Singh, D. 1974. *Seed Sci. Technol.* 2: 443–456. With permission.)

and sides in such cotyledons have moderate and evenly distributed protoplasmic contents. The epidermal cells lining the concavity contain homogeneous cytoplasm and normal interphase nuclei (Figure 9.3D, E).

In chickpea (*Cicer arietinum*), the symptomatic cotyledons have epidermis and one or two subepidermal layers similar to those of the normal cotyledons. However, the mesophyll cells of deeper layers contain comparatively thin and weak cell walls and prominently vacuolated more or less homogeneous cytoplasm. Comparative histochemical studies revealed the occurrence of poor protein contents in the cells of the symptomatic zone, but the starch content was comparable to that of the normal cotyledons.

9.2.1.2 Necrosis in Lettuce Cotyledons

Physiological necrosis, rot, or spotted cotyledons (Figure 9.4A, B), in seedlings of lettuce (*Lactuca sativa*) is a recognized abnormality (Dempsey and Harrington, 1951; Rogers, 1953; Cuddy and Lyall, 1959; Bass, 1970). Cotyledonary necrosis is rarely evident in freshly harvested seeds. It is not evident immediately in imbibed, excised embryos, and it takes several days to manifest. It is influenced by storage conditions, and the process of deterioration is gradual. Initial deterioration causes only delay in midrib greening, but severe effects result in weak growth of the embryonic axis and the cotyledons, which fail to emerge from the seed coat (Dempsey and Harrington, 1951; Bass, 1970). Finley (1959) has experimentally shown that it is caused by physiological drought due to moisture stress at temperature of 32°C or above over an extended period.

Smith (1989) has given a comparative account of ultrastructure of normal and symptomatic cotyledons with particular reference to the mobilization of reserves during germination. The ultrastructure of cotyledonary cells reveals the presence of electron-dense protein bodies with phytin-like deposits, many lipid bodies, plastids, elongated mitochondria, microbodies, and a prominent nucleus with a compact nucleolus (Figure 9.4C). The anatomy of symptomatic cotyledons shows that the necrotic patch, according to size and severity, represents a region of localized cell death. This necrosis is predominantly localized in the midrib region and in the transverse section is seen to be symmetrical (Figure 9.4D). The epidermal cells and the provascular cells are apparently more resistant to deterioration than the palisade or mesophyll cells. The affected mesophyll cells fail to mobilize food reserves and show distortion of cytoplasmic contents. The cells develop irregularity in the plasma-lemma, fretted protein bodies with numerous inclusions and confluence of lipid bodies. While manifestations of membrane damage, such as discontinuity, myelin-like figures adjacent to the plasmalemma, and/or withdrawal of the plasmalemma from the cell wall, are seen (Figure 9.4E), it is speculated that the delayed mobilization of reserves may be the result of membrane lesions (Smith, 1989).

9.2.1.3 Cracking of Cotyledons

Only macro-morphological observations are known. McCollum (1953), Pirson (1966), Pollock, Roos, and Manalo (1969), and Kietreiber (1969) observed seedlings

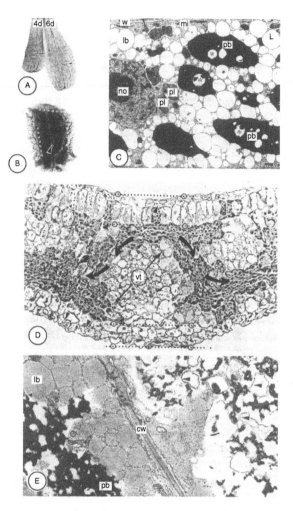

FIGURE 9.4 Morphology and histology of lettuce cotyledons showing physiological necrosis cultivar Great Lakes. A, Cleared unaged (control) cotyledons 4 and 6 days after imbibition of water showing normal vascular supply, development, and growth of margins. B, Cleared cotyledon of a low vigor embryo showing necrosis and marginal development of vascular supply 7 days after inhibition. The necrotic zone is demarcated by dots. C, TEM photograph of a palisade cell from the cotyledon (high vigor embryo) after 2 days imbibition showing a compact nucleus and cell organelles. D, Ts necrotic cotyledon (low vigor) showing normal vascular bundle and surrounding parenchyma cells with incomplete vacuolation. Note symmetrically located mesophyll cells that have failed to mobilize reserves (curve arrows). E, TEM photograph of parts of two adjacent mesophyll cells of a necrotic (low vigor) cotyledon after 5 days of imbibition. Note somewhat irregular plasmalemma along the cell wall, fretted nature of protein bodies, numerous membraneous inclusions (arrow) in other protein bodies, and lipid bodies. (Abbreviations: cw, cell wall; d, days; lb, lipid bodies; mi elongated mitochondria; no, nucleolus; pl, plastid; pb, protein bodies; vt, vascular tissue.) (From Smith, M.T. 1989. *Seed Sci. Technol.* 17: 453–462. With permission.)

of bean and pea showing abnormalities due to drought. The cotyledons showed transverse cracks resembling mechanical damage (Neergaard, 1979). Affected seeds showed poor emergence, and the seedlings were very weak. Dickson (1973) attributes this disorder to a combination of factors, e.g., the genotype, the rate of seed drying at maturation, and imbibition rate at germination. According to Dickson, transverse cracking in cotyledons is common in most *Phaseolus* bean cultivars and can occur as much as 100% in some cultivars.

Jain (1984) observed cracking of cotyledons in chickpea. The disorder was observed both in *desi* and *kabuli* types. Cultivars of desi chickpea had higher incidence of seeds with cracked cotyledons than those of kabuli. Cracks are transverse breaks extending partially or completely across the cotyledons and usually occur in both the cotyledons and rarely in one of them only.

9.2.2 HIGH HUMIDITY EFFECTS

The high air humidity in the field at the time when the crop is close to maturity or high humidity during storage adversely affect the viability of the seed and also cause seedling abnormalities (Neergaard, 1979). In Denmark, such environmental conditions cause gray discoloration in seeds of radish (Neergaard, 1945) and white mustard (Jørgensen, 1967). From a comparative anatomical study of normal and gray discolored white mustard seeds, Jørgenson (1967) demonstrated that the direct cause of the discoloration is swelling of the subepidermal parenchyma of the seed coat. This swelling of subepidermal layers results in the distortion and breaks in the epidermis.

9.3 CONCLUDING REMARKS

The deleterious effects of disorders, such as marsh spot, hollow heart, cracking of cotyledons, and physiological necrosis in cotyledons, are well known. They result in delayed germination, poor root system, reduced size of seedlings, and poor seedling growth (Löhnis, 1936; Pethybridge, 1936; Perry and Howell, 1965; Perry, 1967; Harrison and Perry, 1973; Gane and Biddle, 1973; Jain, 1984; Dempsey and Harrington, 1951; Bass, 1970; Smith, 1989). Effects of marsh spot causing dead plumule tips are more severe than those of hollow heart (Perry and Howell, 1965). The histology of marsh spot-affected cotyledons shows distinctive features of disturbed biochemical or metabolic functioning. Loosening of the cell wall, dissociation and discontinuity of plasma membrane, degradation of cytoplasm, fretted protein bodies, corroded starch grains, and secretion of oily pigmented material, which accumulate in enlarged intercellular air spaces, reveal profound differences in structures in normal cotyledons (Singh and Mathur, 1992). The degradation of the membrane system and macromolecules might account for the poor performance of affected seed. What triggers the new biochemical activity in peas and how is it related to manganese deficiency needs to be understood.

The histology of hollow-hearted cotyledons shows poor cell contents, particularly reserve materials. These cells seem to lose more water during drying and on imbibition and take more time to recover, causing concavity. Don et al. (1984)

concluded that irrespective of germination conditions, hollow heart is related to the quantity of deteriorated tissues near the center of the adaxial surface of cotyledons in ungerminated seeds. The question whether these cells are dead as suggested by Moore (1964) and Perry and Howell (1965) has not been confirmed by Singh (1974). The seed development and maturation, including the development of the embryo, comprise an early phase of cell division, expansion, and differentiation, and a period of physiological maturation in which the deposition of reserve materials takes place. During the last few days or hours, this is a simple drying process, not accompanied by accumulation of nutrient reserves. The deposition of reserve materials, which takes place in the endosperm or embryo, or both, is usually progressive from periphery to center. In the case of pea seed, where the endosperm is negligible, it occurs in the embryo. Premature drying of seeds will imply a cut in the storage phase, which is bound to affect cells of the adaxial central region of the cotyledons more than those of other parts.

The anatomical study of necrotic lettuce cotyledons (Smith, 1989) has also brought out a visible manifestation of membrane damage, such as breaks, myelin-like figures adjacent to the plasmalemma, withdrawal of the plasmalemma from the cell wall, irregular nuclei with dilated membrane of the envelope, fretted nature of protein bodies, and fusion of lipid droplets. Smith (1989) remarked that the subcellular mechanism of deterioration in cotyledons appears no different from that reported in root tips of aging seeds (Villiers, 1973; Smith, 1978). However, according to Smith (1989), it needs to be established if lipid peroxidation is involved in this process.

The histological studies of symptomatic parts of structures showing physiogenic disorders, particularly ultrastructural observations, provide clear evidence of structural changes and disturbed biochemical functioning. Histochemical and experimental studies may provide further evidence on the mechanism of these deteriorations.

REFERENCES

Agrios, G.N. 1988. *Plant Pathology*, 3rd ed. Academic Press, San Diego.
Allen, J.D. 1961. Hollow heart of pea seed. *N.Z. J. Res.* 4: 286–288.
Bass, L.N. 1970. Prevention of physiological necrosis (red cotyledon) in lettuce seed (*Lactuca sativa* L.). *J. Am. Soc. Hort. Sci.* 95: 550–553.
Bennett, W.F. 1993. *Nutrient Deficiencies and Toxicities in Crop Plants*. The American Phytopathological Society, St. Paul, MN.
Cuddy, T.F. 1959. Marsh spot in peas. *Proc. Assoc. Off. Seed Analysts N. Am.* 49: 156–158.
Cuddy, T.F. and Lyall, L.H. 1959. Spotted cotyledons of lettuce. *Proc. Assoc. Off. Seed Analysts N. Am.* 49: 103–106.
Darley, E.F. and Middleton, J.T. 1966. Problems of air pollution in plant pathology. *Ann. Rev. Phytopathol.* 4: 103–118.
De Bruijn, H.L.G. 1933. Kwade harten van der erwten. *Tijdschr. Plant Ziekt.* 39: 281–318.
Dempsey, W.H. and Harrington, J.F. 1951. Red cotyledons of lettuce. *Calif. Agric.* 5: 4.
Dickson, M.H. 1973. Selection of transverse cracking resistance in beans. *Ann. Rep. Bean Improvement Coop.* 16: 21–22.

Don, R., Bustamente, L., Rennie, W.J., and Seddon, M.G. 1984. Hollow heart of pea (*Pisum sativum*). *Seed Sci. Technol.* 12: 707–721.

Eickstein, O., Bruno, A., and Turrentine, J.W. 1937. *Kennzeichen des Kalimangels.* Zweite Auflage. Verlag Gesellschaft für Ackerbau, Berlin.

Evans, L.S. 1984. Acidic precipitation effects on terrestrial vegetation. *Ann. Rev. Phytopathol.* 22: 397–420.

Finley, A.M. 1959. Drought spot of lettuce cotyledons. *Plant Dis. Rep.* 43: 629–632.

Gane, A.J. and Biddle, A.J. 1973. Hollow heart of pea (*Pisum sativum*). *Ann. Appl. Biol.* 74: 239–247.

Gupta, M., Pandey, N., and Sharma, C.P. 1994. Zinc deficiency on seed coat topography of *Vicia faba* Linn. *Phytomorphology* 44: 135–138.

Harris, H.C. and Gilman, R.L. 1957. Effect of boron on peanuts. *Soil Sci.* 84: 233–242.

Heintze, S.G. 1938. Readily soluble manganese of soils and marsh spot of peas. *J. Agric. Sci. Camb.,* 28: 175–186.

Heintze, S.G. 1956. The effects of various soil treatments on the occurrence of marsh spot in pea and on manganese uptake and yield of oats and timothy. *Plant Soil* 7: 218–252.

Heit, C.E. and Crosier, W.F. 1961. Nasturtium seed germination as affected by abnormal seed development. *Proc. Assoc. Off. Seed Analysts N. Am.* 51: 78–81.

Harrison, J.G. and Perry, D.A. 1973. Effect of hollow heart on growth of peas. *Ann. Appl. Biol.* 73: 103–109.

Hoffman, I.C. 1933. Potash starvation in the greenhouse. *Better Crops with Plant Food* 18: 10.

Jacobson, J.S., and Hill, A.C. Eds. 1970. *Recognition of Air Pollution Injury to Vegetation: A Pictorial Atlas.* Air Pollution Control Association, Pittsburgh, PA.

Jain, S.K. 1984. Physical, Physiological and Pathological Disorders of Chickpea (*Cicer arietinum* L.). Ph.D. thesis, University of Rajasthan, Jaipur, India.

Jain, S.K. and Singh, D. 1985. Hollow heart in chickpea seed. *ICN* 12: 30–31.

Jamalainen, E.A. 1936. Herneen Siementen sisainen urmeltuminen. [Internal necrosis of pea seeds.] *Valtion Maatalousk Julk.* 79: 1–8.

Jørgensen, J. 1967. Nogle undersøgelser over årsagerne til grafarvningen af frø af gul sennep (*Sinapsis alba*). In *Statsfrokontrollen Kobenhaven. Beretning for det 96. arbejdsar fra 1. juli 1966 til 30 juni.* 1967: 78–97.

Kietreiber, M. 1969. Abnormale Saprossentwicklung bei Bohnenkeimlingen. *Jahrbuch 1968 der Bundesanstalt für Planzenbau und Samenprufüng in Wien*: 38–45.

Lacasse, N.L. and Treshow, M., Eds. 1976. *Diagnosing Vegetation Injury Caused by Air Pollution.* Applied Science Associates, Inc., Washington, D.C.

Lacey, M.S. 1934. Studies in bacteriosis. XXI. An investigation of marsh spot of peas. *Ann. Appl. Bbiol.* 21: 621–640.

Lawrence, J.A. and Weinstein, L.H. 1981. Effects of air pollutants on plant productivity. *Ann. Rev. Phytopathol.* 19: 257–271.

Levitt, J. 1972. *Responses of Plants to Environmental Stresses.* Academic Press, New York.

Lohnis, M.P. 1936. Wat veroorgaakt kwade harten in erwten. *Tijdschr. Plziekt.* 42: 159–167.

Mansholt, J.H. 1894. Antwoord op vraag in no. 4 van dit blad. [About a disorder in pea.] *Nederlandsch Landbouw, Weekblad* 7: 2–3.

Mattusch, von P. 1973. Die Hohlberzigkeit, eine physiologische Störung von Samen der Gemuseerbse (*Pisum sativum* L.) *Nachrichtenbl. Dtsch. Pflanzenschutzdienst, Braun-schweig* 35: 179–182.

McCollum, J.P. 1953. Factors affecting cotyledonary cracking during the germination of beans (*Phaseolus vulgaris*). *Plant Physiol.* 28: 267–274.

McMurtrey, J.E, Jr. 1953. Environmental nonparasitic injuries. *Yearbook Agriculture*, U.S. Department of Agriculture, Washington, D.C., pp. 94–100.

Moore, R.P. 1964. Garden pea cotyledon cavities. *Newsl. Assoc. Off. Seed Analysts N. Am.* 38: 12–13.

Myers, A. 1947. "Hollow heart": an abnormal condition of the cotyledons of *Pisum sativum* L. *J. Aust. Inst. Agric. Sci.* 13: 76.

Myers, A. 1948. "Hollow heart": an abnormal condition of the cotyledons of *Pisum sativum* L. *Proc. Int. Seed Test Assoc.* 14: 35–37.

Neergaard, P. 1945. *Danish Species of Alternaria and Stemphylium*. Einar Munksgaard, Copenhagen.

Neergaard, P. 1979. *Seed Pathology*. Vols. 1 and 2. Macmillan Press, London.

Noble, M. 1960. Marsh spot and hollow heart in peas. *Proc. Int. Seed Test Assoc.* 25: 536–538.

Pell, E.J. 1979. How air pollutants induce disease. In *Plant Disease: An Advanced Treatise*. Horsfall, J.G. and Cowling, E.B., Eds. Academic Press, New York. Vol. 4, pp. 273–292.

Perry, D.A. 1967. Seed vigour and field establishments of peas. *Proc. Int. Seed. Test. Assoc.* 32: 3–12.

Perry, D.A. and Harrison, J.G. 1973. Causes and development of hollow heart in pea seed. *Ann. Appl. Biol.* 73: 95–101.

Perry, D.A. and Howell, P.J. 1965. Symptoms and nature of hollow heart in pea seed. *Plant Pathol.* 14: 111–116.

Pethybridge, G.H. 1936. Marsh spot in pea seed: is it a deficiency disease? *J. Min. Agric. Fish.* 43: 55–58.

Pirson, H. 1966. Über Trockenschäden an Bohnen und Erbsen. *Saatgutwirtschaft* 18: 240: 242–243.

Pollock, B.M., Roos, E.E., and Manalo, J.R. 1969. Vigor of garden bean seeds and seedlings influenced by initial seed moisture, substrate oxygen and imbibition temperature. *J. Am. Soc. Hort. Sci.* 94: 577–584.

Rogers, C.B.W. 1953. Report of the subcommittee on the evaluation of lettuce seedlings. *Proc. Assoc. Off. Seed Analysts N. Am.* 43: 35.

Scaife, A. and Turner, M. 1984. *Diagnosis of Mineral Disorders in Plants*. Vols. 1 and 2. Chemical Publishing, New York.

Singh, D. 1974. Occurrence and histology of hollow heart and marsh spot in peas. *Seed Sci. Technol.* 2: 443–456.

Singh, D. and Mathur, R. 1992. Comparative anatomy and ultrastructure of normal and marsh spot affected cotyledons of pea. *Phytomorphology* 42: 145–150.

Smith, M.T. 1978. Cytological changes in artificially aged seeds during imbibition. *Proc. Electron Microsc. Soc. Southern Africa* 8: 105–106.

Smith, M.T. 1989. The ultrastructure of physiological necrosis in cotyledons of lettuce seeds (*Lactuca sativa* L.). *Seed Sci. Technol.* 17: 453–462.

Stapel, C. and Bovein, P. 1943. *Mark frøafrødernes Sygdomme og Skadedyr.* Det Kgl. Danske Landhusholdnings Selskab, Copenhagen.

Villiers, T.A. 1973. Ageing and longevity of seed in field conditions. In *Seed Ecology*. Heydecker, W. Ed. Butterworths, London, pp. 265–288.

Wallace, T. 1951. *The Diagnosis of Mineral Deficiencies in Plants by Visual Symptoms*. His Majesty's Stationery Office, London.

10 Microtechniques in Seed Histopathology

Seeds, after dispersal, are autonomous and exposed to the hazards of the environments. They have a strong protective covering with cuticle, waxy coatings, and seed coat and pericarp with thick-walled lignified or suberised cells. The usual histological techniques cannot be applied to cut seeds in a dry state. Seeds need to be softened and cut into small pieces. Immature developing seeds and internal components of seeds, the endosperm and embryo or the embryo after removing the seed coat and pericarp, can be processed like any other soft material.

There are several books on plant microtechnique (Johansen, 1940; Baker, 1958; O'Brien and McCully, 1981; Gerlach, 1984; Neergaard, 1997) and electron microscopy (Glauert, 1974; Aldrich and Todd, 1986; Robards and Wilson, 1993). In this chapter, only some tips and methods, which are useful in the study of seed histopathology, are provided. For detailed information and for SEM and TEM techniques refer to the books cited above.

10.1 CHOICE OF MATERIAL

- Artificially or naturally infected seeds, undamaged or degraded or mechanically injured, have been used (Singh, 1983). Naturally field-infected undamaged seeds should have priority over artificially inoculated seeds or seeds from field-inoculated plants.
- Selection should be made following laboratory screening of seed samples for seed-borne fungi.
- Select samples with single pathogen infection or predominant infection of the pathogen under study. In the latter condition, samples with good infection percentage of the target pathogen and low infection of other fungi, which are readily eliminated after chlorine pretreatment, may be preferred.
- Asymptomatic as well as symptomatic seeds, the latter categorized into weakly, moderately, and heavily infected, should be examined.
- Size of the seed sample examined should be large enough to give correct assessment.

The above procedure is also useful for selecting seed samples infected by bacteria, but for viruses, half-seeds corresponding to those halves that test positive for the virus in serological assay and/or other tests are recommended (Carroll, 1969; Alvarez and Campbell, 1978; Hunter and Bowyer, 1993).

10.2 DETERMINATION OF THE IDENTITY
OF INTERNAL MYCELIUM

The problem of determining the identity of internally seed-borne mycelium of fungi poses some difficulty. D. Singh, while working at the Danish Government Institute of Seed Pathology (DGISP), Copenhagen (1973 to 1974), developed the technique called *component plating*. It has been widely used to detect the internal infection of fungi, which readily sporulate or produce structures, permitting their identification (Maden et al., 1975; Singh, Mathur, and Neergaard, 1977, 1980; Singh, 1983).

10.2.1 PROCEDURE FOR COMPONENT PLATING

- Soak seeds in water at room temperature, just long enough to permit separation of components, namely, seed coat and pericarp, endosperm and embryo.
- Dissect seed aseptically to separate components using sterilized scalpel, needles, and forceps.
- Sterilize each component by washing in 1% chlorine solution.
- Plate on wet blotters (as in the standard blotter test) or potato dextrose agar medium (PDA).
- Incubate for 7 days under near ultra-violet (NUV) or daylight fluorescence tubes.
- Examine the different components under stereobinocular microscope on day 8.

Note: Period for soaking of seeds in water and for incubation of plates may be determined in preliminary tests.

Separated components of seeds can also be used to determine bacteria and viruses using infectivity, serological, and other tests (Schaad, 1988; Lange, Wu, and van Vuurde, 1992).

10.3 SEED SOFTENING

Dry seeds need to be softened for any histological study. Seeds are usually soaked in water, but prolonged soaking of infected seeds at room temperature or temperature congenial for revival and growth of the pathogen may enable the pathogen to spread to new areas. The following treatments are usually followed.

- Hydrofluoric acid treatment: Soak seeds in 5 to 20% hydrofluoric acid until rendered soft, then wash in running water for 24 or more hours until traces of the acid are removed. Store in 70% ethyl alcohol.
- Picric acid treatment: Water-boil seeds for 1 to 2 hours. Cool and transfer to aqueous saturated picric acid solution at room temperature or at 40°C (keep in oven) for 2 to 4 weeks, depending on the hardness of seed. Wash in running water until traces of the acid are removed. Store in 70% ethyl alcohol.

- Water boiling: Water-boil seeds for 2 to 48 hours on a water bath until seed coat becomes soft. Treating seeds in an autoclave has also been suggested but the present authors prefer water boiling on a water bath. Cool and store in 70% ethyl alcohol.
- Seeds with mucilage in seed coat (*Eruca, Linum*) or endosperm (*Cyamopsis, Trigonella*) should be boiled only for a short period (5 to 10 minutes) or kept overnight at 60°C in an oven to soften them.

Seeds boiled in water give satisfactory results because this treatment does not create problems with stains and staining procedures. Sections from acid-treated seeds stain well, but on storage the stains often fade.

Large seeds with a thick seed coat (*Cucurbita, Citrullus, and Hevea*) and pericarp (*Helianthus*) should be boiled and, after cooling, the seed coat and pericarp should be separated from the rest of the seed components. Divide the former into small pieces and the latter into two longitudinal halves before processing further.

10.4 HISTOLOGICAL METHODS

10.4.1 WHOLE-MOUNT METHOD

A general account of preparing whole-mounts of plant parts is given by Johansen (1940), Gardner (1975), and Neergaard (1997). These methods cannot be applied as such for preparing whole-mounts of seed that consist of a number of components and of both hard and soft tissues. Whole-mount preparations are, of course, very useful for determining the location of the pathogen in seed components because the method is quick and gives a total picture of the characteristics of the mycelium and its spread.

The procedure for preparing whole-mounts of seed components in some studies (Maden et al., 1975; Singh, Mathur, and Neergaard, 1977, 1980; Agarwal et al., 1985) is as follows:

- Water-boil seeds for 30 to 45 minutes.
- Cool and separate seed components.
- Boil components of one seed in a test tube in 5 or 10% aqueous KOH or NaOH solution or in 5% HCl for 10 or 20 minutes.
- Wash thoroughly with tap water to remove traces of alkali or acid.
- Stain with cotton blue and mount in lactophenol.

In the case of bulky components or those that do not become transparent in the above treatment, it may be cleared in lactophenol by gently heating the slide on a flame or if prolonged treatment is to be given, in the oven at 80°C (Agarwal et al., 1987).

Acid treatment is found superior for seeds with crystals in the seed coat as in the case of sesame. The calcium oxalate crystals, present in the epidermis, are removed by acid treatment (Singh, Mathur, and Neergaard, 1977).

Many of the chemicals used for clearing tissues, e.g., sodium hydroxide (5%) and lactic acid (9%), are corrosive. Safety precautions must be taken while using them. A lab coat, protective gloves, and protective glasses should be used (Mathur and Kongsdal, 2003).

10.4.2 FREEHAND SECTIONS

Sections, cut by razor or by means of a sliding microtome and handled loosely, not attached to the slide by means of an adhesive, are in this category. The material may be fresh or fixed, but it should be fairly rigid. Hard or woody components of seed are usually cut freehand. Dry seeds are cut after softening.

The difficulty lies in holding the seed tightly to avoid bending when the razor strikes the seed. This can be achieved by holding seed in the pith, any soft but rigid plant material, or by embedding directly (without dehydration and infiltration) in paraffin wax or soap.

Sections may be stained with cotton blue and mounted in lactophenol or stained with safranin and fast green and mounted in glycerin jelly or polyvinyl alcohol (PVA).

10.4.3 MICROTOMY

Usually paraffin-embedded seeds and seed parts are used for histopathological investigations of seeds infected by microorganisms other than viruses. Resin-embedded pieces of components of seeds and ultratrome cut sections are required for TEM for the study of viruses. A brief account of the paraffin method is given below. For ultramicrotomy and TEM and SEM techniques, the readers are advised to consult Glauret (1974), Aldrich and Todd (1986), Robards and Wilson (1993), and Neergaard (1997).

10.4.3.1 Fixing and Storage

Buds, flowers, developing seeds as such, and mature dry seeds after water boiling may be fixed in FAA (formalin-acetic acid-alcohol) for 24 to 48 hours. For better fixation, dehydration, and infiltration, seeds may be cut longitudinally on one side, exposing the internal soft components. Wash and store in 70% ethyl alcohol.

10.4.3.2 Dehydration

The tertiary butyl alcohol (TBA) series is most satisfactory, although other dehydrating series, e.g., alcohol-xylol and chloroform, have also been used. TBA does not cause excessive hardening of the material. The composition of TBA proposed by Johansen (1940) is widely used (Table 10.1).

Keep soft materials for 2 to 4 hours in each solution. Seeds may be kept for 12 hours in each grade.

After a 100% alcohol solution, give three changes with pure TBA. Keep the material for 6 hours in the first two changes and overnight in the last.

TABLE 10.1
Composition of Tertiary Butyl Alcohol Series

Reagents	Alcohol Percentage				
	50	70	85	95	100
Distilled water	50	30	15	—	—
95% ethyl alcohol	40	50	50	45	—
Tertiary butyl alcohol	10	20	35	55	75
100% ethyl alcohol	—	—	—	—	25

10.4.3.3 Infiltration

Infiltration is the process of transfer from TBA to the paraffin wax in tissue. It is an important but slow process. The authors prefer to carry out the process initially under a bulb (60 W). Add flakes of paraffin wax to a vial. Gradually increase the amount by adding more flakes every 2 hours. Continue until the amount of mixture has doubled, leave overnight, drain half of the fluid, add melted paraffin wax, and keep for 6 or more hours. Repeat the process twice and transfer the vials into the oven at 60°C. Drain the fluid and change with melted wax. Make changes every 6 or more hours until traces of TBA are removed.

10.4.3.4 Embedding

Embed the infiltrated material in melted wax. Arrange the material in proper order (for details, see Johansen, 1940).

10.4.3.5 Softening of Embedded Material

This step is very important for microtomy of seeds. The following protocol is recommended:

- Cut block into small individual blocks with one seed per block.
- Trim individual blocks, particularly on the cut side of the seed leaving about 1 mm of the paraffin wax covering it.
- Immerse blocks in 1% aqueous solution of sodium lauryl sulphate for 24 hours.
- Wash with water.
- Transfer to a mixture of glycerin and acetic acid (1:1) for 1 to 4 weeks depending on the size and hardness of seed.
- Wash thoroughly with water and store.

10.4.3.6 Sectioning and Mounting of Ribbons

Microtome sections can be cut using any rotary microtome. For more details consult Johansen (1940).

The paraffin ribbons containing sections are cut in pieces of suitable size and fixed on slides using Haupt's or Meyer's adhesive (Johansen, 1940). For sections of seeds, the double adhesive process has given better results. Smear the slides with Haupt's or Meyer's adhesive, cover the smeared surface with gloy solution, mount the ribbons, spread, drain the extra fluid, and dry.

10.4.3.7 Staining and Mounting

1. Temporary preparations: Deparaffinize sections with xylol, bring to water, stain with cotton blue, and mount in lactophenol. Hyphae are stained blue.
2. Permanent preparations: Several standard staining procedures are described by Johansen (1940), Baker (1958), and Gerlach (1984). Safranin–light green and safranin–fast green are found most suitable for histopathology. Only the main steps are given below. For details consult Johansen (1940).

Deparaffinize slides in xylol, bring down slides to 70% alcohol, stain in safranin solution, wash excess stain with water, destain in picric acid–alcohol solution or 70% ethyl alcohol until the dye is removed from the thin-walled cells. Dehydrate slides through alcohol grades (30, 50, 70 and 90%). Counter stain with light green or fast green, differentiate in clove oil, and wash in a mixture of xylol and 100% alcohol (1:1) to remove traces of clove oil. Transfer to pure xylol and mount in D.P.X., Canada balsam, or Caedax solution.

Slides must be dried before storing in slide boxes.

10.5 PROCEDURES FOR PREPARING SOME REAGENTS AND STAINS

10.5.1 Fixative

Formalin-Acetic Acid-Alcohol (FAA)

70% ethyl alcohol 90 ml
Glacial acetic acid 5 ml
Formalin 5 ml

For delicate materials, 50% ethyl alcohol may be used in place of 70%.

10.5.2 Adhesives

Haupt's Adhesive

Plain, finely divided pure gelatin 1 g
Distilled water 100 ml
Phenol crystals 2 g
Glycerin 15 ml

Dissolve gelatin in water at 30°C. When completely dissolved, add phenol crystals and glycerin.

Meyer's Adhesive

White of fresh egg 50 ml
Glycerin 50 ml
Sodium salicylate 1 g

Shake the mixture and filter through sterile cotton or cheese cloth.

Gloy Solution

Distilled water 10 ml
Gloy (liquid) 2 drops
Potassium dichromate small crystal or a pinch

Shake the mixture vigorously. Do not store for long.

10.5.3 MOUNTING MEDIA

10.5.3.1 Aqueous Mounting Media

Lactophenol

Lactic acid
Glycerin
Phenol crystals or distilled water

Mix the above reagents in a ratio of 1:1:1:1 or 2:1:1:1.
The use of phenol has been found to be hazardous in laboratories in Sweden and Denmark, and use should be avoided. The mixture of lactic acid, glycerol, and water (1:2:1) has been found suitable (Mathur and Kongsdal, 2003). A safe and good medium for temporary mounts is 2% aqueous glycerin.

Lactophenol Cotton Blue

0.1 g aniline blue (cotton blue) in 100 ml lactophenol

Staining, clearing, and mounting are achieved in one step.

Polyvinyl Alcohol (PVA)

Polyvinyl alcohol 1.66 g
Distilled water 10 ml
Lactic acid 10 ml
Glycerol 1 ml

Add polyvinyl alcohol slowly to the water. Stir on magnetic stirrer. Dissolve for 1 to 2 hours. Add lactic acid, stirring vigorously, followed by glycerol.

Glycerin Jelly

> Gelatin (pure and high quality) 1 part by weight
> Distilled water 6 parts by weight
> Glycerin 7 parts by weight
> Phenol crystals 1 g per 100 g mixture

Dissolve gelatin in distilled water for 2 hours or longer. Add glycerin and phenol crystals. Warm for 15 minutes, stirring continuously. Filter through cheese cloth in a wide-mouth bottle. The mixture solidifies on cooling. Take a small piece of jelly onto a slide, gently warm, mount the material, and place the coverglass.

Note: Slides mounted in aqueous mounting media or other temporary liquid media may be sealed with nail polish.

10.5.3.2 Nonaqueous Mounting Media

> Canada balsam
> Caedax
> DPX

The authors prefer to use dilute caedax solution: 2 parts of caedax and 1 part of xylene.

10.5.4 Stains

Safranin O

> *Alcoholic solution*
>   95% ethyl alcohol 100 ml
>   Safranin 1 g

Dilute with equal amount of distilled water when the solution needs to be used.

> *Methyl cellosolve solution*
>   Safranin 2 g
>   Methyl cellosolve 100 ml
>   Ethyl alcohol 95% 50 ml
>   Distilled water 50 ml
>   Sodium acetate 2 g
>   Formalin 4 ml

Dissolve safranin in methyl cellosolve. Add alcohol, distilled water, and other chemicals. Stir and store.

This solution produces a sharp contrast in stained tissues.

Fast Green and Light Green

Alcohol solution
   Ethyl alcohol 95% 100 ml
   Stain 0.2 or 0.5 g (0.2 g preferred)

Clove oil solution
   Stain 0.2 or 0.5 g
   Ethyl alcohol 95% 50 ml
   Clove oil 50 ml

Fast green and light green staining is rapid, and thus the process should be quickly completed.

Mixed Solution of Safranin–Light Green and Safranin–Fast Green

   Conc. HCl 1.25 ml
   Distilled water 75 ml
   Ethyl alcohol (95%) 120 ml
   Safranin 0 1.5 g
   Light green or fast green 0.5 g

Plasma-filled hyphae are stained red while host cells are green (Pedersen, 1956). The authors prefer to use safranin and fast green and light green as separate stains.

REFERENCES

Agarwal, K., Sharma, J., Singh, T., and Singh, D. 1987. Histopathology of *Alternaria tenuis* infected black-pointed kernels of wheat. *Bot. Bull. Academia Sinica* 28: 123–130.

Agarwal, K., Singh, T., Singh, D., and Mathur, S.B. 1985. Studies on glume blotch disease of wheat I. Location of *Septoria nodorum* in seed. *Phytomorphology* 35: 87–91.

Aldrich, H.C. and Todd, W.J. 1986. *Ultrastructure Techniques for Microorganisms*. Plenum Press, New York.

Alvarez, M. and Campbell, R.N. 1978. Transmission and distribution of squash mosaic virus in seeds of cantaloupe. *Phytopathology* 68: 257–263.

Baker, J.R. 1958. *Principles and Biological Microtechnique*. Methuen, London.

Carroll, T.W. 1969. Electron microscopic evidence for the presence of barley stripe mosaic virus in cells of barley embryos. *Virology* 37: 649–657.

Gardner, R.O. 1975. An overview of botanical clearing technique. *Stain Technol.* 50: 99–105.

Gerlach, D. 1984. *Botanische Mikrotechnik*, 3rd ed. Thieme, Stuttgart.

Glauert, A.M. Ed. 1974. *Practical Methods in Electron Microscopy*. Vol. 3, Part 1. North Holland Publishing Company, Amsterdam.

Hunter, D.G. and Bowyer, J.W. 1993. Cytopathology of lettuce mosaic virus — infected lettuce seeds and seedlings. *J. Phytopathol.* 137: 61–72.

Johansen, D.A. 1940. *Plant Microtechnique*. McGraw-Hill, New York.

Lange, L., Wu, W.-S., and van Vuurde, J.W.L. 1992. *Seed Transmitted Virus Diseases: Biology, Detection and Control*. Yi Hsien Publishing Company Ltd., Taipei, Taiwan.

Maden, S., Singh, D., Mathur, S.B., and Neergaard, P. 1975. Detection and location of seed-borne inoculum of *Ascochyta rabiei* and its transmission in chickpea *(Cicer arietinum)*. *Seed Sci. Technol.* 3: 667–681.

Mathur, S.B. and Kongsdal, O. 2003. *Common Laboratory Seed Health Testing Methods for Detecting Fungi*. International Seed Testing Association, Zurich, Switzerland.

Neergaard, E. 1997. *Methods in Botanical Histopathology*. Danish Government Institute of Seed Pathology for Developing Countries, Kandrups Botrykkeri, Copenhagen, Denmark.

O'Brien, T.P. and McCully, M.E. 1981. *The Study of Plant Structure, Principles and Selected Methods*. Termarcarphi, Melbourne.

Pedersen, P. N. 1956. Infection of barley by loose smut, *Ustilago nuda* (Hens.) Rostr. *Friesia* 5: 341–348.

Robards, A.W. and Wilson, A.J., Eds. 1993. *Procedures in Electron Microscopy*. John Wiley & Sons, Chichester, U.K.

Schaad, N.W. Ed. 1988. *Laboratory Guide for Identification of Plant Pathogenic Bacteria*, 2nd ed. American Phytopathological Society, St. Paul, MN.

Singh, D. 1983. Histopathology of some seed-borne infections: a review of recent investigations. *Seed Sci. Technol.* 11: 651–663.

Singh, D., Mathur, S.B., and Neergaard, P. 1977. Histopathology of sunflower seeds infected by *Alternaria tenuis*. *Seed Sci. Technol.* 5: 579–586.

Singh, D., Mathur, S.B., and Neergaard, P. 1980. Histopathological studies of *Alternaria sesamicola* penetration in sesame seed. *Seed Sci. Technol.* 8: 85–93.

Index

A

Abelmoschus esculentus, 57
Abrus precatarious, 48
Acanthaceae, 52
Achyranthes, 69
Acidovorax spp., 3, 169, 185, *186,* 187, 190
Acremonium spp., 150
Acronidiella eschscholtziae, 90
Adhesives, 266–267
Agrobacterium, 169
Albuginaceae, 82
Albugo spp.
    Brassicaceae, 7
    colonization, 107, 111, *112*
    infection, 102, 154
Alfalfa, 188
Alfalfa mosaic virus (AMV), 209, 212–213, 221
Allium, 20, 53, *see also* Onion
Alternanthera, 69
Alternaria spp.
    colonization, 125, 132, *133*
    host–pathogen interactions, 104
    ovule and seed infection, 92
    stigma and style infection, 88
    threshed seeds infection, 93
    vascular supply infection, 83
Amantiferae, 20
Amaranthaceae
    embryo, 53
    integuments, 31
    nucellus, 29
    seeds, 47
    seed structure, 68–70, *71*
Amphitropous ovules, 18, *see also* Ovules
AMV, *see* Alfalfa mosaic virus (AMV)
Anatropous ovules, 18, *see also* Ovules
Anguina spp.
    bacterial infection, 173, 188
    basics, 235–238
    gall formation, 3
    nematodes, 229–235, 241–242
Annonaceae, 31
Anther gall, 235
Anthers, *12,* 170, *see also* Stamen
Anthraxon, 9

Aphelenchoides spp.
    bacterial infection, 173
    nematodes, 229, 231, 239–241
Apiaceae
    endosperm, 25
    ovules, 17
    seeds, 2, 47
    seed structure, 63, *65*
Apiculus, 73
Apple, 9, 171
Aqueous mounting media, 267–268
Arachis spp., 60–61
Aril, 51, 93, 178
Arillode, 37
Artabotrys, 16
Arthrobacter, 169
Asclepiadaceae, 20, 52
Ascochyta spp., 143, *144,* 154
Ascomycetes, 113–119, *114, 116–118*
Ascomycotina, 2
Aspergillus spp., 92–94
Asteraceae
    chalaza, 31
    endosperm, 25
    seeds, 2, 47
    seed structure, 68, *69*
Atlas, 210
Atriplex, 70
Atropous ovules, 18, *see also* Ovules
Aureobasidium lini, 90
Avena spp., 70, *see also* Oats
Avocado, 93
Avocado pear, 110
Axile seeds, 53

B

Bacterial infection
    basics, 3, 169, 191–192
    Clavibacter, 187–188, *189*
    Curtobacterium, 188, 190
    disseminated seeds, 177–178
    histopathology, 178–190
    nematodes, 241–242, *243*
    Pantoea, 190
    penetration, 169–178

Sugar beets, *see also* Beets
    colonization, 110, 124
    nematodes, 244
    ovary and fruit wall infection wall, 90
Sunflower
    colonization, 110, 119, 149
    flowers, 7, 9
    severity of infection, 102
Sunn hemp seeds, 147
Survival, bacterial infections, 190–191
Suspensor, 27–28
Syngamy, 23

T

Tagetes, 68
Tegmic type seeds, 31, 55
TEM, *see* Transmission electron microscopy
    (TEM)
Tenuinucellar ovules, 20, *see also* Ovules
Testal type seeds, 31, 55
Testa nematodes, 239–241
Threshed seeds, 92–93, 177–178
Tiliaceae, 25
Tilletia, 7
TMV, *see* Tobacco mosaic virus (TMV)
Tobacco, 132
Tobacco mosaic virus (TMV), 222
Tobacco ring spot virus (TRSV), 208, 213
Tolyposporium, 7
Tomato
    bacterial infection, 170–171, 173–174, 176,
        188
    colonization, 145
    flowers, 7
    infection penetration, 94
    stigma and style infection, 88
    viral infection, 207–208
Tradescantia, 53
Transmission electron microscopy (TEM), 3, 13,
    241
Transport systems, viral infection, 205–206
Trichothecium spp., 61, 104, 141
Trifolium, 85
Trigonella, 60–61
Triticum spp., 20, 39–40, 70, 73, *see also* Wheat
Tropaelum majus, 253
TRSV, *see* Tobacco ring spot virus (TRSV)
Tube cells, 40
Turnera ulmifolia, 51
Turnip seeds, 111

U

Uniola, 9
Unisexuality, 9
Unitegmic seeds, 55
Urdbean, 217
Uredinales, 82
Ustilaginales, 82
Ustilago spp.
    infection, 101
    ovary and fruit wall infection wall, 90
    ovule and seed infection, 92
    penetration, 93
    seed formation, 7

V

Vandezeia subterranea, 61
Vascularization, 9–10, *10*
Vascular supply, *18,* 174
Verticillum spp., 83, 140–141
Vicia faba, 253
Vigna spp.
    host–pathogen interactions, 104
    ovules, 20
    seed structure, 61
    viral infections, 217
Viola tricolor, 27
Viral infections
    barley stripe mosaic virus, 209–210, 212–213
    basics, 2, 199, *199–203,* 222
    bean common mosaic virus, 217
    cellular contacts, 205–206
    cryptoviruses, 2, 205, 209
    cytopathological effects, 219–221
    inactivation and longevity, 221–222
    infected plants, 206–207
    infection and multiplication, 199, 203–205
    isolation, 205–206
    lettuce mosaic virus, 217, *218*
    localization, 209–219, *213–216*
    movement, 206–209
    ovule and seed, 207–209
    pea seed-borne mosaic virus, 217, 219,
        219–220
    storage, 221–222
    transport systems, 205–206
Viroids, 2

W

Walnut blight, 172
Western wheatgrass nematodes, 238

Wheat, *see also Triticum* spp.
  bacterial infection, 170, 177, 188, 192
  colonization, 120–122, 124, 134, 146
  ear cockle disease, 231–235
  flowers, 7, 9
  host–pathogen interactions, 104–105
  infection penetration, 93
  mineral deficiency, 249
  nematodes, 231, 241
  ovary and fruit wall infection wall, 90
  ovule and seed infection, 92
  Puccinia graminis, 3
  seed coat surface, 49
  severity of infection, 102
White tip nematodes, 239, 242
Whole-mount method, 263–264
Wings, 51, 178
Wounds, bacterial infection, 173

X

Xanthomonas spp., 3, 169–178, 182–185,
    190–191

Z

Zea spp.
  embryo, 27
  infection, 101
  ovules, 20
  seed structure, 70, 73
Zygotes, 27

Milton Keynes UK
Ingram Content Group UK Ltd.
UKHW040446071024
449327UK00020B/1040